Kinetic Theory of Gases in Shear Flows

Fundamental Theories of Physics

An International Book Series on The Fundamental Theories of Physics: Their Clarification, Development and Application

Volume 131

Kinetic Theory of Gases in Shear Flows

Nonlinear Transport

by

Vicente Garzó
Departamento de Física,
Universidad de Extremadura, Badajoz, Spain

and

Andrés Santos
Departamento de Física,
Universidad de Extremadura, Badajoz, Spain

KLUWER ACADEMIC PUBLISHERS
DORDRECHT / BOSTON / LONDON

A C.I.P. Catalogue record for this book is available from the Library of Congress.

ISBN 1-4020-1436-8

Published by Kluwer Academic Publishers,
P.O. Box 17, 3300 AA Dordrecht, The Netherlands.

Sold and distributed in North, Central and South America
by Kluwer Academic Publishers,
101 Philip Drive, Norwell, MA 02061, U.S.A.

In all other countries, sold and distributed
by Kluwer Academic Publishers,
P.O. Box 322, 3300 AH Dordrecht, The Netherlands.

Printed on acid-free paper

Printed in the Netherlands.

Contents

List of Figures

List of Tables

Foreword

The kinetic theory of gases as we know it dates to the paper of Boltzmann in 1872. The justification and context of this equation has been clarified over the past half century to the extent that it comprises one of the most complete examples of many-body analyses exhibiting the contraction from a microscopic to a mesoscopic description. The primary result is that the Boltzmann equation applies to dilute gases with short ranged interatomic forces, on space and time scales large compared to the corresponding atomic scales. Otherwise, there is no a priori limitation on the state of the system. This means it should be applicable even to systems driven very far from its equilibrium state. However, in spite of the physical simplicity of the Boltzmann equation, its mathematical complexity has masked its content except for states near equilibrium. While the latter are very important and the Boltzmann equation has been a resounding success in this case, the full potential of the Boltzmann equation to describe more general nonequilibrium states remains unfulfilled.

An important exception was a study by Ikenberry and Truesdell in 1956 for a gas of Maxwell molecules undergoing shear flow. They provided a formally exact solution to the moment hierarchy that is valid for arbitrarily large shear rates. It was the first example of a fundamental description of rheology far from equilibrium, albeit for an unrealistic system. With rare exceptions, significant progress on nonequilibrium states was made only 20–30 years later. This was done in two complementary ways. The first was the discovery of several exact solutions to kinetic models — simplified versions of the Boltzmann equation preserving its essential properties. The second was the development by Bird of a direct Monte Carlo simulation method to solve numerically the Boltzmann equation. In the last 15 years these tools have been extended and applied extensively. The result is a new field of nonequilibrium phenomena in gases being explored for the first time more than one hundred years after Boltzmann's remarkable insight.

The authors of this book have been at the center of this development, being responsible for some of the most important mathematical and numerical advances in recent years. They have chosen to limit attention in this monograph to shear flows and therefore they are able to cover the subject in exceptional detail. The presentation is self-contained so that the student and research scientist from Physics, Chemistry, or Chemical Engineering now has access to a unified presentation of this exciting field heretofore distributed across a large number of technical papers in many different journals. It will become a primary reference for all who are interested in nonlinear transport, and it will be the springboard for future studies using the methods illustrated here. Equally important, as the results are exact, they will provide important benchmarks with which to test more general but phenomenological studies of shear flows.

The authors have done a remarkable job of complementing technical details of the analysis, necessary to appreciate how exact results are extracted, with a significant number of figures for visual elaboration of the physical properties studied. The bibliography alone will make this an invaluable reference for years to come. I hope that this book will serve as a model for others to follow, synthesizing and unifying under one cover important advances of a field that appear piecewise, unconnected, and often unappreciated by all but a few experts in that field.

JAMES W. DUFTY, UNIVERSITY OF FLORIDA
March 2003

To Leo, Alex, and Ana (V.G.)

To Ángeles, María del Mar,
and David (A.S.)

Acknowledgments

We are very grateful to Profs. David Jou (Universitat Autònoma de Barcelona, Spain), James W. Dufty (University of Florida, USA), and Mariano López de Haro (UNAM, Mexico) for their encouragement and support. They also had the patience of critically reading the manuscript and made detailed and useful comments and suggestions. Special thanks are due to David Jou, who first suggested the possibility of writing this monograph, as well as to James Dufty, who was happy to accept our invitation to write the foreword.

Of course, this book would not have been possible without the work of those colleagues we have had the privilege of collaborating with on some of the covered topics. They are, in alphabetical order, L. Acedo, M. Alaoui, A. V. Bobylev, J. J. Brey, J. W. Dufty, J. Gómez Ordóñez, C. S. Kim, M. Lee, M. López de Haro, J. F. Lutsko, C. Marín, J. M. Montanero, R. F. Rodríguez, and M. Tij.

Finally, we also want to express our gratitude to the members of our research group for their support, help, and friendship: Luis Acedo, Antonio Astillero (who performed new computer simulations), Conchita Marín, José María Montanero, Juan Jesús Ruiz–Lorenzo, Héctor Sánchez–Pajares, José Manuel Vaquero, and Santos Bravo Yuste.

Introduction

In my opinion it would be a great tragedy for science if the theory of gases were temporarily thrown into oblivion because of a momentary hostile attitude toward it, as was for example the wave theory because of Newton's authority.
I am conscious of being only an individual struggling weakly against the stream of time. But it still remains in my power to contribute in such a way that, when the theory of gases is again revived, not too much will have to be rediscovered.

—Ludwig Boltzmann, 1898

The thermodynamical and statistical-mechanical description of macroscopic systems in equilibrium is well established, at least from a formal point of view. Thus, if the system is in thermal equilibrium with a heat bath at a given temperature T, the canonical ensemble assigns to each microstate a probability proportional to the Boltzmann factor. The corresponding normalization condition defines the partition function $Z(T)$, which provides all the thermodynamic information of the system through its relation to the free energy as $A(T) = -k_B T \ln Z(T)$.

On the other hand, the systems surrounding us are usually out of equilibrium. This is especially true in the case of fluid systems, since for instance they tend to flow when subjected to shear stresses and so depart from equilibrium. The classical, hydrodynamic description of nonequilibrium fluids is based on three major stages. In the first stage, one introduces local densities of mass, momentum, and energy, which are conserved quantities. The rate of change of these local densities can then be expressed in terms of the divergence of the corresponding mass, momentum, and energy fluxes. These *balance* equations are exact but do not constitute a closed set of equations. In the second stage, the irreversible parts of the momentum and energy fluxes are assumed to be linear functions of the flow velocity and temperature gradients, respectively. These linear relationships between fluxes and gradients define the *Navier–Stokes constitutive equations*: Newton's friction law for the stress tensor and Fourier's law of heat conduction. The associated coefficients are the transport coefficients of the fluid, namely the shear and bulk viscosities and the thermal conductivity, respectively. When these phenomenological laws for the fluxes are inserted into the balance equations, the resulting set of equations is still not closed because one needs to determine the hydrostatic pressure p and the internal energy density e in terms of the local mass density ρ and the local nonequilibrium temperature T. This is done in the third stage, where the equilibrium equations of state relating p and e to ρ and T are extended to the nonequilibrium local quantities. This extension is called the *local equilibrium hypothesis*. The three-stage process leads to the closed set of Navier–Stokes hydrodynamic equations. This scheme is essentially the same for fluid mixtures, except that the number of transport coefficients is larger than in the case of a simple fluid.

The Navier–Stokes equations are an invaluable tool for the description of gases and liquids under a wide range of conditions, from unsteady to stationary states, from quiescent to frictional situations, from compressible to incompressible fluids, from laminar to turbulent flows. The analysis of the many interesting phenomena encapsulated in the Navier–Stokes hydrodynamic equations is the subject of a scientific field that attracts the attention of mathematicians, physicists, and engineers.

One of the main objectives of nonequilibrium statistical mechanics is the derivation of the phenomenological constitutive equations from first principles, as well as the assessment of their range of validity and the determination of the transport coefficients in terms of the underlying microscopic parameters. Application of linear response theory leads to linear relationships between the fluxes and the hydrodynamic gradients when the latter are small enough so that the wavelength of the spatial variation of the hydrodynamic fields is much larger than a microscopic characteristic length. The most important outcome of the theory is the derivation of the elegant Green–Kubo formulas, expressing the transport coefficients as time integrals of equilibrium correlation functions for fluctuations in the corresponding fluxes. The linear response theory is not limited to fluids but applies to solids as well. However, the price to be paid is that one has to deal with the equations of motion for the many-body system. Therefore, in order to get specific results it is necessary to carry out approximations.

The prototype system that embodies the most relevant physical properties of ordinary fluids consists of a low-density gas of particles interacting via a pair-wise, central short-ranged potential. If the gas is rarefied enough, a simple picture already grasped by the fathers of kinetic theory arises: every particle moves most of the time freely (or under the action of an external force, if it exists), except when eventually its trajectory approaches that of another particle and, as a result of their mutual interaction, both particles abruptly change their respective velocities. Thus, in a dilute gas the interactions among the particles reduce to a sequence of uncorrelated binary collisions. The collisions can be considered as events well localized in space and time because the typical distance ℓ traversed by a particle between two successive collisions is much larger than the range σ of the interactions and, analogously, the typical time elapsed between two collisions is much larger than the duration of a collision.

The simplifications inherent to the low-density limit allow for a huge reduction of the microscopic description of the system: from a general statistical-mechanical treatment to a contracted kinetic theory description, from the N-body phase space probability density to the one-body velocity distribution function, from the reversible Liouville equation to the irreversible Boltzmann equation. The Boltzmann equation is the master equation for a low-density gas and describes the rate of change of the velocity distribution function f as produced by two uncoupled mechanisms: free motion and localized, instantaneous binary collisions. The key point in the Boltzmann equation is the assumption that the velocities of two particles that are about to collide are statistically uncorrelated. The justification of this *molecular chaos* hypothesis requires the interpretation of f as the most *probable* number of particles that at a given time are located near a certain point and move with a certain velocity. This statistical meaning of the Boltzmann equation explains the solution to Loschmidt's paradox on the apparent incompatibility between the irreversible tendency of the distribution function towards equilibrium and the reversibility of the underlying microscopic dynamics.

Perhaps, the most important success of the Boltzmann equation is the derivation of the Navier–Stokes constitutive relations as the first-order approximation in an expansion of f in powers of the hydrodynamic gradients, with expressions for the transport coefficients in terms of the temperature and the parameters of the interaction potential. This method of solution is the celebrated Chapman–Enskog expansion. It allows one to get the so-called *normal* solutions to the Boltzmann equation, a special class of solutions in which the spatial and temporal dependence of the distribution function occurs through a functional dependence (local in time and non-local in space) on the hydrodynamic fields.

Despite the large number of applications of the Navier–Stokes hydrodynamic description, there exist many domestic fluid substances (paint, shampoo, ketchup, tooth paste, grease, yoghurt, ...) that present significant deviations from the linear constitutive equations under ordinary conditions. In particular, when those substances are subjected to shearing, Newton's friction law fails (except for very small shear rates) since the shear stress is no longer proportional to the shear rate and, in addition, normal stress differences appear. This *non-Newtonian* behavior is characterized by material functions such as the generalized or nonlinear shear viscosity and the viscometric functions. The study of non-Newtonian behavior is the subject of an important scientific discipline known as Rheology. The common feature of non-Newtonian fluids, such as colloidal or polymeric solutions made up of very large molecules, is that they are *structurally* complex.

On the other hand, non-Newtonian behavior can also be found in structurally simple fluids that, otherwise, are in "complex" states far from equilibrium. The structurally simplest system able to present rheological effects due to the complexity of the state is again a monatomic dilute gas. As said before, the Boltzmann equation holds in the low-density limit where the mean free path ℓ is much larger than the "size" σ of the particles. However, the kinetic equation remains valid even if the hydrodynamic length h characterizing the spatial variation of the hydrodynamic fields is comparable to or smaller than the mean free path. In this situation the gas is expected to exhibit a non-Newtonian behavior with nonlinear constitutive relations. In principle, the Chapman–Enskog method, which is essentially a perturbation expansion in powers of ℓ/h, accounts for nonlinear transport when worked out beyond the Navier–Stokes order. Nevertheless, since the expansion is generally only *asymptotic*, it is not practical if ℓ/h is not small enough. Consequently, in that case one needs to resort to a non-perturbative approach. This is the point of view that will be generally used along this monograph.

Due to the complicated nature of the nonlinear Boltzmann equation, it is useful to consider model interaction potentials. Already in 1867, Maxwell realized that if the particles repel each other with a force inversely proportional to the fifth-power of the distance, the velocity integration in the collisional rate of change for a given quantity can be carried out without the explicit knowledge of the distribution function. This property leads to an important simplification on the hierarchy of equations for the velocity moments of the distribution function. Although a gas of so-called Maxwell molecules is not a realistic representation of the interaction potential in real gases, the reason to study it is at least two-fold. First, the use of this interaction potential allows one to get exact results in some non-trivial situations. The importance of finding exact solutions cannot be overemphasized. Their knowledge can be exploited to shed light on the physical mechanisms involved in complex situations, as well as to assess the degree of reliability of approximate methods. Second, it is expected that the influence of the specific details of the interaction potential on the nonlinear transport properties is only residual, once the relevant quantities are conveniently nondimensionalized. In this context, the exact results obtained for Maxwell molecules can be extended, at least in a semi-quantitative way, to other more realistic potentials. Quoting Ernst and Brito (2002), let us summarize the relevance of Maxwell molecules by saying that they are for kinetic theory what harmonic oscillators are for quantum mechanics and what dumbbells are for polymer physics.

When exact information can be extracted from the Boltzmann equation, this generally occurs through the first few velocity moments. Nevertheless, the determination of the full explicit knowledge of the velocity distribution function remains an unsurmountable task, even for Maxwell molecules. Of course, the problem is much worse in the case of general interactions. This limitation can be overcome by using model kinetic equations constructed by replacing the Boltzmann collision term by a mathematically simpler term which, otherwise, retains the main physical properties of the true collision term. The simplest choice assumes that the main effect of collisions can be modeled as a tendency of the distribution function to relax towards local equilibrium with an effective relaxation time. This is the well-known Bhatnagar–Gross–Krook kinetic model. An alternative approach to make progress in the understanding of the nonequilibrium properties described by the Boltzmann equation consists of the use of numerical methods to solve it. This is not a simple task since the distribution function depends in general on seven variables (three spatial coordinates, three components of the velocity vector, and time) and the Boltzmann equation is a nonlinear integro-differential equation. The most efficient and fruitful method is Bird's Direct Simulation Monte Carlo (DSMC) method. In this method, the distribution function is represented by the velocities of a certain number of sample particles and collisions are generated stochastically with scattering rates and post-collision velocities determined from the Boltzmann equation.

By a concurrent use of the three tools mentioned above (Boltzmann equation for Maxwell molecules, kinetic models, and DSMC method), we intend in this book to provide an in-depth study about how a dilute gas responds under extreme nonequilibrium conditions. This response is basically characterized by *nonlinear transport* equations relating fluxes and hydrodynamic gradients through generalized transport coefficients that depend on the strength of the gradients. In general, these transport coefficients deviate significantly from the Navier–Stokes ones, reducing to the latter in the limit of zero gradients. The hydrodynamic fluxes are given by low-degree velocity moments and so probe the distribution of velocities of the order of the thermal velocity. However, many interesting phenomena (e.g. chemical reactions or other processes with a high activation energy) are strongly influenced by the population of particles with an energy much larger than the thermal energy. This motivates the analysis of high-degree moments and the high-velocity tail of the distribution function.

The above ambitious goals cannot be achieved for general nonequilibrium states and so one must consider specific situations. Here we have chosen to focus on laminar *shear flows* with the simplest geometry. The gas is enclosed between two infinite, parallel plates in relative motion along the x-direction and the gradients only appear along the y-direction orthogonal to the plates. The paradigmatic example of shear flow is characterized by a linear velocity profile and uniform density and temperature. We will refer to this state as uniform shear flow (USF) across this book. Despite its apparent simplicity, the USF can be sustained by the application of appropriate boundary conditions (namely generalized periodic boundary conditions) and displays a wealth of quite interesting properties for large shear rates. The price to be paid by the absence of gradients different from the constant shear rate is that the viscous heating is not counterbalanced by energy dissipation and, consequently, the temperature grows with time, unless a non-conservative external force is applied. A more realistic shear flow problem corresponds to a steady state in which the viscous heating is compensated for by a non-uniform heat flux. As a consequence, neither the density nor the temperature are uniform and, moreover, the velocity profile is not strictly linear. This second shear flow state will be referred to as planar Couette flow to avoid confusion with USF. The major difference between both shear flow problems lies in the presence of a coupling between energy and momentum transport in the Couette flow, whereas only momentum transport is present in the USF.

The non-Newtonian transport properties are determined from the shear-rate dependence of the stress tensor P_{ij}. The ratio between the shear stress P_{xy} and the shear rate defines the nonlinear shear viscosity, while the normal stress differences are measured by the viscometric functions. They are different in the USF and Couette flow states but exhibit qualitatively similar effects. Thus, the nonlinear shear viscosity decreases with the shear rate (shear-thinning behavior) and the normal stress P_{xx} along the flow direction is larger than the normal stress P_{yy} orthogonal to the shearing plane. In general, the expansion of the transport properties in powers of the shear rate, which is not but the Chapman–Enskog expansion specialized to the geometry of both shearing problems, is not convergent but asymptotic. This result may serve as an illustrative example to clarify the debated issue about the convergence of the Chapman–Enskog expansion. A physically interesting quantity not directly related to the rheological properties is the nonequilibrium entropy. We devote some effort to describe its nonlinear dependence on the shear rate. The results show that it remains larger than the extrapolation to finite shear rates of the entropy given by the Navier–Stokes approximation.

Apart from nonlinear transport, we provide the exact nonlinear shear-rate dependence of the fourth-degree velocity moments of a Maxwell gas described by the Boltzmann equation under USF. Quite interestingly, those moments *diverge* for shear rates larger than a certain critical value. Analogous singular behaviors are found for higher-degree moments, the associated critical shear rates decreasing as the degree of the moments grows. These singularities are consequences of an algebraic high-velocity tail of the velocity distribution function. On the other hand, this effect tends to be less pronounced as the interaction potential becomes more repulsive, due to the enhancement of the collision frequency of particles with high energy.

A feature of the Couette flow not shared by the USF is the existence of a nonzero heat flux $\mathbf{q}$. The component q_y of the heat flux parallel to the thermal gradient is simply proportional to this gradient, as in the classical Fourier's law. However, the proportionality constant is not the familiar thermal conductivity coefficient but a shear-rate dependent function accounting for the influence of shear flow on the energy transport. Additionally, there exists a component q_x of the heat flux parallel to the flow direction and hence orthogonal to the thermal gradient. This component, which is absent in Fourier's law, is proportional to the thermal gradient, the associated transport coefficient being again a nonlinear function of the shear rate. The physical origin of the component q_x lies in the nonlinear coupling between momentum and energy transport that appears far from equilibrium. For sufficiently large shear rates, the magnitude of q_x can even be larger than that of q_y.

The study of gas mixtures is important not only from a fundamental point of view but also from a practical point of view. Needless to say, most of the gases present in nature and industrial processes are constituted by particles of different species. The extension from the one-component to the multi-component case is not trivial at all since the number of physical quantities characterizing the nonequilibrium state in a mixture is higher than in a single gas and, besides, they depend on a larger number of parameters. At a kinetic theory level, one has to deal with a coupled set of nonlinear Boltzmann equations, one for each species. An important part of this book addresses nonlinear transport in shear flows for multi-component gases. Some of the findings obtained for a single gas remain qualitatively valid for mixtures, except that besides the shear-rate dependence of the transport coefficients, they are nonlinear functions of the parameters of the mixture (masses, composition, and sizes). However, there are new effects due to the larger complexity of the system. For instance, we find that for sufficiently mass-disparate mixtures, the nonlinear shear viscosity starts increasing with the shear rate (shear-thickening effect), reaches a maximum value, and decreases thereafter (shear-thinning effect). Another interesting and unexpected behavior takes place in a binary mixture under USF in the tracer limit, i.e. when the concentration of one of the species is much smaller than that of the other one. An exact solution of the Boltzmann equation shows that there exists a critical value of the shear rate beyond which the contribution of the tracer species to the total energy of the mixture is not negligible and in fact can even be larger than that of the solvent species. This phenomenon can be interpreted as a nonequilibrium phase transition in velocity space. Apart from the transport properties related to the momentum and energy fluxes, a mixture can exhibit diffusion processes when concentration gradients are present. In fact, diffusion (including self-diffusion) is the most characteristic transport problem in a mixture since it cannot appear if all the particles are absolutely identical. When the mixture is sheared, diffusion is affected by the shear flow and, as a consequence, Fick's law is generalized in the sense that the diffusion coefficient is replaced by a shear-rate diffusion tensor. The same applies to the Dufour effect, i.e. the heat flow induced by a concentration gradient.

The above few paragraphs summarize the main topics addressed in this monograph. We have attempted to provide a comprehensive and unified presentation of those topics, most of them scattered so far in many different papers. With the information explosion across journals, important results are getting lost to most of us and the connection between those results is even more difficult. Our work tries to mitigate this obstacle, at least in the specific field covered in this book. Besides, the book offers some aspects not previously reported, such as a more detailed treatment of the nonequilibrium entropy, some new comparisons with computer simulations, the description of a peculiar non-monotonic behavior of the shear stress, or the analysis of shear thickening in mixtures.

Our aim has been to develop the contents of the book with an approach somewhat intermediate between an extensive review article and a text. Similarities with the former are due to its exhaustive treatment of a seemingly narrow subject. However, it is more like the latter in that we have made every effort to present the results in a pedagogical and self-contained way and make connection with a broader context. Another feature of the book is the choice of topic. Every field has its continual advances to be reported, but nonlinear transport has been a field with few results until recently (except at the phenomenological level). Access to predictions from the Boltzmann equation far from equilibrium has been provided in the last two decades by theoretical and computational tools for specific but interesting states such as shear flow. From that point of view, it is important to emphasize that this book does not present a new field but a review of recent advances in an otherwise well-developed field.

The scope of the book pursues two major objectives. One is to understand quantitatively the rheology of a physical system, albeit one not closely related to real complex fluids. The second objective is to use this specific class of nonequilibrium states to explore and test conceptual issues that transcend nonlinear transport more generally. To that end, our analysis has intended to involve complementary and reinforcing methods: analytic, numerical, and simulational. Consequently, the results obtained may be considered as carefully controlled and unambiguous. This distinguishes our presentation from others that might emphasize only mathematical methods or realistic phenomenology. In this sense, our book does not supersede many excellent texts in the literature devoted to kinetic theory but may serve as a complement to them, especially in what concerns nonlinear transport in gas mixtures.

We hope that our monograph succeeds in providing an up-to-date overview of the exciting field of non-Newtonian behavior of rarefied gases. It can be read as a whole or can be used as a resource for selected topics from specific chapters. We expect that the text may be useful to students and researchers in nonequilibrium statistical-mechanics, kinetic theory of rarefied gases, irreversible thermodynamics, physical chemistry, chemical engineering, fluid mechanics, or applied mathematics. An attempt has been made to provide the bibliographical sources that have a more direct connection with our treatment. Of course, this is unavoidably a biased and subjective choice; we apologize for the omissions that may exist.

Chapter 1

KINETIC THEORY OF DILUTE GASES

1. Introduction

Kinetic theory of gases can be considered as a branch of statistical mechanics, although it preceded it from a historical point of view (Brush, 1965; Brush, 1976; Cercignani, 1998). The aim of kinetic theory is to explain, by using a statistical approach, the macroscopic properties of gases out of equilibrium from the knowledge of the laws governing the dynamics of the microscopic constituents (atoms or molecules). The first serious advances in kinetic theory began in the 18th Century with the work of Daniel Bernoulli (1700–1782), continued along the 19th Century with John Herapath (1790–1868), John J. Waterston (1811-1883), James P. Joule (1818–1889), August K. Krönig (1822–1879), and, especially, Rudolf Clausius (1822–1888). The theory reached its maturity with James C. Maxwell (1831–1879) and Ludwig Boltzmann (1844–1906). The former derived the velocity distribution function of a gas in equilibrium, obtained the transfer equations for mass, momentum, and energy in a dilute gas, and showed that the coefficients of shear viscosity and thermal conductivity of a dilute gas are independent of the density. Maxwell also analyzed a class of interaction models where the particles repel each other with a potential inversely proportional to the w-th power of the distance and found that the collisional rate of change of a given quantity involved the relative velocity raised to the power $(w-4)/w$; in the special case of $w=4$ the relative velocity drops out and the velocity integration can be carried out without the explicit knowledge of the velocity distribution function. This simple interaction model with $w=4$ defines the so-called Maxwell molecules.

The most fundamental contribution to the kinetic theory was provided by Boltzmann. In 1872 he derived his reputed equation for the rate of change of the velocity distribution function for a dilute gas arbitrarily far from equilibrium.

The essential assumption underlying the Boltzmann equation is the absence of statistical correlations between the pre-collisional velocities of two colliding particles ("molecular chaos" hypothesis or *Stosszahlansatz*). Boltzmann also introduced a functional of the distribution function that has the properties of a nonequilibrium *entropy*: its production is non-negative and vanishes only at equilibrium. This consequence of the Boltzmann equation is commonly known as the H-theorem and explains how molecular collisions make the system tend irreversibly towards equilibrium. The statistical nature of the H-theorem was not fully recognized by some of Boltzmann's contemporaries, who confronted it with paradoxes such as Loschmidt's and Zermelo's. For an enjoyable account of this exciting episode in the history of physics we refer the reader to the excellent books by Brush (1976) and Cercignani (1998).

The Boltzmann equation is a nonlinear integro-differential equation involving seven independent variables (the three components of position, the three components of velocity, and time). Its complexity prevented any significant advance until 1912, when the mathematician David Hilbert (1862–1943) proposed an approximate method to solve the Boltzmann equation by a series expansion around equilibrium in powers of the Knudsen number, the latter being defined as the ratio of the mean free path to a hydrodynamic or macroscopic length. A closely related method was developed independently in the period 1916–1917 by Sydney Chapman (1888–1970) and, especially, David Enskog (1884–1947). The idea behind the Chapman–Enskog method is to take the *local* equilibrium state as the reference state and expand around it in powers of the hydrodynamic gradients. Both methods allow to derive the Navier–Stokes equations and obtain expressions for the transport coefficients in terms of the interaction potential. The standard reference for the Chapman–Enskog theory, its applications, as well as the evaluation of collisional integrals is the book by Chapman and Cowling (1970).

For states not necessarily close to equilibrium, the Boltzmann equation is in general extremely difficult to solve. Most of the known exact solutions correspond to uniform states, especially in the case of Maxwell molecules (Ernst, 1981; Bobylev, 1988). An important breakthrough was the discovery by Bobylev (1975a, 1975b) and, independently, by Krook and Wu (1976, 1977) of a special *exact* solution to the Boltzmann equation for a uniform, isotropic gas of Maxwell molecules. This solution is known as the BKW-mode. On the other hand, a closed explicit solution is not known for any non-trivial inhomogeneous state. For Maxwell molecules, however, the hierarchy of moment equations can be recursively solved exactly in some specific inhomogeneous situations. This includes the so-called homo-energetic affine flows (Ikenberry and Truesdell, 1956; Truesdell, 1956; Truesdell and Muncaster, 1980; Galkin, 1995), in which the flow velocity is a linear function of the spatial coordinates and the fluxes are uniform, and the planar Fourier flow (Asmolov et al., 1979).

The crucial difficulties for solving the Boltzmann equation lie in the intricacy of the collision term, which incorporates the detailed mechanism of binary collisions. These mathematical difficulties have led to the proposal of *model* kinetic equations, in which the Boltzmann collision integral operator is replaced by a simpler term that, otherwise, retains the most important physical properties of the actual collision term. The simplest and more widely used kinetic model is the one proposed by Bhatnagar et al. (1954) and, independently, by Welander (1954). It is usually referred to as the BGK model and consists of replacing the Boltzmann collision term by a single relaxation-time term to the local equilibrium distribution. The mathematical tractability of the BGK model allows the possibility of deriving the explicit form of the velocity distribution function in some interesting physical situations, as will be shown along the book. From the knowledge of the distribution function one can compute the velocity moments and compare them with those obtained exactly from the Boltzmann equation, when available. The general good agreement found between both routes shows the reliability of the BGK model to study transport properties even far from equilibrium.

An alternative route to kinetic modeling is to look for numerical solutions to the Boltzmann equation. The method that has been proven to be more efficient and fruitful is Bird's Direct Simulation Monte Carlo (DSMC) method. For a detailed description of the method we refer the reader to the author's monograph (Bird, 1994). Simulation results complement the scarce exact results of the Boltzmann equation for Maxwell molecules. In general, one finds that the results obtained for Maxwell molecules can be extended in a semi-quantitative way to more general potentials by using appropriate reduced quantities.

In this chapter we will give an overview of the derivation and fundamental properties of the Boltzmann equation. Its content will be rather standard but somewhat biased towards the objectives of this monograph. Our main aim is to provide the definitions, notation, and basic information that will be needed to follow the remaining chapters, which constitute the core of the book. First, the heuristic derivation of the Boltzmann equation is presented and its basic properties (such as the conservation laws and the H-theorem) are analyzed. Then, the solution of the Boltzmann equation by the Chapman–Enskog expansion in the hydrodynamic gradients is worked out to the Navier–Stokes order. The generalization of the Boltzmann equation and the Navier–Stokes transport coefficients to a multi-component system is also carried out. Finally, we describe some of the kinetic models proposed to simplify the Boltzmann collision term, both for a single gas and for a gas mixture.

2. Derivation of the Boltzmann equation

In this Section we derive the Boltzmann equation by using classical heuristic arguments similar to those originally made by Boltzmann himself (Brush,

1965; Boltzmann, 1995). A more detailed exposition can be found in many excellent textbooks (Kogan, 1969; Chapman and Cowling, 1970; Ferziger and Kaper, 1972; Résibois and de Leener, 1977; Cercignani, 1988; McLennan, 1989; Woods, 1993).

Let us consider a dilute monatomic gas of $\mathcal{N}$ identical particles of mass m interacting through a pair-wise central interaction potential of finite range. At a kinetic level the relevant information on the state of the system is contained in the one-body velocity distribution function $f(\mathbf{r}, \mathbf{v}; t)$, defined so that $f(\mathbf{r}, \mathbf{v}; t) d\mathbf{r} d\mathbf{v}$ is the *most probable* (or average) number of particles which at time t lie in the volume element $d\mathbf{r}$ centered at the point $\mathbf{r}$ and moving with velocities in the range $d\mathbf{v}$ about $\mathbf{v}$. The average number of particles per unit volume is the local number density $n(\mathbf{r}, t)$, related to f through

$$n(\mathbf{r}, t) = \int d\mathbf{v}\, f(\mathbf{r}, \mathbf{v}; t). \tag{1.1}$$

The average velocity of those particles located around the point $\mathbf{r}$ at time t defines the local flow velocity $\mathbf{u}(\mathbf{r}, t)$ as

$$\mathbf{u}(\mathbf{r}, t) = \frac{1}{n(\mathbf{r}, t)} \int d\mathbf{v}\, \mathbf{v} f(\mathbf{r}, \mathbf{v}; t). \tag{1.2}$$

The average kinetic energy measured in the frame moving with the flow velocity defines a nonequilibrium local temperature $T(\mathbf{r}, t)$ as

$$\frac{3}{2} n(\mathbf{r}, t) k_B T(\mathbf{r}, t) = \frac{m}{2} \int d\mathbf{v}\, \left[\mathbf{v} - \mathbf{u}(\mathbf{r}, t)\right]^2 f(\mathbf{r}, \mathbf{v}; t), \tag{1.3}$$

where $k_B = 1.38054 \times 10^{-23}$ J/K is the Boltzmann constant. We assume that the density n, the temperature T, and the mass m are such that $n\lambda_B^3 \ll 1$, where $\lambda_B = \hbar / \sqrt{m k_B T / 2\pi}$ is the thermal de Broglie wavelength, $\hbar$ being the Planck constant divided by 2π. Therefore, a description based on classical mechanics is sufficient.

Our aim is to get the equation for the rate of change of f with time. In general, there are two sources of temporal change of f: free streaming and interactions, namely

$$\frac{\partial f}{\partial t} = \left.\frac{\partial f}{\partial t}\right|_{\text{str}} + \left.\frac{\partial f}{\partial t}\right|_{\text{int}}. \tag{1.4}$$

The first term on the right-hand-side represents the change of f due to the free motion of the particles under the action, in general, of an *external* force $\mathbf{F}$. The second term refers to the change due to the mutual interactions among the particles. To get the free streaming contribution, let us assume for the moment that the interactions are absent. Such a system is usually called a Knudsen gas. In that case, the particles that at time t are located at the point $\mathbf{r}$ and moving with velocity $\mathbf{v}$ will be located at $\mathbf{r}' = \mathbf{r}+\mathbf{v}\delta t$ having a velocity $\mathbf{v}' = \mathbf{v}+m^{-1}\mathbf{F}\delta t$ at a later time $t' = t+\delta t$, where δt is an infinitesimal time interval. Consequently,

$$f(\mathbf{r}, \mathbf{v}; t)d\mathbf{r}d\mathbf{v} = f(\mathbf{r}', \mathbf{v}'; t')d\mathbf{r}'d\mathbf{v}'. \tag{1.5}$$

Taking into account that

$$f(\mathbf{r}', \mathbf{v}'; t') = f(\mathbf{r}, \mathbf{v}; t) + \mathbf{v} \cdot \frac{\partial f}{\partial \mathbf{r}}\delta t + \frac{\mathbf{F}}{m} \cdot \frac{\partial f}{\partial \mathbf{v}}\delta t + \frac{\partial f}{\partial t}\delta t + \mathcal{O}(\delta t^2) \tag{1.6}$$

and that the Jacobian of the transformation $(\mathbf{r}, \mathbf{v}) \to (\mathbf{r}', \mathbf{v}')$ yields

$$d\mathbf{r}'d\mathbf{v}' = \left[1 + \frac{\partial}{\partial \mathbf{v}} \cdot \frac{\mathbf{F}}{m}\delta t + \mathcal{O}(\delta t^2)\right] d\mathbf{r}d\mathbf{v}, \tag{1.7}$$

Eq. (1.5) leads to

$$\left.\frac{\partial f}{\partial t}\right|_{\text{str}} = -\mathbf{v} \cdot \frac{\partial f}{\partial \mathbf{r}} - \frac{\partial}{\partial \mathbf{v}} \cdot \left(\frac{\mathbf{F}}{m} f\right). \tag{1.8}$$

In Eqs. (1.7) and (1.8) we have considered the possibility that the external force $\mathbf{F}$ may depend on the velocity (non-conservative force). Inserting Eq. (1.8) into Eq. (1.4), one gets

$$\frac{\partial f}{\partial t} + \mathbf{v} \cdot \nabla f + \frac{\partial}{\partial \mathbf{v}} \cdot \left(\frac{\mathbf{F}}{m} f\right) = \left.\frac{\partial f}{\partial t}\right|_{\text{int}}. \tag{1.9}$$

The interaction term is much more difficult to deal with than the free streaming term. From the exact Liouville equation one can derive a formal expression for $\partial f/\partial t|_{\text{int}}$ that involves the *two-body* distribution function (Ferziger and Kaper, 1972; Balescu, 1975; McLennan, 1989). In fact this is just the first equation of a hierarchy of equations (BBGKY hierarchy) where the s-body distribution function is coupled to the $(s+1)$-body distribution. This coupling makes Eq. (1.9) be a non closed equation for the velocity distribution function f. On the other hand, a closed equation can be obtained under some *statistical* hypotheses that are justified in the low-density limit. The resulting closed equation is the Boltzmann equation.

If the density of the gas is low enough, the average separation $n^{-1/3}$ between particles is much larger than the range σ of the interaction potential (i.e. $n\sigma^3 \ll 1$), and so any particle moves freely most of the time. However, eventually two particles come close enough to interact with each other and then they change their velocities. A simultaneous interaction of more than two particles is highly unlikely and can be neglected. Consequently, in a dilute gas the interactions among the particles reduce to a succession of elastic *binary collisions*. A collision takes place in a region of size of the order of σ and lasts a time of the order of $\tau_c \approx \sigma/v_0$, where $v_0 = \sqrt{2k_BT/m}$ is the thermal velocity.[1] The typical distance traversed by a given particle between two successive collisions is the *mean free path* $\ell \sim (n\sigma^2)^{-1}$ and the typical time elapsed is the *mean free time* $\tau \sim \ell/v_0$ (Reif, 1965). The condition $n\sigma^3 \ll 1$ then implies $\sigma \ll \ell$ and $\tau_c \ll \tau$, so the collisions are events well localized in space and time. The space and time scales over which the velocity distribution function varies are expected to be, at least, of the order of ℓ and τ, respectively. Then, when evaluating the rate of change of f due to the interactions, the collisions will be considered as instantaneous and spatially localized. To show up this reduction of the interactions to binary collisions, we will adopt the change of notation

$$\left.\frac{\partial f}{\partial t}\right|_{\text{int}} \rightarrow \left.\frac{\partial f}{\partial t}\right|_{\text{coll}}. \tag{1.10}$$

Before delving into the derivation of the Boltzmann collision term, it is instructive to give a brief account of the dynamics of binary collisions. Let us consider two colliding particles having velocities $\mathbf{v}$ and $\mathbf{v}_1$ before the collision. After the collision, the velocities change to $\mathbf{v}'$ and $\mathbf{v}_1'$, respectively, where the conservation laws of momentum and energy imply

$$\mathbf{v}' = \mathbf{v} - (\mathbf{g}\cdot\widehat{\boldsymbol{\sigma}})\widehat{\boldsymbol{\sigma}}, \quad \mathbf{v}_1' = \mathbf{v}_1 + (\mathbf{g}\cdot\widehat{\boldsymbol{\sigma}})\widehat{\boldsymbol{\sigma}}, \tag{1.11}$$

where $\mathbf{g} = \mathbf{v} - \mathbf{v}_1$ is the pre-collisional relative velocity and $\widehat{\boldsymbol{\sigma}}$ is a unit vector pointing on the apse line, i.e. the line joining the centers of the two particles at their closest approach. The geometry of the collision in a frame of reference moving with the particle with pre-collisional velocity $\mathbf{v}_1$ is sketched in Fig. 1.1. That particle acts as a target while the other one acts as the incident particle. In the figure we have taken the convention that $\widehat{\boldsymbol{\sigma}}$ points from the incident particle to the target particle at the closest approach. Note that $|\mathbf{g}'| = |\mathbf{g}|$ and $\mathbf{g}'\cdot\widehat{\boldsymbol{\sigma}} = -\mathbf{g}\cdot\widehat{\boldsymbol{\sigma}}$, where $\mathbf{g}' = \mathbf{v}' - \mathbf{v}_1'$ is the post-collisional relative velocity. The orientation of the vector $\mathbf{g}'$ can be specified by the polar angle χ and the azimuthal angle φ with the direction of $\mathbf{g}$ taken as the z-axis. The angle θ that the unit vector $\widehat{\boldsymbol{\sigma}}$ makes with $\mathbf{g}$ is related to the *scattering* angle χ by $2\theta = \pi - \chi$, so $2\sin\theta\cos\theta = \sin\chi$. Therefore, if $d\Omega = \sin\chi d\chi d\varphi$ is an element of solid angle about $\mathbf{g}'$ and $d\widehat{\boldsymbol{\sigma}} = \sin\theta d\theta d\varphi$ is an element of solid angle about $\widehat{\boldsymbol{\sigma}}$, one has

$$g d\Omega = 4|(\mathbf{g}\cdot\widehat{\boldsymbol{\sigma}})| d\widehat{\boldsymbol{\sigma}}. \tag{1.12}$$

[1] Note that $\tau_c \approx \sigma/v_0$ is only a rough estimate in some cases. For instance, for hard spheres of diameter σ the collisions are strictly instantaneous and so $\tau_c = 0$.

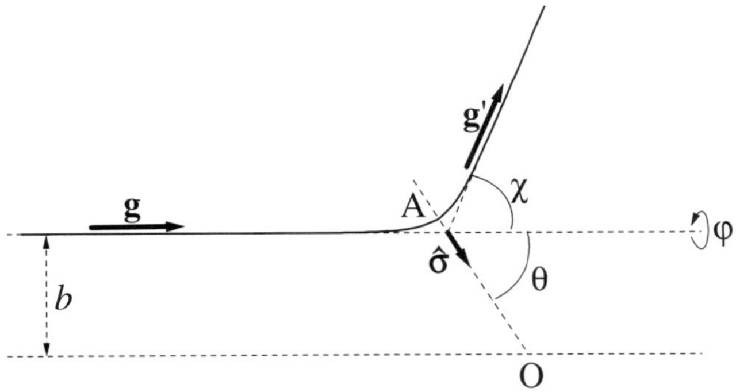

Figure 1.1. Sketch of the geometry of a binary collision. The point O is the center of the target particle. The incident particle approaches the target particle with a relative velocity **g** and an impact parameter b. The line joining the point A of closest approach and the point O is the apse line. After the interaction, the incident particle moves with a relative velocity $\mathbf{g}'$. The angle χ between the incident and deflected directions is the scattering angle. The trajectory depicted in the figure corresponds to a repulsive interaction potential $\phi(r) = Kr^{-8}$ with a reduced impact parameter $\beta \equiv (mg^2/16K)^{1/8}b = 0.7$. In this case, the distance of closest approach is $r_0 = 1.332b$ and the scattering angle is $\chi = 66.58°$.

Instead of the scattering angle χ, one can alternatively use the *impact parameter* b to specify a collision, as illustrated in Fig. 1.1. For a given interaction potential $\phi(r)$ and a given value of the magnitude of the relative velocity g, there exists a well defined relationship between χ and b (Goldstein, 1959). For instance, in the case of hard spheres of diameter σ, one simply has $b/\sigma = \sin\theta = \cos(\chi/2)$, regardless of the relative velocity. More in general,

$$\chi(b) = \pi - 2\int_{r_0}^{\infty} dr \frac{b/r^2}{[1-(b/r)^2 - 4\phi(r)/mg^2]^{1/2}}, \tag{1.13}$$

where r_0 is the distance at closest approach, which is given as the root of $1-(b/r)^2 - 4\phi(r)/mg^2$. In the special case of power-law repulsive potentials of the form $\phi(r) = Kr^{-w}$, the scattering angle χ depends on both b and g through the scaled dimensionless parameter $\beta = (mg^2/2wK)^{1/w}b$, namely

$$\chi(\beta) = \pi - 2\int_0^{\beta_0} d\beta' \left[1-\beta'^2 - \frac{2}{w}\left(\frac{\beta'}{\beta}\right)^w\right]^{-1/2}, \tag{1.14}$$

where $\beta_0(\beta) = b/r_0$ is the root of the quantity enclosed by brackets. Figure 1.2 shows $\cos(\chi/2)$ as a function of the reduced impact parameter β for several values of w. The value $w = 4$ defines the so-called Maxwell interaction, while the limit $w \to \infty$ corresponds to hard spheres. Except for hard spheres (in which case $\chi = 0$ for $\beta = b/\sigma > 1$, σ being the diameter of a sphere), the scattering angle χ vanishes in the limit $\beta \to \infty$ only. In some practical

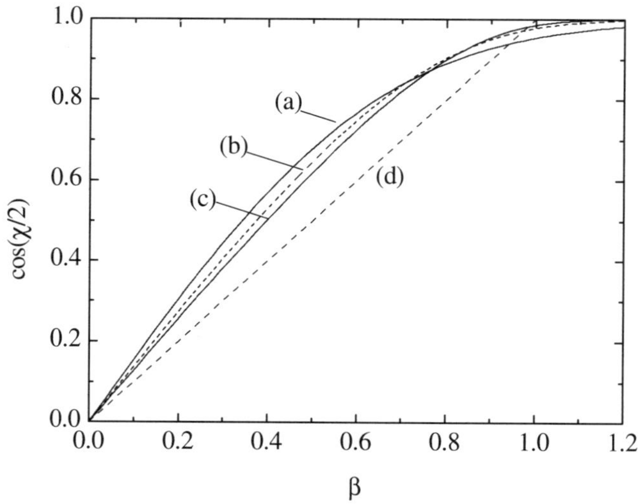

Figure 1.2. Plot of $\cos(\chi/2)$, where χ is the scattering angle, versus the reduced impact parameter β for repulsive potentials of the form $\phi(r) = Kr^{-w}$ with (a) $w = 4$ (Maxwell molecules), (b) $w = 8$, (c) $w = 12$, and (d) $w \to \infty$ (hard spheres).

applications, however, it is customary to introduce an angular cutoff, which corresponds to a certain maximum value of the reduced impact parameter β.

In the case that, instead of only one incident particle, we have a beam of incident particles, all moving with the same relative velocity g, the number of particles scattered per unit time into the element of solid angle $d\Omega$ about a direction characterized by the angles (χ, φ) is

$$\Phi B(g, \chi) d\Omega, \tag{1.15}$$

where Φ denotes the flux of incident particles (i.e. the number of incident particles per unit area and unit time) and

$$B(g, \chi) = \frac{b}{\sin\chi} \left| \frac{db}{d\chi} \right| \tag{1.16}$$

is the differential cross section (Goldstein, 1959). For hard spheres, $B(g, \chi) = \frac{1}{4}\sigma^2$ is independent of the relative velocity and the scattering angle. In the more general case of r^{-w}-potentials, the scaling in Eq. (1.14) implies that

$$B(g, \chi) \propto g^{-4/w}, \tag{1.17}$$

where the proportionality factor depends on the scattering angle. In particular, for Maxwell molecules ($w = 4$), the collision rate $gB(g, \chi)$ is independent of the relative velocity g. Note that the total cross section $\int d\Omega\, B(g, \chi)$ diverges if $w \leq 2$. Henceforth we will consider that $w > 2$.

The *collisional* change of the number of particles,

$$f(\mathbf{r}, \mathbf{v}; t)d\mathbf{r}d\mathbf{v} \to f(\mathbf{r}, \mathbf{v}; t + \delta t)d\mathbf{r}d\mathbf{v}, \tag{1.18}$$

during the time interval $(t, t + \delta t)$ is due to two separate processes. On the one hand, those particles having a velocity $\mathbf{v}$ at time t and suffering a collision during the time step δt will change their velocities. Consequently, those particles will not contribute to $f(\mathbf{r}, \mathbf{v}; t + \delta t)$. On the other hand, there are particles with a velocity $\mathbf{v}'$ different from $\mathbf{v}$ at time t which may suffer collisions and have a post-collisional velocity $\mathbf{v}$ at time $t + \delta t$. Consequently, those particles will contribute to $f(\mathbf{r}, \mathbf{v}; t + \delta t)$. The first process represents the *loss* term L, while the second one represents the *gain* term G:

$$\left.\frac{\partial f}{\partial t}\right|_{\text{coll}} = G - L. \tag{1.19}$$

Let us begin with the loss term. We consider a target particle with velocity $\mathbf{v}_1$. The flux of incident particles with a velocity in the range $d\mathbf{v}$ about $\mathbf{v}$ is

$$\Phi = g f(\mathbf{v})d\mathbf{v}. \tag{1.20}$$

According to Eq. (1.15), the number of particles scattered into the element $d\Omega$ by the target during a time interval δt is $gf(\mathbf{v})B(g, \chi)d\Omega d\mathbf{v}\delta t$. Since there is not only one target particle with velocity $\mathbf{v}_1$, but there are $f(\mathbf{v}_1)d\mathbf{v}_1$ particles per unit volume with velocity between $\mathbf{v}_1$ and $\mathbf{v}_1 + d\mathbf{v}_1$, then the number of particles scattered due to encounters of type $(\mathbf{v}, \mathbf{v}_1)$ per unit volume is

$$gf(\mathbf{v})f(\mathbf{v}_1)B(g, \chi)d\Omega d\mathbf{v}_1 d\mathbf{v}\delta t. \tag{1.21}$$

The *net* number of scattered particles is obtained by integrating over all the scattering directions and over all the velocities of the target particles. The rate of loss (per unit volume) of particles with velocity between $\mathbf{v}$ and $\mathbf{v} + d\mathbf{v}$ is therefore

$$L d\mathbf{v} = \int d\mathbf{v}_1 \int d\Omega\, g B(g, \chi) f(\mathbf{v}) f(\mathbf{v}_1) d\mathbf{v}. \tag{1.22}$$

The rate of gain can be evaluated in a similar way. Consider an inverse collision such that, for a given unit vector $-\widehat{\boldsymbol{\sigma}}$, the post-collisional velocities are $(\mathbf{v}, \mathbf{v}_1)$. The corresponding pre-collisional velocities $(\mathbf{v}', \mathbf{v}_1')$ are given by Eq. (1.11). Thus, following similar steps as those made in the case of Eq. (1.22), we get

$$G d\mathbf{v}' = \int d\mathbf{v}_1' \int d\Omega\, g' B(g', \chi) f(\mathbf{v}') f(\mathbf{v}_1') d\mathbf{v}'. \tag{1.23}$$

Taking into account that $g' = g$ and that the Jacobian of the transformation $(\mathbf{v}', \mathbf{v}_1') \to (\mathbf{v}, \mathbf{v}_1)$ is equal to 1, we get

$$\begin{aligned} \left.\frac{\partial f}{\partial t}\right|_{\text{coll}} &= \int d\mathbf{v}_1 \int d\Omega\, g B(g,\chi) \left[f(\mathbf{v}')f(\mathbf{v}_1') - f(\mathbf{v})f(\mathbf{v}_1)\right] \\ &\equiv J[\mathbf{v}|f,f]. \end{aligned} \tag{1.24}$$

Insertion of this into Eq. (1.9) finally yields the *Boltzmann equation*

$$\frac{\partial f}{\partial t} + \mathbf{v}\cdot\nabla f + \frac{\partial}{\partial \mathbf{v}}\cdot\left(\frac{\mathbf{F}}{m}f\right) = J[\mathbf{v}|f,f]. \tag{1.25}$$

The conventional notation $J[\mathbf{v}|f,f]$ means that the Boltzmann collision term is a bilinear operator acting on the velocity distribution function f. An equivalent form for the Boltzmann collision operator $J[\mathbf{v}|f,f]$ can be obtained by using Eq. (1.12), in which case the integration over $\widehat{\boldsymbol{\sigma}}$ is restricted to a hemisphere (for instance the one with $\mathbf{g}\cdot\widehat{\boldsymbol{\sigma}} \geq 0$). In particular, for hard spheres of diameter σ, one has $B(g,\chi) = \frac{1}{4}\sigma^2$ and therefore

$$J[\mathbf{v}|f,f] = \sigma^2 \int d\mathbf{v}_1 \int d\widehat{\boldsymbol{\sigma}}\, \Theta(\mathbf{g}\cdot\widehat{\boldsymbol{\sigma}})(\mathbf{g}\cdot\widehat{\boldsymbol{\sigma}}) \left[f(\mathbf{v}')f(\mathbf{v}_1') - f(\mathbf{v})f(\mathbf{v}_1)\right], \tag{1.26}$$

where $\Theta(x)$ is Heaviside's step function. Alternatively, we can also use Eq. (1.16) to replace $B(g,\chi)d\Omega$ by $b\,db\,d\varphi$ in Eq. (1.24). The Boltzmann equation is a nonlinear integro-differential equation for f as a function of the three spatial coordinates, the three velocity components, and time. It must be solved subjected to a given initial condition $f(\mathbf{r},\mathbf{v};0)$ and given boundary conditions.

Upon deriving the Boltzmann equation, we have implicitly made use of a number of hypotheses. First, it has been assumed that the time interval δt is such that $\tau_c \ll \delta t \ll \tau$, where, as said before, τ_c is the typical duration of a collision and τ is the mean free time. The condition $\delta t \ll \tau$ partially justifies the steps from Eq. (1.20) to Eq. (1.21), since one can assume that the collision cylinder contains at most one particle with velocity $\mathbf{v}$ and that multiple scattering can be ignored. On the other hand, the condition $\delta t \gg \tau_c$ allows us to consider the collisions as instantaneous events, so that all the functions f appearing in the Boltzmann collision term are evaluated at the same time t. This makes the Boltzmann equation to describe a Markovian process. In addition, it has been assumed that the mean free path ℓ is much larger than the range of the interaction σ and, consequently, all the functions f appearing in the Boltzmann collision term are evaluated at the same point $\mathbf{r}$. These assumptions are supported by the low-density condition $n\sigma^3 \ll 1$ and so they are not problematic. In fact, they are not in contradiction with mechanics at this stage. On the other hand, there is a crucial assumption in Eq. (1.21) which departs from the laws of mechanics and has a *statistical* nature. Strictly speaking, the product $f(\mathbf{r},\mathbf{v})f(\mathbf{r},\mathbf{v}_1)$ in

Eq. (1.21) should be replaced by $f_2(\mathbf{r}, \mathbf{v}; \mathbf{r}, \mathbf{v}_1)$, where $f_2(\mathbf{r}, \mathbf{v}; \mathbf{r}_1, \mathbf{v}_1)$ is the *two-body* distribution function, representing the number of pairs of particles located simultaneously at positions $(\mathbf{r}, \mathbf{r}_1)$ and moving with velocities $(\mathbf{v}, \mathbf{v}_1)$. Therefore, in Eq. (1.21) it has been assumed that

$$f_2(\mathbf{r}, \mathbf{v}; \mathbf{r}, \mathbf{v}_1) = f(\mathbf{r}, \mathbf{v}) f(\mathbf{r}, \mathbf{v}_1) \tag{1.27}$$

for any pair of *pre-collisional* velocities $\mathbf{v}$ and $\mathbf{v}_1$. Equation (1.27) implies that there are no correlations between the velocities of two particles that are about to collide. This is the famous Boltzmann's hypothesis of molecular chaos (or *Stosszahlansatz*), which allows the Boltzmann equation (1.25) to be a closed equation for the one-body distribution function. While the Stosszahlansatz is in contradiction with mechanics (since interactions create velocity correlations), it must be interpreted in terms of probabilistic arguments. The search for a rigorous justification of the molecular chaos hypothesis has received a great deal of attention (Balescu, 1975; Dorfman and van Beijeren, 1977; Cercignani et al., 1994; Cercignani, 1998). At an intuitive level, one can realize that, in a very dilute gas, when two particles collide it is very unlikely that they have met before, either directly or indirectly through a common set of other particles, so their velocities can be considered as uncorrelated. It is important to stress that the molecular chaos assumption (1.27) is needed only for particles that are about to collide; after the collision, the scattered velocities are strongly correlated.

For a dense gas, the Boltzmann equation is not expected to be appropriate. On the other hand, Enskog proposed in 1922 (Brush, 1965) a heuristic modification of the Boltzmann equation for a system of hard spheres accounting for finite-density effects. First, the Enskog equation takes into account that the centers of the two colliding spheres are separated by a distance equal to the diameter σ. This incorporates the transport of momentum and energy occurring across finite distances in collisions, which is important in dense gases and liquids. Second, the collision frequency is increased by a factor χ associated with the spatial correlations due to excluded volume effects. Otherwise, the assumption of molecular chaos is maintained. The Enskog collision operator is (Chapman and Cowling, 1970; Ferziger and Kaper, 1972; van Beijeren and Ernst, 1973; Dorfman and van Beijeren, 1977)

$$\begin{aligned} J_E[\mathbf{r}, \mathbf{v}|f, f] &= \sigma^2 \int d\mathbf{v}_1 \int d\widehat{\boldsymbol{\sigma}}\, \Theta(\mathbf{g} \cdot \widehat{\boldsymbol{\sigma}})(\mathbf{g} \cdot \widehat{\boldsymbol{\sigma}}) \left\{ \chi[\mathbf{r}, \mathbf{r} - \boldsymbol{\sigma}|n] f(\mathbf{r}, \mathbf{v}') \right. \\ & \quad \left. \times f(\mathbf{r} - \boldsymbol{\sigma}, \mathbf{v}_1') - \chi[\mathbf{r}, \mathbf{r} + \boldsymbol{\sigma}|n] f(\mathbf{r}, \mathbf{v}) f(\mathbf{r} + \boldsymbol{\sigma}, \mathbf{v}_1) \right\}, \end{aligned} \tag{1.28}$$

where $\boldsymbol{\sigma} = \sigma\widehat{\boldsymbol{\sigma}}$, and $\chi[\mathbf{r}, \mathbf{r} \pm \boldsymbol{\sigma}|n]$ is the equilibrium pair correlation function at contact corresponding to the density field $n(\mathbf{r})$.

3. General properties of the Boltzmann equation. H-theorem

Let $\psi(\mathbf{v})$ be an arbitrary function of the velocity. Its associated (local) density is

$$\Psi(\mathbf{r},t)=\int d\mathbf{v}\,\psi(\mathbf{v})f(\mathbf{r},\mathbf{v};t). \tag{1.29}$$

For instance, if $\psi(\mathbf{v})=\{1,\mathbf{v},v^2\}$, then the corresponding densities are the hydrodynamic fields $\Psi=\{n,n\mathbf{u},3nk_BT/m+nu^2\}$. The evolution equation for Ψ can be obtained from the Boltzmann equation (1.25) by multiplying both sides by $\psi(\mathbf{v})$ and integrating over the velocity. The result can be written as

$$\frac{\partial\Psi}{\partial t}+\nabla\cdot\mathbf{\Phi}_\psi=\sigma_\psi^{(F)}+\sigma_\psi^{(c)}, \tag{1.30}$$

where

$$\mathbf{\Phi}_\psi=\int d\mathbf{v}\,\mathbf{v}\psi(\mathbf{v})f(\mathbf{v}) \tag{1.31}$$

is the *flux* of the quantity ψ,

$$\sigma_\psi^{(F)}=\int d\mathbf{v}\,\frac{\partial\psi}{\partial\mathbf{v}}\cdot\frac{\mathbf{F}}{m}f(\mathbf{v}) \tag{1.32}$$

is a production term associated with the external force, and

$$\begin{aligned}\sigma_\psi^{(c)} &= \int d\mathbf{v}\,\psi(\mathbf{v})J[\mathbf{v}|f,f]\\ &= \int d\mathbf{v}\int d\mathbf{v}_1\int d\Omega\,gB(g,\chi)\psi(\mathbf{v})\left[f(\mathbf{v}')f(\mathbf{v}_1')-f(\mathbf{v})f(\mathbf{v}_1)\right]\end{aligned} \tag{1.33}$$

is the production term due to the collisions. Equation (1.30) has the well-known structure of a local *balance* equation. If the total source $\sigma_\psi^{(F)}+\sigma_\psi^{(c)}$ vanishes, Eq. (1.30) expresses the conservation of the quantity $\psi(\mathbf{v})$. Upon obtaining Eq. (1.30) we have considered that $\psi(\mathbf{v})$ does not depend explicitly on time and position.

The production term $\sigma_\psi^{(c)}$ can be expressed in a more symmetrical form than in Eq. (1.33). First, we make the change of variables $\mathbf{v}\to\mathbf{v}_1$ and $\mathbf{v}_1\to\mathbf{v}$. The resulting expression is unchanged, except that $\psi(\mathbf{v})\to\psi(\mathbf{v}_1)$. Taking the arithmetic mean of both equivalent expressions, one has

$$\begin{aligned}\sigma_\psi^{(c)} &= \frac{1}{2}\int d\mathbf{v}\int d\mathbf{v}_1\int d\Omega\,gB(g,\chi)\left[\psi(\mathbf{v})+\psi(\mathbf{v}_1)\right]\\ &\quad\times\left[f(\mathbf{v}')f(\mathbf{v}_1')-f(\mathbf{v})f(\mathbf{v}_1)\right].\end{aligned} \tag{1.34}$$

Next, we change $\mathbf{v} \to \mathbf{v}'$ and $\mathbf{v}_1 \to \mathbf{v}_1'$, which implies that $\mathbf{v}' \to \mathbf{v}$ and $\mathbf{v}_1' \to \mathbf{v}_1$. Since $d\mathbf{v}'d\mathbf{v}_1' = d\mathbf{v}d\mathbf{v}_1$ and $g' = g$, it turns out that we arrive to an expression similar to Eq. (1.34), except that $\psi(\mathbf{v}) + \psi(\mathbf{v}_1) \to \psi(\mathbf{v}') + \psi(\mathbf{v}_1')$ and there is a change of sign. Taking again the arithmetic mean, we finally get

$$\begin{aligned} \sigma_\psi^{(c)} &= \frac{1}{4}\int d\mathbf{v}\int d\mathbf{v}_1\int d\Omega\, gB(g,\chi)\left[\psi(\mathbf{v})+\psi(\mathbf{v}_1)-\psi(\mathbf{v}')-\psi(\mathbf{v}_1')\right] \\ &\quad\times\left[f(\mathbf{v}')f(\mathbf{v}_1')-f(\mathbf{v})f(\mathbf{v}_1)\right]. \end{aligned} \tag{1.35}$$

An important consequence of Eq. (1.35) is that $\sigma_\psi^{(c)} = 0$ for any function $\psi(\mathbf{v})$ such that

$$\psi(\mathbf{v}) + \psi(\mathbf{v}_1) - \psi(\mathbf{v}') - \psi(\mathbf{v}_1') = 0 \tag{1.36}$$

for any pair of velocities $(\mathbf{v}, \mathbf{v}_1)$ and any unit vector $\widehat{\boldsymbol{\sigma}}$. A quantity satisfying this condition is called a *collisional invariant*. Obviously, $\psi = 1$, $\psi = v_i$ $(i = x, y, z)$, and $\psi = v^2$ are collisional invariants. In general, it can be proved (Grad, 1949; Balescu, 1975; Cercignani et al., 1994) that all the collisional invariants are linear combinations of the above five functions, i.e.

$$\psi(\mathbf{v}) = A_0 + \mathbf{A}_1 \cdot \mathbf{v} + A_2 v^2. \tag{1.37}$$

Taking $\psi = 1$ in Eq. (1.30), one obtains the continuity equation

$$D_t n + n\nabla \cdot \mathbf{u} = 0, \tag{1.38}$$

where $D_t \equiv \partial_t + \mathbf{u}\cdot\nabla$ is the material time derivative, i.e. the time derivative in the Lagrangian frame of reference moving with the fluid. In the case $\psi = mv_i$, one gets the balance equation for momentum,

$$\rho D_t \mathbf{u} + \nabla \cdot \mathsf{P} = \boldsymbol{\sigma}_u, \tag{1.39}$$

where $\rho = mn$ is the mass density,

$$\mathsf{P} = m\int d\mathbf{v}\, \mathbf{V}\mathbf{V} f(\mathbf{v}) \tag{1.40}$$

is the pressure or stress tensor (i.e. the momentum flux measured in the Lagrangian frame), $\mathbf{V} \equiv \mathbf{v} - \mathbf{u}$ being the so-called peculiar velocity, and

$$\boldsymbol{\sigma}_u = \int d\mathbf{v}\, \mathbf{F} f(\mathbf{v}) \tag{1.41}$$

is the production of momentum due to the external force. The hydrostatic pressure p is defined as one third of the trace of the stress tensor, $p = \frac{1}{3}\mathrm{tr}\,\mathsf{P}$. Then, Eq. (1.3) implies that

$$p(\mathbf{r}, t) = n(\mathbf{r}, t) k_B T(\mathbf{r}, t). \tag{1.42}$$

Equation (1.42) is a generalization to nonequilibrium states of the equation of state of an ideal gas. The traceless tensor

$$\Pi_{ij} = P_{ij} - p\,\delta_{ij} \tag{1.43}$$

is the *irreversible* part of the pressure tensor. Here, δ_{ij} is Kronecker's delta. Finally, if $\psi = \frac{1}{2}mv^2$, one arrives at the balance equation for energy, namely

$$\frac{3}{2}nk_B D_t T + \nabla\cdot\mathbf{q} + \mathsf{P}:\nabla\mathbf{u} = \sigma_T, \tag{1.44}$$

where

$$\mathbf{q} = \frac{m}{2}\int d\mathbf{v}\, V^2\mathbf{V} f(\mathbf{v}) \tag{1.45}$$

is the heat flux (or the energy flux in the Lagrangian frame) and

$$\sigma_T = \int d\mathbf{v}\, \mathbf{V}\cdot\mathbf{F} f(\mathbf{v}) \tag{1.46}$$

is the production of energy due to the external force. In the particular case of an external force independent of the velocity, $\boldsymbol{\sigma}_u = n\mathbf{F}$ and $\sigma_T = 0$, while for a non-conservative force proportional to the peculiar velocity, $\mathbf{F} = \zeta m\mathbf{V}$, one has $\boldsymbol{\sigma}_u = \mathbf{0}$ and $\sigma_T = 3\zeta nk_B T$. Equations (1.38), (1.39), and (1.44) constitute the fundamental conservation equations of classical fluid dynamics. They provide the time variation of the local densities of the conserved quantities (mass, momentum, and energy) as the divergence of the corresponding fluxes plus the sources produced by the action of the external field. The conservation equations are not closed equations, unless the momentum and heat fluxes are expressed in terms of the hydrodynamic fields n, $\mathbf{u}$, and T. This point will be addressed in the next Section. It must be noted that the balance equations (1.38), (1.39), and (1.44) actually apply for any range of fluid densities. In that case, apart from the kinetic contributions (1.40) and (1.45) to the fluxes P and $\mathbf{q}$, there exist collisional transfer contributions which require the knowledge of the two-body distribution function (Résibois and de Leener, 1977; McLennan, 1989). However, in a dilute gas the collisional transfer contributions can be neglected against their kinetic contributions.

Now we are going to focus on the most important property of the Boltzmann equation: the *irreversible* tendency to equilibrium of an isolated rarefied gas. The key point is the introduction of the Boltzmann entropy (per unit volume) s as a functional of the velocity distribution function:

$$s(\mathbf{r},t) = -k_B\int d\mathbf{v}\, f(\mathbf{r},\mathbf{v};t)\ln f(\mathbf{r},\mathbf{v};t) + C(\mathbf{r},t), \tag{1.47}$$

where

$$C(\mathbf{r},t) = n(\mathbf{r},t)k_B\left(3\ln\frac{m}{2\pi\hbar} + 1\right), \tag{1.48}$$

$\hbar$ being the Planck constant divided by 2π. The additive term C has been included to recover the equilibrium expression of the entropy of an ideal gas,

$$s_0(n, e) = nk_B \left[\ln \frac{(me/3\pi n)^{3/2}}{n\hbar^3} + \frac{5}{2} \right], \tag{1.49}$$

where $e = \frac{3}{2}nk_BT$ is the internal energy per unit volume, when f is the Maxwell–Boltzmann equilibrium distribution

$$f_{\text{eq}}(\mathbf{v}) = n \left(\frac{m}{2\pi k_B T} \right)^{3/2} e^{-mv^2/2k_BT}. \tag{1.50}$$

The total entropy $S(t)$ is obtained as

$$S(t) = \int d\mathbf{r}\, s(\mathbf{r}, t). \tag{1.51}$$

The balance equation for the entropy density s can be easily obtained by following steps similar to those made in Eq. (1.30). The result is

$$D_t s + s\nabla \cdot \mathbf{u} + \nabla \cdot \mathbf{\Phi}_s = \sigma_s^{(F)} + \sigma_s^{(c)}, \tag{1.52}$$

where

$$\mathbf{\Phi}_s = -k_B \int d\mathbf{v}\, \mathbf{V} f \ln f \tag{1.53}$$

is the entropy flux in the Lagrangian frame. The entropy production due to the external force can be obtained from Eq. (1.32) with the result

$$\sigma_s^{(F)} = k_B \int d\mathbf{v} f \frac{\partial}{\partial \mathbf{v}} \cdot \frac{\mathbf{F}}{m}. \tag{1.54}$$

It vanishes for forces independent of the velocity. In the remainder of this Section, we will consider that the system is isolated and so $\mathbf{F} = \mathbf{0}$, although the subsequent analysis is essentially unchanged if $\mathbf{F}$ is independent of the velocity or in the special case of a "drag" force of the form $\mathbf{F} \propto \mathbf{V}$ (Garzó et al., 1990). Therefore, the evolution equation for the total entropy is

$$\frac{\partial S}{\partial t} = \int d\mathbf{r}\, \sigma_s^{(c)}(\mathbf{r}, t), \tag{1.55}$$

where it has been assumed that, since the system is isolated, no entropy flux exists across the boundaries.

The entropy production due to collisions is given by Eq. (1.35) with $\psi(\mathbf{v}) = -k_B \ln f(\mathbf{v})$:

$$\begin{aligned} \sigma_s^{(c)} &= \frac{k_B}{4} \int d\mathbf{v} \int d\mathbf{v}_1 \int d\Omega\, gB(g, \chi) \\ &\quad \times \left[f(\mathbf{v}')f(\mathbf{v}_1') - f(\mathbf{v})f(\mathbf{v}_1) \right] \ln \frac{f(\mathbf{v}')f(\mathbf{v}_1')}{f(\mathbf{v})f(\mathbf{v}_1)}. \end{aligned} \tag{1.56}$$

Since for any $x > 0$ and $y > 0$ one has the inequality $(x - y)\ln(x/y) \geq 0$, where the equality only holds if $x = y$, and given that $g > 0$, $B > 0$, and $f \geq 0$, it follows that

$$\sigma_s^{(c)} \geq 0. \tag{1.57}$$

The equality in the above relation is possible if and only if

$$\ln f(\mathbf{v}') + \ln f(\mathbf{v}_1') = \ln f(\mathbf{v}) + \ln f(\mathbf{v}_1). \tag{1.58}$$

That means that $\ln f(\mathbf{v})$ is a collisional invariant, and therefore, must have the structure of Eq. (1.37). The coefficients A_0, $\mathbf{A}_1$, and A_2 can be expressed in terms of the hydrodynamic fields n, $\mathbf{u}$, and T through Eqs. (1.1)–(1.3). As a consequence, the only distribution function for which the collisional production of entropy $\sigma_s^{(c)}$ vanishes has the form

$$f_L(\mathbf{r}, \mathbf{v}; t) = n(\mathbf{r}, t) \left[\frac{m}{2\pi k_B T(\mathbf{r}, t)}\right]^{3/2} \exp\left\{-\frac{m\left[\mathbf{v} - \mathbf{u}(\mathbf{r}, t)\right]^2}{2k_B T(\mathbf{r}, t)}\right\}. \tag{1.59}$$

This distribution is called the *local equilibrium* distribution function. According to Eq. (1.58), it has the property $J[f_L, f_L] = 0$. In general, however, the local equilibrium distribution f_L with non-uniform fields is not a solution of the Boltzmann equation (1.25).[2] On the other hand, Eq. (1.59) is a physically relevant solution when the hydrodynamic fields are stationary and uniform, in which case f_L becomes the Maxwell–Boltzmann *equilibrium* distribution (1.50), where by a Galilean change of frame we have taken $\mathbf{u} = \mathbf{0}$. For any initial condition $f(\mathbf{r}, \mathbf{v}; 0)$, the solution of the Boltzmann equation is such that the total entropy monotonically increases with time until reaching a maximum plateau corresponding to the equilibrium state. Even if the system is initially prepared with a local equilibrium distribution (and so the initial entropy production is zero), the time derivative $\partial_t f = -\mathbf{v} \cdot \nabla f \neq 0$ at $t = 0^+$, the system leaves the local equilibrium state, and the entropy production becomes positive.

Equation (1.57) is a formulation of the celebrated Boltzmann's H-theorem, which shows that any nonequilibrium state of an isolated gas evolves irreversibly towards the equilibrium state, monotonically increasing its entropy. This provides a statistical or probabilistic interpretation of the second principle of thermodynamics. Of course, the hypothesis of molecular chaos plays a crucial role in its derivation from the Boltzmann equation.

---

[2]It can be proved that the local equilibrium distribution is a solution of the (collisionless) Boltzmann equation in special nonequilibrium states where the temperature T is uniform, the flow velocity $\mathbf{u}$ is a linear function of $\mathbf{r}$, and $\ln n$ is a quadratic function of $\mathbf{r}$ (Grad, 1949). The time evolution is such that T^{-1} is a quadratic function of t, $\mathbf{u}/T$ is a linear function of t, and $\ln\left(nT^{-3/2}e^{-mu^2/2k_BT}\right)$ is stationary. This represents situations, such as a rigid body rotation or a radial free expansion, that are not of physical interest.

To close this Section, it is appropriate to say a few words about the moment equations stemming from the Boltzmann equation. When $\psi(\mathbf{v})$ in Eqs. (1.29), (1.31), and (1.33) is a polynomial of degree k, then the integral Ψ is a kth-degree moment of the distribution, $\mathbf{\Phi}_\psi$ is a $(k+1)$-th degree moment, and $\sigma_\psi^{(c)}$ is the corresponding collisional moment. In general, the determination of $\sigma_\psi^{(c)}$ requires the knowledge of the whole distribution function or, equivalently, of its infinite set of moments. However, for Maxwell molecules the collision rate $gB(g,\chi)$ is independent of the relative velocity g and, as a consequence, a collisional moment of degree k involves only moments of degree equal to or smaller than k. The collisional moments of degrees 2, 3, and 4 for a single gas and of degrees 1 and 2 for a multi-component gas are given in Appendices 1.A and 1.B, respectively. Even in the case of Maxwell molecules, the equation for a moment of degree k involves moments of degree $k+1$ due to the free streaming term $\mathbf{v}\cdot\nabla$ of the Boltzmann equation. Therefore, in general one has to deal with an infinite *hierarchy* of moment equations. To illustrate this point, let us assume, for simplicity, that the spatial dependence of the velocity distribution is restricted to the variable x. Then, the evolution equation for the moment

$$M_k = \int d\mathbf{v}\, V_x^k f(\mathbf{v}) \tag{1.60}$$

reads (in the absence of an external force)

$$\begin{aligned}&\frac{\partial}{\partial t}M_k + \frac{\partial}{\partial x}\left(M_{k+1}+u_x M_k\right)\\&+k\left(M_k+u_x M_{k-1}\right)\frac{\partial u_x}{\partial x} = \int d\mathbf{v}\, V_x^k J[\mathbf{v}|f,f].\end{aligned} \tag{1.61}$$

On the other hand, if the moments M_k are spatially uniform, i.e. all the spatial dependence of f occurs through the flow velocity, then the left-hand side of Eq. (1.61) only involves moments of degrees k and $k-1$, so that the hierarchy can be recursively solved, at least for Maxwell molecules.

In order to get an approximate solution of the hierarchy of moment equations, Grad (1949) devised a widely used method. The idea of Grad's method is to expand the distribution function f in a complete set of orthogonal polynomials (essentially Hermite polynomials), the coefficients being the corresponding velocity moments. Next, this expansion is truncated by retaining terms up to a given order k, and so the (orthogonal) moments of degree higher than k are neglected. When this truncated expansion is used to compute the collisional moments, one finally gets a *closed* set of moment equations up to degree k. In the usual thirteen-moment approximation, the expansion includes the density, the three components of the flow velocity, the six elements of the pressure tensor, and the three components of the heat flux:

$$f(\mathbf{v}) = f_L(\mathbf{v}) \left\{ 1 + \frac{m}{n(k_B T)^2} \left[\left(\frac{mV^2}{5k_B T} - 1 \right) \mathbf{V} \cdot \mathbf{q} + \frac{1}{2} \Pi_{ij} V_i V_j \right] \right\}. \tag{1.62}$$

This approximation leaves out the seven third-degree moments not included in the heat flux, as well as all the moments of degree higher than or equal to four.

4. Chapman–Enskog expansion

As said in the previous Section, the derivation of the Boltzmann equation requires the existence of two well separated length scales: $\sigma \ll \ell$, where σ is the range of the interaction potential and ℓ is the mean free path. In addition, in an inhomogeneous gas, one can introduce a characteristic *hydrodynamic* length h, which is the typical range over which the distribution function changes appreciably. Of course, h cannot be larger than the size of the system L. Associated with σ, ℓ, and h there are three time scales: the duration of a collision $\tau_c = \sigma/v_0$, the mean free time $\tau = \ell/v_0$, and the hydrodynamic time $\tau_h = h/v_0$, where $v_0 = \sqrt{2k_B T/m}$ is the thermal velocity. Let us now suppose that $\ell \ll h$ or, equivalently, $\tau \ll \tau_h$. Then the following picture of the time evolution of the gas arises (Balescu, 1975; Dorfman and van Beijeren, 1977). Consider a time t after the initial state such that $\tau \ll t \ll \tau_h$, so in a small region (of size much smaller than h) around a point $\mathbf{r}$ the particles have had time to suffer a number of collisions. Due to these collisions, it is expected that the velocity distribution function in that region reaches a state close to that of local equilibrium, characterized by the local values of n, $\mathbf{u}$, and T, which may vary from one region to another. For much longer times, $t \gg \tau_h$, the particles have travelled distances much larger than the hydrodynamic length and, consequently, the variation of the hydrodynamic fields across the system have smoothed out and the gas finally achieves the total equilibrium state.

In this picture, we can distinguish two separate processes. In a *fast* first stage (for times of the order of the mean free time τ), the gas tends to relax quickly towards a local equilibrium state, due to collisions. Then, in a *slow* second stage (for times of the order of the hydrodynamic time τ_h), the time evolution of the gas is governed by hydrodynamics and, for $t \gg \tau_h$, the total equilibrium state is reached. The first stage defines the so-called *kinetic* regime, which is sensitive to the initial state. In contrast, in the second stage the gas has forgotten the details of the initial condition and the state is entirely characterized by the five hydrodynamic fields n, $\mathbf{u}$, and T. This stage defines the *hydrodynamic* regime. The origin of the privileged role played by n, $\mathbf{u}$, and T lies in the fact that these quantities are related to the densities of the conserved quantities (mass, momentum, and energy). Since they do not change in time because of collisions but only due to the free motion of the particles, they evolve much more slowly than other moments of the velocity distribution function, such as the irreversible part of the pressure tensor and the heat flux. According to this scenario, the distribution function in the hydrodynamic regime is expected to depend on space and time only through a *functional* dependence on the hydrodynamic fields:

$$f(\mathbf{r}, \mathbf{v}; t) = f[\mathbf{v}|n, \mathbf{u}, T]. \tag{1.63}$$

This functional dependence means that, in order to determine f at a given point $\mathbf{r}$, we need the knowledge of the hydrodynamic fields, not only at that point, but also at the remaining points of the system. This is formally equivalent to the knowledge of the fields and all their spatial derivatives at $\mathbf{r}$. A solution to the Boltzmann equation of the form (1.63) is called a Hilbert-class or *normal* solution. Obviously, the normal solution is actually a class of solutions but not the most general class. It is restricted to times much larger than the mean free time (i.e. outside the so-called *initial layer*) and applies to points separated from the boundaries distances much larger than the mean free path (i.e. outside the *boundary layer*). When the normal form (1.63) is inserted into Eqs. (1.40) and (1.45) for the fluxes and into Eqs. (1.41) and (1.46) for the production terms, the conservation equations (1.38), (1.39), and (1.44) constitute a closed set of equations for the hydrodynamic fields.

The Chapman–Enskog method is a constructive way of obtaining the normal solution as an expansion in powers of the ratio $\ell/h = \tau/\tau_h$. Since $h^{-1} \sim |\nabla \ln f|$, the expansion in powers of ℓ/h is actually equivalent to an expansion in powers of hydrodynamic gradients. More explicitly, we look for normal solutions in the form

$$f = f^{(0)} + \varepsilon f^{(1)} + \varepsilon^2 f^{(2)} + \cdots, \tag{1.64}$$

where ε is a formal expansion parameter, usually referred to as the *uniformity parameter*, measuring the dependence on spatial gradients. This bookkeeping parameter will be used to label the relative orders of magnitude of the different terms and will be set $\varepsilon = 1$ at the end of the calculations. For the sake of simplicity, in the remaining of this Section we will assume the absence of an external force, $\mathbf{F} = \mathbf{0}$, and so the production terms $\boldsymbol{\sigma}_u$ and σ_T vanish. By construction, the zeroth-order approximation $f^{(0)}$ is the local equilibrium distribution and therefore has the same first five velocity moments as the actual distribution f:

$$\int d\mathbf{v}\, \{1, \mathbf{v}, v^2\}(f - f^{(0)}) = \{0, \mathbf{0}, 0\}. \tag{1.65}$$

Since this must hold to any order in ε, one has

$$\int d\mathbf{v}\, \{1, \mathbf{v}, v^2\} f^{(k)} = \{0, \mathbf{0}, 0\}, \quad k \geq 1. \tag{1.66}$$

The expansion (1.64) generates a similar expansion for the momentum and heat fluxes. To obtain the hierarchy of equations for the successive approximations $f^{(k)}$ from the Boltzmann equation (1.25), we formally replace ∇ by $\varepsilon\nabla$ and expand the time derivative operator as[3]

$$\partial_t = \partial_t^{(0)} + \varepsilon\partial_t^{(1)} + \varepsilon^2\partial_t^{(2)} + \cdots . \tag{1.67}$$

Since $f^{(0)}, f^{(1)}, f^{(2)} \ldots$ depend on time only through the hydrodynamic fields, the action of the operator $\partial_t^{(k)}$ is

$$\partial_t^{(k)} = \left(\partial_t^{(k)} n\right)\frac{\partial}{\partial n} + \left(\partial_t^{(k)}\mathbf{u}\right)\cdot\frac{\partial}{\partial \mathbf{u}} + \left(\partial_t^{(k)} T\right)\frac{\partial}{\partial T}. \tag{1.68}$$

The operators $\partial_t^{(k)} n, \partial_t^{(k)}\mathbf{u}, \partial_t^{(k)} T$ are obtained from the macroscopic conservation equations (1.38), (1.39), and (1.44) by expanding the fluxes, setting $\nabla \to \varepsilon\nabla$, and collecting terms of the same order in ε. They are given as

$$\partial_t^{(0)} n = \partial_t^{(0)}\mathbf{u} = \partial_t^{(0)} T = 0, \tag{1.69}$$

$$\partial_t^{(1)} n = -\nabla\cdot(n\mathbf{u}), \tag{1.70}$$

$$\rho\partial_t^{(1)}\mathbf{u} = -\rho\mathbf{u}\cdot\nabla\mathbf{u} - \nabla p, \tag{1.71}$$

$$\frac{3}{2}nk_B\partial_t^{(1)} T = -\frac{3}{2}nk_B\mathbf{u}\cdot\nabla T - p\nabla\cdot\mathbf{u}, \tag{1.72}$$

$$\partial_t^{(k)} n = 0, \quad k \geq 2, \tag{1.73}$$

$$\rho\partial_t^{(k)}\mathbf{u} = -\nabla\cdot\mathsf{P}^{(k-1)}, \quad k \geq 2, \tag{1.74}$$

$$\frac{3}{2}nk_B\partial_t^{(k)} T = -\nabla\cdot\mathbf{q}^{(k-1)} - \mathsf{P}^{(k-1)} : \nabla\mathbf{u}, \quad k \geq 2. \tag{1.75}$$

Upon writing Eqs. (1.71) and (1.72) we have taken into account that $P_{ij}^{(0)} = p\delta_{ij}$ and $\mathbf{q}^{(0)} = \mathbf{0}$. Note that $\partial_t^{(0)} = 0$ because $\mathbf{F} = \mathbf{0}$; if, however, $\mathbf{F} = \zeta m\mathbf{V}$, then $\partial_t^{(0)} T = 2\zeta T$. Equations (1.70)–(1.72) are the conservation equations for an *inviscid* (or ideal) fluid, i.e. a fluid with zero viscosity and thermal conductivity. Those equations are known as the Euler hydrodynamic equations.

Inserting the expansions in powers of ε into the Boltzmann equation (1.25) (with $\mathbf{F} = \mathbf{0}$) and equating terms of the same order in ε, we get a hierarchy of equations for each $f^{(k)}$. The zeroth-order equation is simply

$$0 = J[f^{(0)}, f^{(0)}], \tag{1.76}$$

[3]In standard textbooks (Chapman and Cowling, 1970) one usually assigns a factor ε^{-1} to the Boltzmann collision operator J and so the operators ∂_t and ∇ are formally of order ε^0. In this formulation, the operator $\partial_t^{(0)}$ is the same as our operator $\partial_t^{(1)}$ of Eq. (1.67). Of course the results are fully equivalent in both schemes.

whose solution is the local equilibrium distribution (1.59), as expected. The first-order equation reads

$$\mathcal{L}\Phi^{(1)} = -\left(\partial_t^{(1)} + \mathbf{v}\cdot\nabla\right)\ln f_L, \tag{1.77}$$

where $\Phi^{(1)} \equiv f^{(1)}/f_L$ and $\mathcal{L}$ is the *linearized* collision operator

$$\begin{aligned}\mathcal{L}X(\mathbf{v}) &= -\frac{1}{f_L(\mathbf{v})}\left(J[\mathbf{v}|f_L, f_L X] + J[\mathbf{v}|f_L X, f_L]\right) \\ &= \int d\mathbf{v}_1 \int d\Omega\, gB(g,\chi) f_L(\mathbf{v}_1) \\ &\quad \times \left[X(\mathbf{v}) + X(\mathbf{v}_1) - X(\mathbf{v}') - X(\mathbf{v}_1')\right],\end{aligned} \tag{1.78}$$

where in the last step use has been made of the property $f_L(\mathbf{v})f_L(\mathbf{v}_1) = f_L(\mathbf{v}')f_L(\mathbf{v}_1')$. The right-hand side of Eq. (1.77) can be written more explicitly by using the Euler equations (1.70)–(1.72). After some algebra, one gets

$$\left(\partial_t^{(1)} + \mathbf{v}\cdot\nabla\right)\ln f_L = \frac{1}{k_B T}\mathbf{S}(\mathbf{V})\cdot\nabla\ln T + \frac{1}{k_B T}\mathsf{C}(\mathbf{V}) : \nabla\mathbf{u}, \tag{1.79}$$

where

$$\mathbf{S}(\mathbf{V}) = \left(\frac{mV^2}{2} - \frac{5}{2}k_B T\right)\mathbf{V}, \tag{1.80}$$

$$C_{ij}(\mathbf{V}) = m\left(V_i V_j - \frac{1}{3}V^2\delta_{ij}\right). \tag{1.81}$$

Since C is a symmetric traceless tensor, the rate-of-strain tensor $\nabla\mathbf{u}$ can be replaced in Eq. (1.79) by its symmetric traceless part

$$D_{ij} = \frac{1}{2}\left(\nabla_i u_j + \nabla_j u_i - \frac{2}{3}\nabla\cdot\mathbf{u}\,\delta_{ij}\right). \tag{1.82}$$

In order to solve the first-order equation (1.77), it is convenient to consider some properties of the linear operator $\mathcal{L}$. First, we introduce the inner product of two velocity functions $X_1(\mathbf{v})$ and $X_2(\mathbf{v})$ as

$$\langle X_1|X_2\rangle = \frac{1}{n}\int d\mathbf{v}\, f_L(\mathbf{v})X_1^\dagger(\mathbf{v})X_2(\mathbf{v}), \tag{1.83}$$

where the dagger denotes the complex conjugate and we have considered that X_1 and X_2 are, in general, complex functions. By following steps similar to those made in deriving Eq. (1.35), it is straightforward to get

$$\begin{aligned}\langle X_1|\mathcal{L}X_2\rangle &= \frac{1}{4n}\int d\mathbf{v}\int d\mathbf{v}_1\int d\Omega\, gB(g,\chi)f_L(\mathbf{v}_1)f_L(\mathbf{v}_1)\\ &\quad\times\left[X_1^\dagger(\mathbf{v})+X_1^\dagger(\mathbf{v}_1)-X_1^\dagger(\mathbf{v}')-X_1^\dagger(\mathbf{v}_1')\right]\\ &\quad\times\left[X_2(\mathbf{v})+X_2(\mathbf{v}_1)-X_2(\mathbf{v}')-X_2(\mathbf{v}_1')\right]\\ &= \langle X_2|\mathcal{L}X_1\rangle^\dagger. \end{aligned}\tag{1.84}$$

Therefore, $\mathcal{L}$ is a Hermitian operator, so its eigenvalues are real numbers. Moreover, we have $\langle X|\mathcal{L}X\rangle \geq 0$ because the integrand in Eq. (1.84) is positive semi-definite for $X_1 = X_2$. Consequently, the eigenvalues of $\mathcal{L}$ are non-negative. The null eigenvalue is five-fold degenerate and its corresponding eigenfunctions are the collisional invariants. It can be proved (Dorfman and van Beijeren, 1977; Cercignani, 1990) that for repulsive interaction potentials of the form $\phi(r) = Kr^{-w}$ with $2 < w < \infty$ the spectrum of $\mathcal{L}$ is discrete, while for hard spheres ($w \to \infty$), the spectrum has a discrete and a continuum part, the latter having a nonzero lower bound. When an angular cutoff is considered, the spectrum also has a continuum part for $2 < w < 4$ and $4 < w < \infty$, having again a nonzero lower bound if $w > 4$. In conclusion, if $2 < w < 4$ without angular cutoff or $w \geq 4$ with and without cutoff, a gap exists between the null eigenvalue and the first nonzero eigenvalue. The existence of a gap is important because the inverse of its width provides a measure of the mean free time τ. An important property of the operator $\mathcal{L}$ is that it is *isotropic*, i.e. it commutes with the rotation operators in velocity space (Résibois and de Leener, 1977). Consequently,

$$\mathcal{L}X(V)Y_\ell^m(\widehat{\mathbf{V}}) = X'(V)Y_\ell^m(\widehat{\mathbf{V}}),\tag{1.85}$$

where X and X' are functions depending only on the magnitude of the velocity and $Y_\ell^m(\widehat{\mathbf{V}})$ are spherical harmonics, depending on the unit vector $\widehat{\mathbf{V}}$. This implies that the eigenfunctions of $\mathcal{L}$ can be written as

$$\Psi_{r\ell m}(\mathbf{V}) = \psi_{r\ell}(V)Y_\ell^m(\widehat{\mathbf{V}})\tag{1.86}$$

and the eigenvalues $\lambda_{r\ell}^0$ only depend on r and ℓ. In particular, $\lambda_{00}^0 = \lambda_{01}^0 = \lambda_{10}^0 = 0$ and $\psi_{00} = 1$, $\psi_{01} \propto V$, and $\psi_{10} \propto V^2 - 3k_BT/m$ are related to the collisional invariants. The explicit form for the functions $\psi_{r\ell}$ is exactly known for Maxwell molecules only. In this case $\psi_{r\ell}$ is (Grad, 1949; Résibois and de Leener, 1977; McLennan, 1989; Cercignani, 1990)

$$\psi_{r\ell}(V) = N_{r\ell}\xi^\ell L_r^{(\ell+\frac{1}{2})}(\xi^2),\tag{1.87}$$

$$\boldsymbol{\xi} = (m/2k_BT)^{1/2}\,\mathbf{V},\tag{1.88}$$

where $\boldsymbol{\xi}$ is the peculiar velocity normalized with respect to the thermal velocity, $L_r^{(\ell+\frac{1}{2})}(\xi^2)$ are generalized Laguerre polynomials (also known as Sonine polynomials) (Abramowitz and Stegun, 1972; Gradshteyn and Ryzhik, 1980), and

$$N_{r\ell} = \left[2\pi^{3/2} \frac{r!}{\Gamma(r+\ell+\frac{3}{2})} \right]^{1/2} \tag{1.89}$$

are normalization constants, $\Gamma(x)$ being the gamma function. The corresponding eigenvalues are

$$\begin{aligned} \lambda_{r\ell}^0 &= n \int d\Omega\, gB(g,\chi) \Big[1 + \delta_{r0}\delta_{\ell 0} - \cos^{2r+\ell}\left(\frac{\chi}{2}\right) P_\ell\left(\cos\frac{\chi}{2}\right) \\ &\quad - \sin^{2r+\ell}\left(\frac{\chi}{2}\right) P_\ell\left(\sin\frac{\chi}{2}\right), \Big] \end{aligned} \tag{1.90}$$

where $gB(g,\chi)$ is independent of the relative velocity g and $P_\ell(x)$ are Legendre polynomials (Abramowitz and Stegun, 1972; Gradshteyn and Ryzhik, 1980). It is easy to check that $\lambda_{r,1}^0 = \lambda_{r+1,0}^0$ and $\lambda_{11}^0/\lambda_{02}^0 = \frac{2}{3}$, $\lambda_{03}^0/\lambda_{02}^0 = \frac{3}{2}$, $\lambda_{12}^0/\lambda_{02}^0 = \frac{7}{6}$, $\lambda_{21}^0/\lambda_{02}^0 = 1$. A rather extensive table of eigenvalues for Maxwell molecules, as well as for hard spheres (in the latter case for $0 \leq r \leq 13$ and $r + \ell = 13$) can be found in a work by Alterman et al. (1962). In some cases it is convenient to define a simpler version of Maxwell molecules by setting the collision rate $gB(g,\chi)$ independent of the scattering angle (Hendriks and Nieuwenhuizen, 1982). This scattering model will be referred to as the *isotropic* Maxwell model. Of course, the eigenvalues are in general different in both Maxwell models. For instance, $\lambda_{04}^0/\lambda_{02}^0 = \frac{7}{4}$ in the isotropic Maxwell model, while $\lambda_{04}^0/\lambda_{02}^0 = 1.87313$ for actual Maxwell molecules.

Now we turn to the linear integral equation (1.77). According to the Fredholm alternative (Margeneau and Murphy, 1956), the inhomogeneous integral equation (1.77) has a solution if and only if the inhomogeneous term is orthogonal to the collisional invariants. This condition is satisfied since the polynomials $\mathbf{S}$ and C appearing in Eq. (1.79) are indeed orthogonal to the collisional invariants. The polynomials $\mathbf{S}$ and C are linear combinations of the spherical harmonics Y_1^m and Y_2^m, respectively. More explicitly,

$$\mathbf{S}(\mathbf{V}) = -\frac{m}{2}\left(\frac{2k_BT}{m}\right)^{3/2} \xi L_1^{3/2}(\xi^2)\widehat{\boldsymbol{\xi}}, \tag{1.91}$$

$$C_{ij}(\mathbf{V}) = 2k_BT\xi^2 L_0^{5/2}(\xi^2)\left(\widehat{\xi}_i\widehat{\xi}_j - \frac{1}{3}\delta_{ij}\right). \tag{1.92}$$

Using the isotropy property of $\mathcal{L}$, we can write the solution to Eq. (1.77) as

$$\Phi^{(1)}(\mathbf{V}) = -\frac{1}{k_BT}A(V^2)\mathbf{V}\cdot\nabla\ln T - \frac{1}{k_BT}B(V^2)\mathsf{C}(\mathsf{V}) : \mathsf{D}, \tag{1.93}$$

where $A(V^2)$ and $B(V^2)$ are scalar functions verifying the equations

$$\mathcal{L}A(V^2)\mathbf{V} = \mathbf{S}(\mathbf{V}), \tag{1.94}$$

$$\mathcal{L}B(V^2)\mathsf{C}(\mathbf{V}) = \mathsf{C}(\mathbf{V}). \tag{1.95}$$

Conditions (1.66) imply that $\Phi^{(1)}$ must be orthogonal to 1, $\mathbf{V}$, and V^2. The first and third conditions are automatically satisfied by the form (1.93), while the second one requires that

$$\langle A(V^2)|V^2\rangle = \frac{1}{n}\int d\mathbf{v}\, f_L(\mathbf{v})A(V^2)V^2 = 0. \tag{1.96}$$

In the particular case of Maxwell molecules, the right-hand sides of (1.94) and (1.95) are linear combinations of the eigenfunctions Ψ_{11m} and Ψ_{02m}, respectively, so that $\mathcal{L}\mathbf{S}(\mathbf{V}) = \lambda_{11}^0\mathbf{S}(\mathbf{V})$, $\mathcal{L}\mathsf{C}(\mathbf{V}) = \lambda_{02}^0\mathsf{C}(\mathbf{V})$. Therefore, the solutions are

$$A(V^2)\mathbf{V} = \frac{1}{\lambda_{11}^0}\mathbf{S}(\mathbf{V}), \tag{1.97}$$

$$B(V^2)\mathsf{C}(\mathbf{V}) = \frac{1}{\lambda_{02}^0}\mathsf{C}(\mathbf{V}). \tag{1.98}$$

In general, however, the solutions to the integral equations (1.94) and (1.95) are not known. By isotropy reasons, it is convenient to expand $A(V^2)$ as a linear combination of generalized Laguerre polynomials with $\ell = 1$:

$$A(V^2) = \sum_{r=0}^{\infty} a_r L_r^{(3/2)}(\xi^2). \tag{1.99}$$

This guarantees that $\mathcal{L}A(V^2)\mathbf{V}$ can be expressed as $\mathbf{V}$ times a linear combination of $L_r^{(3/2)}(\xi^2)$. Analogously,

$$B(V^2) = \sum_{r=0}^{\infty} b_r L_r^{(5/2)}(\xi^2). \tag{1.100}$$

Inserting the expansions (1.99) and (1.100) into the integral equations (1.94) and (1.95), one gets two independent sets of equations for the coefficients a_r and b_r. The requirement (1.96) yields $a_0 = 0$. For practical purposes, one can get an approximate solution by neglecting the coefficients beyond a given order r. This procedure is usually referred to as the Sonine approximation.

Once the functions $A(V^2)$ and $B(V^2)$ are determined, we are in conditions of computing the first-order approximation to the momentum and heat fluxes. The pressure tensor is

$$\begin{aligned} P_{ij}^{(1)} &= \int d\mathbf{v}\, f_L(\mathbf{V})C_{ij}(\mathbf{V})\Phi^{(1)}(\mathbf{V}) \\ &= -\frac{1}{k_B T}\int d\mathbf{V} f_L(\mathbf{V})B(V^2)C_{ij}(\mathbf{V})C_{k\ell}(\mathbf{V})D_{k\ell}, \end{aligned} \tag{1.101}$$

where use has been made of the fact that $A(V^2)\mathbf{V}$ and $\mathsf{C}(\mathbf{V})$ are mutually orthogonal. In the last line of Eq. (1.101) Einstein's convention of repeated indices has been adopted. Note that the off-diagonal element $B(V^2)C_{ij}(\mathbf{V})$ (with $i \neq j$) is orthogonal to $C_{k\ell}(\mathbf{V})$, except if $(k,\ell) = (i,j)$ or $(k,\ell) = (j,i)$; the diagonal element $B(V^2)C_{ij}(\mathbf{V})$ (with $i = j$) is orthogonal to $C_{k\ell}(\mathbf{V})$, except if $k = \ell$. In addition, since D is a traceless tensor, then $D_{zz} + D_{yy} = -D_{xx}$. Therefore, the only scalar products needed in (1.101) are

$$\begin{aligned}\int d\mathbf{V}\, f_L(\mathbf{V}) B(V^2) C_{xy}^2(\mathbf{V}) &= \frac{1}{2}\int d\mathbf{V}\, f_L(\mathbf{V}) B(V^2) C_{xx}(\mathbf{V}) \\ &\quad \times \left[C_{xx}(\mathbf{V}) - C_{yy}(\mathbf{V})\right] \\ &= \frac{m^2}{15}\int d\mathbf{V}\, f_L(\mathbf{V}) V^4 B(V^2). \qquad (1.102)\end{aligned}$$

As a consequence, Eq. (1.101) can be rewritten as

$$P_{ij}^{(1)} = -2\eta_0 D_{ij}, \qquad (1.103)$$

where

$$\eta_0 = \frac{m^2}{15 k_B T}\int d\mathbf{V}\, f_L(\mathbf{V}) V^4 B(V^2) \qquad (1.104)$$

is the *shear viscosity* coefficient. Substituting the expansion (1.100) into Eq. (1.104) and taking into account the orthogonality properties of the generalized Laguerre polynomials (Abramowitz and Stegun, 1972; Gradshteyn and Ryzhik, 1980) one simply gets

$$\eta_0 = n k_B T b_0. \qquad (1.105)$$

Therefore, the pressure tensor up to the first order in the gradients becomes

$$P_{ij} = p\,\delta_{ij} - 2\eta_0 D_{ij}. \qquad (1.106)$$

Equation (1.106) is Newton's law of friction and establishes a linear relationship between the irreversible part of the pressure tensor and the rate-of-strain tensor (1.82).[4] In fluid mechanics, Newton's law is introduced in a phenomenological way. The Chapman–Enskog method provides a justification of Newton's law, as well as a microscopic expression for the shear viscosity.

Similarly, the heat flux is

$$\begin{aligned}\mathbf{q}^{(1)} &= \int d\mathbf{v}\, f_L(\mathbf{V})\mathbf{S}(\mathbf{V})\Phi^{(1)}(\mathbf{V}) \\ &= -\frac{1}{k_B T}\int d\mathbf{V}\, f_L(\mathbf{V}) A(V^2)\mathbf{S}(\mathbf{V})\cdot\mathbf{V}\nabla\ln T, \qquad (1.107)\end{aligned}$$

[4] In general, there is an additional contribution to the pressure tensor of the form $-\zeta_0(\nabla\cdot\mathbf{u})\delta_{ij}$, ζ_0 being the *bulk* viscosity. For a dilute gas, $\zeta_0 = 0$ and this contribution is absent.

where we have taken into account that $B(V^2)\mathsf{C}(\mathbf{V})$ and $\mathbf{S}(\mathbf{V})$ are mutually orthogonal. In addition, $A(V^2)S_i(\mathbf{V})$ is orthogonal to V_j if $i \neq j$. Thus, the heat flux becomes

$$\mathbf{q}^{(1)} = -\kappa_0 \nabla T, \tag{1.108}$$

where κ_0 is the *thermal conductivity* coefficient, given as

$$\begin{aligned}\kappa_0 &= \frac{1}{3k_B T^2}\int d\mathbf{V}\, f_L(\mathbf{V})A(V^2)\mathbf{S}(\mathbf{V})\cdot\mathbf{V} \\ &= \frac{m}{6k_B T^2}\int d\mathbf{V}\, f_L(\mathbf{V})V^4 A(V^2) \\ &= -\frac{5}{2}\frac{nk_B}{m}a_1, \end{aligned}\tag{1.109}$$

where in the second equality Eq. (1.96) has been used and in the third equality the expansion (1.99) has been considered. Equation (1.108) is Fourier's law, establishing a linear relationship between the heat flux and the thermal gradient. Again, the Chapman–Enskog method gives a microscopic expression for the associated transport coefficient κ_0. When the constitutive equations (1.103) and (1.108) are substituted into the conservation equations (1.38), (1.39), and (1.44), the well-known Navier–Stokes hydrodynamic equations are obtained. They are the fundamental equations of fluid mechanics and provide a good description of the gas when the hydrodynamic variables are practically constant over a distance of the order of the mean free path. An interesting dimensionless measure of the viscosity relative to the thermal conductivity is the *Prandtl number* Pr defined as

$$\text{Pr} = c_p \frac{\eta_0}{\kappa_0}, \tag{1.110}$$

where c_p is the specific heat at constant pressure, which in a monatomic ideal gas is $c_p = \frac{5}{2}k_B/m$.

Note that, according to Eq. (1.78), the operator $\mathcal{L}$ is proportional to the density n, so Eqs. (1.94) and (1.95) imply that $A(V^2)$ and $B(V^2)$ are inversely proportional to n. Consequently, the transport coefficients are independent of the density. This is an interesting general result for low-density gases that was first discovered by Maxwell (Brush, 1965; Brush, 1976). On the other hand, the transport coefficients depend on the temperature, the dependence being different for each interaction potential. For repulsive potentials of the form $\phi(r) = Kr^{-w}$, the scaling $gB(g,\chi) \propto g^{1-(4/w)}$ implies that $\mathcal{L} \propto T^q$ with $q \equiv \frac{1}{2} - \frac{2}{w}$. In this case, from Eqs. (1.94) and (1.95) one has $A(V^2) \propto T^{1-q}$ and $B(V^2) \propto T^{-q}$. Thus, the transport coefficients η_0 and κ_0 scale as $T^{1-q} = T^{\frac{1}{2}+\frac{2}{w}}$. For hard spheres, $w \to \infty$ and so $\eta_0, \kappa_0 \propto T^{1/2}$. The complete determination of the transport coefficients requires the solution of the integral equations (1.94) and (1.95). In the case of Maxwell molecules the exact result is given by Eqs. (1.97) and (1.98), so

$$\eta_0 = \frac{nk_BT}{\lambda_{02}^0}, \tag{1.111}$$

$$\kappa_0 = \frac{5}{2}\frac{nk_B^2T}{m\lambda_{11}^0}. \tag{1.112}$$

From Eq. (1.90) we have $\lambda_{11}^0 = \frac{2}{3}\lambda_{02}^0$ and $\lambda_{02}^0 = 3n\mathrm{A}_2$, where

$$\mathrm{A}_2 = 1.3703\sqrt{2K/m}. \tag{1.113}$$

The Prandtl number for Maxwell molecules is $\mathrm{Pr} = \frac{2}{3}$. Except for Maxwell molecules, the Navier–Stokes transport coefficients cannot be determined exactly. A good estimate can be obtained by means of the so-called first Sonine approximation, consisting of taking the leading terms $r = 1$ and $r = 0$ in the expansions (1.99) and (1.100), respectively. In this approximation,

$$a_1 = -\frac{15}{4}\frac{k_BT}{\langle L_1^{(3/2)}(\xi^2)\xi_i|\mathcal{L}L_1^{(3/2)}(\xi^2)\xi_i\rangle}, \tag{1.114}$$

$$b_0 = \frac{5}{2}\frac{1}{\langle \xi_i\xi_j - \frac{1}{3}\xi^2\delta_{ij}|\mathcal{L}(\xi_i\xi_j - \frac{1}{3}\xi^2\delta_{ij})\rangle}. \tag{1.115}$$

It can be proved (Chapman and Cowling, 1970) that, whatever the nature of the interaction between the particles, the denominators in Eqs. (1.114) and (1.115) are equal, so $a_1 = -\frac{3}{2}k_BTb_0$ and $\mathrm{Pr} = \frac{2}{3}$ in the first Sonine approximation. In this approximation, Eq. (1.93) for the distribution function $\Phi^{(1)}$ becomes

$$\Phi^{(1)}(\mathbf{V}) = -\frac{2}{5}\frac{m}{nk_B^3T^2}\kappa_0\mathbf{S}(\mathbf{V})\cdot\nabla\ln T - \frac{2}{nk_B^2T^2}\eta_0\mathsf{C}(\mathbf{V}):\mathsf{D}. \tag{1.116}$$

As an illustration of the first Sonine approximation, we take again repulsive potentials $\phi(r) = Kr^{-w}$. In this case, the shear viscosity is approximated by (Chapman and Cowling, 1970)

$$\eta_0 = \frac{5}{8}\sqrt{\frac{mk_BT}{\pi}}\frac{(2k_BT/wK)^{2/w}}{\Gamma(4-2/w)\,\mathrm{A}^{(2)}(w)}, \tag{1.117}$$

where the constants $\mathrm{A}^{(s)}(w)$ are defined as

$$\mathrm{A}^{(s)}(w) = \int_0^\infty d\beta\,\beta\,[1-\cos^s(\chi)] \tag{1.118}$$

and are tabulated for $s = 1$ and $s = 2$ on p. 172 of Chapman and Cowling's book (Chapman and Cowling, 1970). For Maxwell molecules, one has $\mathrm{A}^{(2)}(4) =$

$1.3703/\pi \simeq 0.436$ and the exact result (1.111) is recovered. In the opposite limit of hard spheres of diameter σ one must take $w \to \infty$ with K/σ^w finite. The result is

$$\eta_0 = \frac{5}{16} \frac{\sqrt{\pi m k_B T}}{\pi \sigma^2}, \tag{1.119}$$

where we have made use of the value $\mathrm{A}^{(2)}(\infty) = \frac{1}{3}$ (Chapman and Cowling, 1970).

The Sonine approximation can be improved by including higher polynomial terms in the expansions (1.99) and (1.100), the convergence of the method being quite good (Chapman and Cowling, 1970). In fact, in the case of hard spheres, the fourth Sonine approximation modifies the first Sonine approximation (1.119) by a factor 1.0160. Analogously the value $\frac{2}{3}$ for the Prandtl number is modified by a factor 0.9911.

Once the first-order distribution $f^{(1)} = f_L \Phi^{(1)}$ has been determined, one can get the corresponding equations for the successive distributions $f^{(k)}$ with $k \geq 2$. The constitutive equations associated with $k = 2$, $k = 3$, ... are known as the Burnett, super-Burnett, super-super-Burnett, ... equations. The corresponding hydrodynamic equations involve the third, fourth, fifth, ... spatial derivatives of the hydrodynamic fields. A detailed account of the Burnett equations can be found in Chapter 15 of (Chapman and Cowling, 1970). Nevertheless, the practical usefulness of the Burnett and super-Burnett equations has a limited scope. First, there is no reason to expect that the Chapman–Enskog expansion is convergent, although it is at least asymptotic (Grad, 1963). As a consequence, if in a given problem the Navier–Stokes equations are not sufficient, it is unlikely that the description is improved by including the next order terms. The shear flow problems studied in this book provide examples of the divergence of the Chapman–Enskog expansion. As a second reason, it is not clear how to determine the additional boundary conditions needed to solve the Burnett and super-Burnett equations. Finally, it has been shown that the inclusion of the Burnett and super-Burnett terms in the hydrodynamic equations make the equilibrium solution to be unstable against small perturbations with a wavelength smaller than a critical value (Bobylev, 1981).

Apart from the fluxes, an important physical quantity is the entropy, defined by Eq. (1.47). The Chapman–Enskog expansion (1.64) generates an expansion of the entropy in powers of the gradients:

$$s = s^{(0)} + \varepsilon s^{(1)} + \varepsilon^2 s^{(2)} + \cdots, \tag{1.120}$$

where $s^{(0)}$ is the local equilibrium entropy. The expansion of $\ln f$ is

$$\ln f = \ln f_L + \varepsilon \Phi^{(1)} + \varepsilon^2 \left[\Phi^{(2)} - \frac{1}{2} \Phi^{(1)^2} \right] + \cdots, \tag{1.121}$$

where $\Phi^{(k)} = f^{(k)}/f_L$. Therefore,

$$s^{(1)} = -k_B \int d\mathbf{v} f_L \Phi^{(1)} \left(1 + \ln f_L\right) = 0, \tag{1.122}$$

according to Eq. (1.66). The second-order contribution to the entropy is

$$s^{(2)} = -\frac{1}{2} k_B \int d\mathbf{v} f_L {\Phi^{(1)}}^2, \tag{1.123}$$

where Eq. (1.66) has been applied again. Thus, we see that the knowledge of the distribution function to Navier–Stokes order allows one to evaluate the entropy to second order in the gradients. In particular, for Maxwell molecules (or for any potential in the first Sonine approximation) $\Phi^{(1)}$ is given explicitly by Eq. (1.116) and so Eq. (1.123) becomes

$$\begin{aligned} s^{(2)} &= -\frac{\eta_0^2}{pT} \mathsf{D} : \mathsf{D} - \frac{1}{5} \frac{m\kappa_0^2}{k_B p} (\nabla \ln T)^2 \\ &= -\frac{1}{4pT} \mathbf{\Pi} : \mathbf{\Pi} - \frac{m}{5 k_B p T^2} \mathbf{q} \cdot \mathbf{q}, \end{aligned} \tag{1.124}$$

where the tensor $\mathbf{\Pi}$ is defined by Eq. (1.43).

5. Boltzmann equation for gas mixtures

It is easy to extend the Boltzmann equation to the case of a mixture of N monatomic gases. First, one has a velocity distribution function $f_s(\mathbf{r}, \mathbf{v}; t)$ for particles of species $s = 1, \ldots, N$. From f_s one can define the number density n_s of species s as

$$n_s(\mathbf{r}, t) = \int d\mathbf{v}\, f_s(\mathbf{r}, \mathbf{v}; t), \tag{1.125}$$

the (barycentric) flow velocity $\mathbf{u}$ of the mixture as

$$\mathbf{u}(\mathbf{r}, t) = \frac{1}{\rho(\mathbf{r}, t)} \sum_{s=1}^{N} m_s \int d\mathbf{v}\, \mathbf{v} f_s(\mathbf{r}, \mathbf{v}; t), \tag{1.126}$$

and the temperature T of the mixture as

$$\frac{3}{2} n(\mathbf{r}, t) k_B T(\mathbf{r}, t) = \sum_{s=1}^{N} \frac{m_s}{2} \int d\mathbf{v}\, \left[\mathbf{v} - \mathbf{u}(\mathbf{r}, t)\right]^2 f_s(\mathbf{r}, \mathbf{v}; t). \tag{1.127}$$

In the above equations, $\rho = \sum_s \rho_s$ is the total mass density, $\rho_s = m_s n_s$ is the mass density of species s, m_s is the mass of a particle of species s, and

$n = \sum_s n_s$ is the total number density. Now, instead of a single Boltzmann equation, we have a coupled set of N nonlinear equations:

$$\frac{\partial f_s}{\partial t} + \mathbf{v} \cdot \nabla f_s + \frac{\partial}{\partial \mathbf{v}} \cdot \left(\frac{\mathbf{F}_s}{m_s} f_s \right) = \sum_{r=1}^{N} J_{sr}[\mathbf{v}|f_s, f_r], \tag{1.128}$$

where the Boltzmann collision integral J_{sr} accounts for the rate of change of f_s due to collisions with particles of species r. Its expression is

$$J_{sr}[\mathbf{v}|f_s, f_r] = \int d\mathbf{v}_1 \int d\Omega \, g B_{sr}(g, \chi) \left[f_s(\mathbf{v}') f_r(\mathbf{v}_1') - f_s(\mathbf{v}) f_r(\mathbf{v}_1) \right]. \tag{1.129}$$

Here, the post-collisional velocities $\mathbf{v}'$ and $\mathbf{v}_1'$ are given by

$$\mathbf{v}' = \mathbf{v} - 2\frac{m_s}{m_s + m_r}(\mathbf{g} \cdot \widehat{\boldsymbol{\sigma}})\widehat{\boldsymbol{\sigma}}, \quad \mathbf{v}_1' = \mathbf{v}_1 + 2\frac{m_r}{m_s + m_r}(\mathbf{g} \cdot \widehat{\boldsymbol{\sigma}})\widehat{\boldsymbol{\sigma}}. \tag{1.130}$$

It is worth remarking that in the special case of mechanically equivalent particles [i.e. $m_s = m$, $B_{sr}(g, \chi) = B(g, \chi)$, and $\mathbf{F}_s = \mathbf{F}$, so the particles of each species only differ by a “tag” or label] the *total* distribution function $f = \sum_s f_s$ obeys the closed Boltzmann equation for a single gas, Eq. (1.25). This physical property follows from the bilinear character of the collision operator, what implies

$$J[f, f] = \sum_{s=1}^{N} \sum_{r=1}^{N} J_{sr}[f_s, f_r], \tag{1.131}$$

in the case of mechanically equivalent particles. This situation corresponds, for instance, to the self-diffusion problem or to a mixture of a vapor and a non-condensable gas of mechanically identical particles.

Let us obtain the macroscopic balance equations for the mixture. By following similar steps as in the single-gas case, it is easy to prove the properties

$$\int d\mathbf{v} \, J_{sr} \left[\mathbf{v}|f_s, f_r \right] = 0, \tag{1.132}$$

$$\int d\mathbf{v} \, m_s \mathbf{v} J_{sr} \left[\mathbf{v}|f_s, f_r \right] + \int d\mathbf{v} \, m_r \mathbf{v} J_{rs} \left[\mathbf{v}|f_r, f_s \right] = \mathbf{0}, \tag{1.133}$$

$$\int d\mathbf{v} \, m_s v^2 J_{sr} \left[\mathbf{v}|f_s, f_r \right] + \int d\mathbf{v} \, m_r v^2 J_{rs} \left[\mathbf{v}|f_r, f_s \right] = 0. \tag{1.134}$$

Equation (1.132) expresses the conservation of the number of particles of each species in any binary collision, while Eqs. (1.133) and (1.134) represent the conservation of the total momentum and energy in any collision. The corresponding macroscopic balance equations follow from the Boltzmann equation

(1.128) and the collisional invariance conditions (1.132)–(1.134). They read

$$D_t n_s + n_s \nabla \cdot \mathbf{u} + \frac{1}{m_s} \nabla \cdot \mathbf{j}_s = 0, \tag{1.135}$$

$$\rho D_t \mathbf{u} + \nabla \cdot \mathsf{P} = \boldsymbol{\sigma}_u, \tag{1.136}$$

$$\frac{3}{2} n k_B D_t T + \nabla \cdot \mathbf{q} + \mathsf{P} : \nabla \mathbf{u} - \frac{3}{2} k_B T \sum_s \frac{1}{m_s} \nabla \cdot \mathbf{j}_s = \sigma_T. \tag{1.137}$$

In the above equations,

$$\mathbf{j}_s = m_s \int d\mathbf{v}\, \mathbf{V} f_s(\mathbf{v}) \tag{1.138}$$

is the mass flux of species s relative to the local flow,

$$\mathsf{P} = \sum_s m_s \int d\mathbf{v}\, \mathbf{V}\mathbf{V} f_s(\mathbf{v}) \tag{1.139}$$

is the total pressure or stress tensor,

$$\boldsymbol{\sigma}_u = \sum_s \int d\mathbf{v}\, \mathbf{F}_s f_s(\mathbf{v}) \tag{1.140}$$

is the production of total momentum due to the external forces,

$$\mathbf{q} = \sum_s \frac{m_s}{2} \int d\mathbf{v}\, V^2 \mathbf{V} f_s(\mathbf{v}) \tag{1.141}$$

is the total heat flux, and

$$\sigma_T = \sum_s \int d\mathbf{v}\, \mathbf{V} \cdot \mathbf{F}_s f_s(\mathbf{v}) \tag{1.142}$$

is the production of total energy due to the external forces. In Eqs. (1.138)–(1.142), as in the one-component case, $\mathbf{V} = \mathbf{v} - \mathbf{u}$ is the peculiar velocity. The H-theorem can be proved by following essentially the same procedure as in the single-gas case (Chapman and Cowling, 1970).

In the case of multi-component gases, the normal solution to the set of Boltzmann equations (1.128) can be obtained by means of the Chapman–Enskog method for small spatial gradients (Chapman and Cowling, 1970; Ferziger and Kaper, 1972). Of course the calculations are much more involved than in the single-gas case since not only the number of transport coefficients is larger but those coefficients also depend on a larger number of parameters. In the Navier–Stokes order, the structure of the constitutive equations is (de Groot and Mazur, 1984)

$$\mathbf{j}_s = -\sum_{r=1}^{N} L_{sr} \left(\frac{\nabla \mu_r}{T} \right)_T - L_{sq} \frac{\nabla T}{T^2}, \tag{1.143}$$

$$\mathbf{J}_q \equiv \mathbf{q} - \frac{5}{2}k_B T \sum_{s=1}^{N} \frac{\mathbf{j}_s}{m_s} = -\kappa_0 \nabla T - \sum_{s=1}^{N} L_{qs} \left(\frac{\nabla \mu_s}{T}\right)_T, \tag{1.144}$$

$$\mathsf{P} = p\mathsf{I} - 2\eta_0 \mathsf{D}, \tag{1.145}$$

where I is the 3×3 unit tensor. In Eqs. (1.143) and (1.144),

$$\mu_s = -\frac{k_B T}{m_s}\left[\ln \frac{(m_s k_B T/2\pi)^{3/2}}{\hbar^3} - \ln \frac{x_s p}{k_B T}\right] \tag{1.146}$$

is the chemical potential per unit mass, $x_s = n_s/n$ being the mole fraction of species s, and $p = \frac{1}{3}\mathrm{tr}\,\mathsf{P} = nk_B T$ is the total hydrostatic pressure. Therefore, $(\nabla \mu_s)_T = (k_B T/m_s)\nabla \ln(x_s p)$. Equation (1.143) shows that there are two different contributions to the mass flux of each species. The first one is proportional to the gradients of the chemical potentials and is referred to as Fick's law, defining the phenomenological coefficients L_{sr}. The second term gives the contribution to the mass transport due to the presence of a thermal gradient (thermal diffusion or the Soret effect). In a mixture, the vector $\mathbf{J}_q$ in Eq. (1.144) is the flux conjugate to the temperature gradient (de Groot and Mazur, 1984). At a kinetic level it is defined as in Eq. (1.141), but replacing $m_s V^2 \mathbf{V}$ by $(m_s V^2 - 5k_B T/2)\mathbf{V}$. The first term on the right-hand side of Eq. (1.144) defines the thermal conductivity coefficient κ_0 of the mixture, while the second one gives the heat flux associated with the concentration gradients (diffusion thermo-effect or the Dufour effect). Finally, Eq. (1.146) is the familiar Newton's law, defining the shear viscosity η_0 of the mixture. Onsager's reciprocal relations (de Groot and Mazur, 1984) state the symmetry properties $L_{sr} = L_{rs}$ and $L_{sq} = L_{qs}$. In addition, according to Eqs. (1.126) and (1.138), $\sum_s \mathbf{j}_s = \mathbf{0}$, what implies $\sum_s L_{sr} = \sum_s L_{sq} = 0$. In the special case of a binary mixture ($N = 2$) with uniform density and temperature, Eqs. (1.143) and (1.144) become

$$\mathbf{j}_s = -\frac{m_1 m_2 n^2}{\rho} D_0 \nabla x_s, \tag{1.147}$$

$$\mathbf{J}_q = -k_B \frac{n\rho}{\rho_1 \rho_2} L_0 \nabla x_1, \tag{1.148}$$

where

$$D_0 = \frac{k_B \rho^2}{m_1^2 m_2^2 n_1 n_2 n} L_{11} \tag{1.149}$$

is the mutual diffusion coefficient, and $L_0 = L_{q1}$ is the Dufour coefficient.

For a binary mixture of Maxwell molecules, one has

$$D_0 = \frac{k_B T}{n \lambda_{12}}, \tag{1.150}$$

λ_{sr} being given by Eq. (1.B.3). Moreover, the Soret and Dufour coefficients vanish, i.e. $L_{sq} = L_{qs} = 0$. This is a peculiarity of the Maxwell interaction. The expressions for the shear viscosity and the thermal conductivity in terms of the mole fraction (x_1), the masses (m_1, m_2), and the force constants (K_{11}, K_{12}, K_{22}) are much more involved (Chapman and Cowling, 1970). Here we explicitly display the expression of the shear viscosity:

$$\begin{aligned}\eta_0 &= \frac{k_BT}{\lambda'_{12}}\left[(m_1x_1 - m_2x_2)^2 + x_1x_2(m_1+m_2)(m_1\gamma_{22}+m_2\gamma_{11})\right.\\ &\left.+2m_1m_2\varpi\right]\left[x_1^2\gamma_{11}(m_1+2m_2\varpi)\right.\\ &\left.+x_1x_2(m_1+m_2)(\gamma_{11}\gamma_{22}+2\varpi)+x_2^2\gamma_{22}(m_2+2m_1\varpi)\right]^{-1},\end{aligned} \tag{1.151}$$

where $\gamma_{11} \equiv \lambda_{11}/\lambda_{12} = \lambda'_{11}/\lambda'_{12}$, $\gamma_{22} \equiv \lambda_{22}/\lambda_{12} = \lambda'_{22}/\lambda'_{12}$, and $\varpi = \lambda_{sr}/\lambda'_{sr} \equiv 0.645$ is a pure number, λ'_{sr} being given by Eq. (1.B.5).

6. Kinetic models for a single gas

Needless to say, the main difficulty in solving the Boltzmann equation (1.25) lies in the intricate mathematical structure of the collision term (1.24). Because of this, the known explicit exact solutions to the Boltzmann equations, apart from the trivial equilibrium solutions, are scarce (Ernst, 1981). For practical purposes, it is then convenient to replace the fine structure of the collision term $J[f, f]$ by a simpler term $Q[f]$ which retains only the most important qualitative properties of the true collision term. The minimum requirements of $Q[f]$ are: (i) it must have the same collisional invariants as $J[f, f]$, namely

$$\int d\mathbf{v}\,\{1, \mathbf{v}, v^2\}Q[f] = \{0, \mathbf{0}, 0\}, \tag{1.152}$$

and (ii) it must vanish at (local) equilibrium, i.e.

$$Q[f_L] = 0. \tag{1.153}$$

The notation $Q[f]$ denotes that Q is in general a nonlinear functional of f, but not necessarily a quadratic one. The kinetic equation obtained by the replacement $J[f, f] \to Q[f]$ is usually referred to as a model kinetic equation or a kinetic model. The simplest choice of $Q[f]$ consistent with the requirements (1.152) and (1.153) is the model proposed by Bhatnagar et al. (1954) and, independently, by Welander (1954). This model is commonly known as the BGK model. In order to construct the BGK model, let us start from the Boltzmann collision operator for hard spheres, Eq. (1.26). It can be rewritten as

$$J[\mathbf{v}|f, f] = -\nu_{\text{HS}}[\mathbf{v}|f]\left(f(\mathbf{v}) - f_R[\mathbf{v}|f]\right), \tag{1.154}$$

where

$$\nu_{\text{HS}}[\mathbf{v}|f] = \sigma^2 \int d\mathbf{v}_1 \int d\widehat{\boldsymbol{\sigma}}\, \Theta(\mathbf{g}\cdot\widehat{\boldsymbol{\sigma}})(\mathbf{g}\cdot\widehat{\boldsymbol{\sigma}}) f(\mathbf{v}_1) \tag{1.155}$$

is the collision frequency of particles with velocity $\mathbf{v}$ and

$$f_R[\mathbf{v}|f] = \nu_{\text{HS}}^{-1}[\mathbf{v}|f]\, \sigma^2 \int d\mathbf{v}_1 \int d\widehat{\boldsymbol{\sigma}}\, \Theta(\mathbf{g}\cdot\widehat{\boldsymbol{\sigma}})(\mathbf{g}\cdot\widehat{\boldsymbol{\sigma}}) f(\mathbf{v}') f(\mathbf{v}_1') \tag{1.156}$$

is a *reference* distribution function. The physical interpretation of Eq. (1.154) is that the effect of collisions is to make the distribution function f to relax towards the reference function f_R with a relaxation time ν_{HS}^{-1}. Note that ν_{HS} is a linear functional of f, while f_R is a nonlinear (and not quadratic) functional of f. The angular integration in Eq. (1.155) can be performed to yield

$$\nu_{\text{HS}}[\mathbf{v}|f] = \pi\sigma^2 \int d\mathbf{v}_1 |\mathbf{v}-\mathbf{v}_1| f(\mathbf{v}_1). \tag{1.157}$$

So far, we have not made any approximation to the Boltzmann collision term. A general class of kinetic models consists of replacing the operator $J[f,f]$ in Eq. (1.154) by a term $Q[f]$ in which $\nu_{\text{HS}}(\mathbf{v})$ and $f_R(\mathbf{v})$ are not necessarily given by Eqs. (1.157) and (1.156), respectively, but still they preserve the conditions (1.152), namely

$$\int d\mathbf{v}\, \{1, \mathbf{v}, v^2\} \nu_{\text{HS}}(\mathbf{v})\, [f(\mathbf{v}) - f_R(\mathbf{v})] = \{0, \mathbf{0}, 0\}. \tag{1.158}$$

As a guide to propose a form for f_R, we require that an H-theorem can be proved. In the kinetic model the entropy production due to collisions is

$$\begin{aligned} \sigma_s^{(c)} &= -k_B \int d\mathbf{v}\, Q[f] \ln f \\ &= k_B \int d\mathbf{v}\, \nu_{\text{HS}}(\mathbf{v})\, [f(\mathbf{v}) - f_R(\mathbf{v})] \ln \frac{f(\mathbf{v})}{f_R(\mathbf{v})} \\ &\quad + k_B \int d\mathbf{v}\, \nu_{\text{HS}}(\mathbf{v})\, [f(\mathbf{v}) - f_R(\mathbf{v})] \ln f_R(\mathbf{v}). \end{aligned} \tag{1.159}$$

A *sufficient* condition for $\sigma_s^{(c)} \geq 0$ is that $\ln f_R$ be a linear combination of the collisional invariants, i.e. f_R is a Maxwellian

$$f_R(\mathbf{v}) = a \exp\left(\mathbf{b}\cdot\mathbf{V} - cV^2\right) \tag{1.160}$$

defined in terms of five local quantities $a(\mathbf{r},t)$, $\mathbf{b}(\mathbf{r},t)$, and $c(\mathbf{r},t)$, such that Eq. (1.158) is fulfilled.[5]

[5]Equation (1.160) is not a *necessary* condition, as exemplified by the fact that in the original Boltzmann equation, although an H-theorem is verified, $\ln f_R$ is not a linear combination of collisional invariants.

In order to get the explicit form of the BGK model, we take the following three simplifying steps. First, we replace $\nu_{\text{HS}}[\mathbf{v}|f]$ by the local equilibrium velocity-dependent collision frequency $\nu_L(\mathbf{v}) = \nu_{\text{HS}}[\mathbf{v}|f_L]$. The result is

$$\nu_L(\mathbf{v}) = n\sigma^2 \left(\frac{2\pi k_B T}{m}\right)^{1/2} \left[e^{-\xi^2} + \frac{\sqrt{\pi}(1+2\xi^2)}{2\xi}\text{erf}(\xi)\right], \tag{1.161}$$

where ξ is the reduced peculiar velocity defined in Eq. (1.88) and $\text{erf}(x)$ is the error function (Abramowitz and Stegun, 1972). For small velocities, $\nu_L(\mathbf{v}) \approx n\sigma^2 (2\pi k_B T/m)^{1/2}\, 2\left(1+\xi^2/3\right)$, while for large velocities one has $\nu_L(\mathbf{v}) \approx n\sigma^2 (2\pi k_B T/m)^{1/2} \sqrt{\pi}\xi \left(1+1/2\xi^2\right)$. Note that, in contrast to the linear dependence of $\nu_{\text{HS}}(\mathbf{v})$ on f, $\nu_L(\mathbf{v})$ is a nonlinear functional of f through n, $\mathbf{u}$, and T. In agreement with the approximation $\nu_{\text{HS}}[\mathbf{v}|f] \to \nu_{\text{HS}}[\mathbf{v}|f_L]$, it might seem natural to make the parallel replacement $f_R[\mathbf{v}|f] \to f_R[\mathbf{v}|f_L] = f_L(\mathbf{v})$. However, the resulting kinetic model violates the conservation laws (1.152), due to the velocity dependence of $\nu_L(\mathbf{v})$. This problem can be remedied if one replaces $f_R[\mathbf{v}|f]$ by a Maxwellian of the form (1.160) (Brey and Santos, 1984; Cercignani, 1990), with the parameters a, $\mathbf{b}$, and c determined by the constraints (1.158). Nevertheless, for practical applications this model with a velocity dependent collision frequency turns out to be still too complicated. Therefore, as a second step, we take an *average* collision frequency $\overline{\nu}_L$ defined as

$$\begin{aligned} \overline{\nu}_L &= \frac{1}{n}\int d\mathbf{v}\, f_L(\mathbf{v})\nu_L(\mathbf{v}) \\ &= 4n\sigma^2\left(\frac{\pi k_B T}{m}\right)^{1/2}. \end{aligned} \tag{1.162}$$

According to elementary kinetic theory arguments (Reif, 1965) the mean free path for hard spheres is

$$\ell = \frac{1}{\sqrt{2}\pi n\sigma^2}, \tag{1.163}$$

so $\overline{\nu}_L = \langle|\mathbf{V}|\rangle/\ell$, where $\langle|\mathbf{V}|\rangle = \sqrt{8k_BT/\pi m}$ is the local equilibrium average of the magnitude of the velocity. Finally, the third step consists of setting $f_R[\mathbf{v}|f] \to f_R[\mathbf{v}|f_L] = f_L(\mathbf{v})$. In summary, these three approximations lead to the kinetic model

$$Q[f] = -\overline{\nu}_L\,(f - f_L)\,. \tag{1.164}$$

This is the BGK model for hard spheres. It describes the collisions in a statistical way without paying attention to the details of the two-body interactions. The physical idea is that the main effect of collisions is to make the distribution function tend to the local equilibrium distribution with a local relaxation time $\overline{\nu}_L^{-1}$. A BGK-like model of the *Enskog equation* Enskog equation (1.28) for a dense system of hard spheres has also been proposed (Dufty et al., 1996; Lutsko, 1997; Santos et al., 1998; Lutsko, 1998).

Although our heuristic derivation has focused on hard spheres, the BGK model can easily be extended to any interaction potential. In that case, the role of $\overline{\nu}_L$ is played by a velocity-independent local collision frequency ν, which is proportional to the density n and depends on the temperature T according to the interaction potential. For instance, dimensional analysis shows that $\nu \propto nT^q$, with $q = 1/2 - 2/w$, for r^w-potentials. The explicit form of ν can be chosen to optimize the agreement with the results obtained from the Boltzmann equation. In particular, for hard spheres it can be convenient to take a collision frequency ν different from $\overline{\nu}_L$, provided that $\nu \propto nT^{1/2}$. Replacing $J[f, f]$ by $Q[f]$ into the Boltzmann equation (1.25), one gets the BGK model kinetic equation:

$$\frac{\partial f}{\partial t} + \mathbf{v} \cdot \nabla f + \frac{\partial}{\partial \mathbf{v}} \cdot \left(\frac{\mathbf{F}}{m} f \right) = -\nu \left(f - f_L \right). \tag{1.165}$$

It is not self-evident to what extent the BGK model is mathematically more tractable than the original Boltzmann equation. In fact, Eq. (1.165) is still a highly nonlinear integro-differential equation since the collision frequency ν and the local equilibrium distribution f_L are functionals of f through their dependence on the hydrodynamic fields n, $\mathbf{u}$, and T. The essential advantage of Eq. (1.165) over the Boltzmann equation is that the velocity dependence of ν and f_L is explicitly known: ν is independent of the velocity and f_L is a Gaussian. Consequently, for given initial and boundary conditions, an implicit solution to the kinetic equation can be obtained displaying the exact velocity dependence but parameterized with the space and time dependence of the hydrodynamic fields. Use of this implicit solution in the definitions of the fields provides velocity-independent integral equations for n, $\mathbf{u}$, and T, and, hence, the full solution. Thus, the original problem in six-dimensional phase space is reduced to one in three-dimensional coordinate space. The problem remains quite difficult in general, but becomes tractable in many cases where additional simplifications such as symmetries can be exploited. Some of those cases are considered along this book.

To illustrate the above reduction, consider the initial-value problem in the absence of an external force. In that case, the *formal* solution to the BGK equation can be written as the following path integral:

$$\begin{aligned} f(\mathbf{r}, \mathbf{v}; t) &= e^{-t\mathbf{v}\cdot\nabla} U(\mathbf{r}, \mathbf{v}; t) \Bigg[f(\mathbf{r}, \mathbf{v}; 0) + \int_0^t dt' \, U^{-1}(\mathbf{r}, \mathbf{v}; t') \\ &\quad \times e^{t'\mathbf{v}\cdot\nabla} \nu(\mathbf{r}, t') f_L(\mathbf{r}, \mathbf{v}; t') \Bigg], \end{aligned} \tag{1.166}$$

where $f(\mathbf{r}, \mathbf{v}; 0)$ is the initial distribution and

$$U(\mathbf{r}, \mathbf{v}; t) \equiv \exp\left[-\int_0^t dt'\, e^{t'\mathbf{v}\cdot\nabla}\nu(\mathbf{r}, t')\right]. \tag{1.167}$$

In the above two equations, the action of the shift operator $\exp(t\mathbf{v}\cdot\nabla)$ is

$$e^{t\mathbf{v}\cdot\nabla}X(\mathbf{r}) = X(\mathbf{r} + t\mathbf{v}). \tag{1.168}$$

Equation (1.166) represents a formal solution because the right-hand side is a functional of the unknown distribution function f through the dependence of the collision frequency ν and the local equilibrium distribution f_L on the hydrodynamic fields. On the other hand, the velocity dependence of the right-hand side is explicit. Therefore, by multiplying both sides of Eq. (1.166) by 1, $\mathbf{V}$, and $\mathbf{V}^2$ one gets a *closed* set of equations for the hydrodynamic fields.

According to Eq. (1.158), the macroscopic balance equations derived from the kinetic model are still given by Eqs. (1.38), (1.39), and (1.44). In the hydrodynamic regime, as discussed in the case of the Boltzmann equation, the BGK equation admits a normal solution. For small spatial gradients, one can construct the normal solution by means of the Chapman–Enskog expansion (1.64). The equation for the first-order contribution $f^{(1)} = f_L\Phi^{(1)}$ is given by Eq. (1.77), except that the linear operator $\mathcal{L}$ is replaced by the constant ν. The solution is simply Eq. (1.93) with $A(V^2)\mathbf{V} \to \nu^{-1}\mathbf{S}(\mathbf{V})$ and $B(V^2) \to \nu^{-1}$. Consequently, the expressions for the Navier–Stokes transport coefficients are

$$\eta_0 = \frac{p}{\nu}, \tag{1.169}$$

$$\kappa_0 = \frac{5}{2}\frac{nk_B^2T}{m\nu}. \tag{1.170}$$

The value of the Prandtl number defined by Eq. (1.110) in the BGK model is $\mathrm{Pr} = 1$, which differs from the value $\mathrm{Pr} \simeq \frac{2}{3}$ given by the Boltzmann equation. This is one of the most important drawbacks of the BGK kinetic model. It is due to the simplicity of the model, where all the relaxation processes are accounted for only by a single relaxation time ν^{-1}. For instance, in the case of Maxwell molecules, Eqs. (1.111) and (1.112) indicate that the relaxation times associated with the shear and heat diffusion modes are $1/\lambda_{02}^0$ and $1/\lambda_{11}^0 = 3/2\lambda_{02}^0$, respectively. In other words, in the BGK model all the nonzero eigenvalues of the linearized collision operator $\mathcal{L}$ are collapsed into a single eigenvalue ν. To make contact with the Boltzmann results for the transport coefficients, one can adjust ν as to reproduce either the shear viscosity or the thermal conductivity. For instance, for Maxwell molecules one would take $\nu = \nu_\eta = \lambda_{02}^0$ in the first case and $\nu = \nu_\kappa = \lambda_{11}^0$ in the second case. For hard spheres (in the first Sonine approximation), Eq. (1.119) implies that

$$\nu_\eta = \frac{16}{5} n\sigma^2 \left(\frac{\pi k_B T}{m}\right)^{1/2}, \quad \nu_\kappa = \frac{2}{3}\nu_\eta. \tag{1.171}$$

None of these values coincide with the local equilibrium collision frequency (1.162), namely $\overline{\nu}_L = \frac{5}{4}\nu_\eta$. This means that the effective collision frequency associated with momentum transport is smaller than the actual average collision frequency. This discrepancy is even larger in the case of ν_κ, which is the effective collision frequency associated with energy transport.

In order to overcome the insufficiency of the BGK model to give the correct Prandtl number, other more sophisticated kinetic models have been proposed in the literature (Cercignani, 1988; Cercignani, 1990). The price to be paid is that they are mathematically more complicated than the BGK model, but they still allow one to get explicit results. Here we will analyze the so-called ellipsoidal statistical (ES) model (Holway, 1966; Cercignani, 1988) and the Liu model (Liu, 1990). Both have in common that the Prandtl number Pr appears explicitly as a free parameter, so the BGK model is recovered in the special case Pr $= 1$. The collision term in the ES and the Liu models share with the BGK model the structure of a single relaxation-time term:

$$Q[f] = -\zeta(f - f_R), \tag{1.172}$$

where ζ is an effective relaxation rate and f_R is a reference function whose choice depends on the model considered. In the case of the ES model, f_R is

$$f_R(\mathbf{v}) = n\left(\frac{mn}{2\pi}\right)^{3/2} (\det \mathsf{R})^{-1/2} \exp\left(-\frac{mn}{2}\mathsf{R}^{-1} : \mathbf{VV}\right), \tag{1.173}$$

where the tensor R is

$$\mathsf{R} = \frac{1}{\mathrm{Pr}}\left[p\mathsf{I} - (1 - \mathrm{Pr})\,\mathsf{P}\right]. \tag{1.174}$$

The ES choice of f_R is based on information theory arguments, i.e. it is the distribution function that maximizes the entropy subjected to given constraints (Holway, 1966). The reference distribution f_R has a finite norm provided that R is a positive definite matrix, i.e. its eigenvalues r_i must be non-negative. From Eq. (1.174), $r_i = \mathrm{Pr}^{-1}[p - (1 - \mathrm{Pr})p_i]$, where p_i are the eigenvalues of the pressure tensor P. Since $p_1 + p_2 + p_3 = 3p$, then $p_i \leq 3p$ and, consequently, the positiveness of r_i implies that Pr $\geq \frac{2}{3}$. The lower bound coincides with the physical value of the Prandtl number Pr $= \frac{2}{3}$. The first few moments of f_R are

$$\int d\mathbf{v}\, \{1, \mathbf{V}, m\mathbf{VV}\} f_R(\mathbf{v}) = \{n, \mathbf{0}, \mathsf{R}\}. \tag{1.175}$$

While in the BGK equation the reference function (namely, the local equilibrium distribution) is a functional of f through its hydrodynamic fields n, $\mathbf{u}$, and T, in the ES model f_R depends also on the dissipative momentum flux. In addition, since $\ln f_R$ is not a linear combination of collisional invariants, the proof of the H-theorem based on Eq. (1.159) does not hold. However, an H-theorem for the ES model has been proved (Andries et al., 2000). A sketch of the proof is presented in Appendix 1.C.

In the application of the Chapman–Enskog method to the ES model one has to expand the reference function f_R in powers of the uniformity parameter ε. To first order,

$$\mathsf{R} = p\mathsf{I} - \varepsilon\frac{1-\mathrm{Pr}}{\mathrm{Pr}}\mathsf{P}^{(1)} + \cdots, \tag{1.176}$$

so $\det \mathsf{R} = 1 + \mathcal{O}(\varepsilon^2)$. Consequently,

$$f_R^{(1)} = -f_L\frac{1-\mathrm{Pr}}{\mathrm{Pr}}p^{-1}\mathsf{P}^{(1)} : \left(\boldsymbol{\xi}\boldsymbol{\xi} - \frac{1}{3}\xi^2\mathsf{I}\right), \tag{1.177}$$

where we have used the property $\mathsf{P}^{(1)} : \mathsf{I} = \mathrm{tr}\,\mathsf{P}^{(1)} = 0$. The first-order distribution $\Phi^{(1)}$ obeys Eq. (1.77), except that

$$\mathcal{L}\Phi^{(1)} \to \zeta\left[\Phi^{(1)} + \frac{1-\mathrm{Pr}}{\mathrm{Pr}}p^{-1}\mathsf{P}^{(1)} : \left(\boldsymbol{\xi}\boldsymbol{\xi} - \frac{1}{3}\xi^2\mathsf{I}\right)\right]. \tag{1.178}$$

Equation (1.178) implies that the elements of the tensor $\mathsf{C}(\mathsf{V})$ defined by Eq. (1.81) are eigenfunctions of $\mathcal{L}$ with the corresponding eigenvalue $\nu = \zeta/\mathrm{Pr}$. The remaining nonzero eigenvalues are collapsed into the collision rate ζ. A class of linearized kinetic models consisting of retaining some eigenvalues of the linearized Boltzmann collision operator and collapsing the rest of the spectrum into a single eigenvalue was proposed by Gross and Jackson (Gross and Jackson, 1959; Sirovich, 1962; Cercignani, 1990).

Multiplying both sides of Eq. (1.77) by mV_iV_j with the replacement (1.178), one gets (1.103) with the shear viscosity given by

$$\eta_0 = p/\nu. \tag{1.179}$$

Thus, $\nu = \zeta/\mathrm{Pr}$ represents the effective collision frequency ν_η associated with the momentum transport. Similarly, multiplying both sides of Eq. (1.77) by $\mathbf{S}(\mathbf{V})$, one obtains (1.108) with the thermal conductivity given by

$$\kappa_0 = \frac{5}{2}\frac{nk_B^2T}{m\zeta}, \tag{1.180}$$

so ζ plays the role of the effective collision frequency ν_κ associated with the energy transport. This shows that the parameter Pr appearing in Eq. (1.174) is actually the Prandtl number. The physical choice is $\mathrm{Pr} = \frac{2}{3}$. The explicit expression of $\Phi^{(1)}$ is given by Eq. (1.116).

In the Liu model (Liu, 1990), the reference function is chosen as having the same functional form as the Chapman–Enskog solution to the linearized Boltzmann equation, i.e.

$$\begin{aligned} f_R(\mathbf{V}) &= f_L(\mathbf{V}) \left[1 + \frac{\zeta^{-1}}{k_B T} \left(1 - \frac{\zeta}{\nu} \right) \mathsf{C}(\mathbf{V}) : \nabla \mathbf{u} \right. \\ & \left. + \frac{\zeta^{-1}}{k_B T} \left(1 - \frac{\zeta}{\nu \mathrm{Pr}} \right) \mathbf{S}(\mathbf{V}) \cdot \nabla \ln T \right]. \end{aligned} \tag{1.181}$$

The transport coefficients obtained from this model are again given by Eqs. (1.179) and (1.180), but now the ratio ζ/ν is an arbitrary parameter of the model. In particular, the choice $\zeta/\nu = 1$ yields the same Burnett transport coefficients appearing in the pressure tensor (Garzó, 1991) as those obtained from the Boltzmann equation (Chapman and Cowling, 1970), while the choice $\zeta/\nu = \mathrm{Pr} = \frac{2}{3}$ gives the same Burnett coefficients appearing in the heat flux. In the linearized case, it is possible to prove an H-theorem for the Liu model (Liu, 1990).

7. Kinetic models for gas mixtures

Of course, all the mathematical difficulties of the Boltzmann collision operator of a single gas are inherited and augmented in the multi-component case. Not only the collision term (1.129) involves two different distribution functions, but also the collision rules (1.130) include the mass ratio of the two colliding particles and the cross section $B_{sr}(g, \chi)$ depends on the parameters of the interaction $\phi_{sr}(r)$. Thus it has been natural the extension to gas mixtures of kinetic models. In the same spirit as in the BGK model for a single gas, most of the kinetic models for mixtures proposed in the literature approximate the Boltzmann collision operator $J_{sr}[f_s, f_r]$ by a single relaxation-time term of the form

$$J_{sr}[\mathbf{v}|f_s, f_r] \to -\nu_{sr} \left[f_s(\mathbf{v}) - f_{sr}(\mathbf{v}) \right], \tag{1.182}$$

where ν_{sr} is an effective collision frequency of a particle of species s with particles of species r. It has the properties $\nu_{sr} \propto n_r$ and $n_s \nu_{sr} = n_r \nu_{rs}$. The net collision frequency for particles of species s is

$$\nu_s = \sum_r \nu_{sr}. \tag{1.183}$$

Equation (1.182) means that the main effect on the distribution function f_s due to collisions with particles of species r is to produce a tendency towards the *reference* distribution function f_{sr}. The velocity dependence of f_{sr} is different in each kinetic model considered. This explicit dependence involves a number of fields to be determined by imposing that Eq. (1.182) retains the main properties of the Boltzmann collision operator J_{sr}. The most important requirements are those related to the conservation of mass, momentum, and energy, Eqs. (1.132)–(1.134). However, the latter conditions are not enough to determine the fields defining f_{sr} and one has to resort to additional properties of the original Boltzmann collision term. Usually, one requires the collisional transfers of momentum and energy obtained from the model to be the same as those of the Boltzmann equation for Maxwell molecules, Eqs. (1.B.1) and (1.B.7).

The most natural way of extending the BGK model to mixtures is to assume a Gaussian form for the reference distribution function f_{sr}:

$$f_{sr}(\mathbf{v}) = n_s \left(\frac{m_s}{2\pi k_B T_{sr}}\right)^{3/2} \exp\left[-\frac{m_s}{2k_B T_{sr}} (\mathbf{v} - \mathbf{u}_{sr})^2\right], \tag{1.184}$$

where $\mathbf{u}_{sr}$ and T_{sr} are fields to be determined. In general, $\mathbf{u}_{sr} \neq \mathbf{u}_{rs}$ and $T_{sr} \neq T_{rs}$, so the number of unknowns is $4N^2$, N being the number of species. According to (1.184) the effect of collisions of particles of species s with particles of species r is to make the distribution f_s tend to a sort of local equilibrium distribution characterized by an effective flow velocity $\mathbf{u}_{sr}$ and an effective temperature T_{sr}. It is straightforward to get the relations

$$m_s \int d\mathbf{v}\, \mathbf{V} \left[f_s(\mathbf{v}) - f_{sr}(\mathbf{v})\right] = \rho_s \left(\mathbf{u}_s - \mathbf{u}_{sr}\right), \tag{1.185}$$

$$\frac{m_s}{2} \int d\mathbf{v}\, V^2 \left[f_s(\mathbf{v}) - f_{sr}(\mathbf{v})\right] = \frac{3}{2} n_s k_B \left(T_s - T_{sr}\right) - \frac{1}{2}\rho_s \left(\mathbf{u}_{sr} - \mathbf{u}\right)^2, \tag{1.186}$$

where $\mathbf{V} = \mathbf{v} - \mathbf{u}$,

$$\mathbf{u}_s = \frac{1}{n_s} \int d\mathbf{v}\, \mathbf{v} f_s(\mathbf{v}) \tag{1.187}$$

is the mean flow velocity of species s and

$$T_s = \frac{m_s}{3 n_s k_B} \int d\mathbf{v}\, V^2 f_s(\mathbf{v}) \tag{1.188}$$

is the partial temperature of species s, measuring the mean kinetic energy of particles of species s. Note that the mass flux of species s can be written as $\mathbf{j}_s = \rho_s(\mathbf{u}_s - \mathbf{u})$, where $\rho_s = m_s n_s$ is the mass density of species s. The condition $\sum_s \mathbf{j}_s = \mathbf{0}$ then yields

$$\mathbf{u} = \frac{1}{\rho} \sum_{s=1}^{N} \rho_s \mathbf{u}_s. \tag{1.189}$$

Analogously, Eq. (1.127) implies that

$$T = \sum_{s=1}^{N} x_s T_s, \tag{1.190}$$

$x_s = n_s/n$ being the mole fraction of species s. It is important to note that the partial temperatures T_s are not hydrodynamic quantities since they do not enter into the energy balance equation because the conserved quantity is the total energy (which is directly related to the total temperature T) and not the energy of each species.

The conservation of the number of particles of each species is automatically satisfied by the collision term (1.182) with the choice (1.184). The momentum and energy conservation equations (1.133) and (1.134) lead to the conditions

$$m_s\left(\mathbf{u}_s - \mathbf{u}_{sr}\right) + m_r\left(\mathbf{u}_r - \mathbf{u}_{rs}\right) = \mathbf{0}, \tag{1.191}$$

$$\frac{3}{2}k_B\left(T_s - T_{sr}\right) - \frac{1}{2}m_s\left(\mathbf{u}_{sr} - \mathbf{u}\right)^2 + \frac{3}{2}k_B\left(T_r - T_{rs}\right) - \frac{1}{2}m_r\left(\mathbf{u}_{rs} - \mathbf{u}\right)^2 = 0, \tag{1.192}$$

where we have taken into account that $n_s\nu_{sr} = n_r\nu_{rs}$. This provides $2N(N+1)$ conditions, so that we still need $2N(N-1)$ additional conditions to completely specify the parameters of the kinetic model. One possibility might be to take the symmetry relations $\mathbf{u}_{sr} = \mathbf{u}_{rs}$ and $T_{sr} = T_{rs}$. However, this choice is too simplistic for T_{sr}. For instance, in the absence of mutual diffusion ($\mathbf{u}_{sr} = \mathbf{u}$) one would have $T_{sr} = T_{rs} = (T_s + T_r)/2$ independently of the masses of both species. Moreover, this simplification prevents the model to reproduce simultaneously the collisional transfer of momentum and energy of the Boltzmann equation for Maxwell molecules, Eqs. (1.B.1) and (1.B.7), regardless of the choice of the collision frequencies ν_{sr}. Therefore, we discard the simple prescription $T_{sr} = T_{rs}$. On the other hand, one can keep the symmetry relation $\mathbf{u}_{sr} = \mathbf{u}_{rs}$ and determine ν_{sr} from Eq. (1.B.1). According to Eq. (1.191), the symmetry relation for $\mathbf{u}_{sr}$ yields

$$\mathbf{u}_{sr} = \frac{m_s\mathbf{u}_s + m_r\mathbf{u}_r}{m_s + m_r}. \tag{1.193}$$

Next, Eq. (1.B.1) allows one to identify ν_{sr} as

$$\nu_{sr} = \frac{m_s + m_r}{m_s m_r}\lambda_{sr} n_r, \tag{1.194}$$

where λ_{sr} is defined by Eq. (1.B.3). Finally, the temperatures T_{sr} are obtained from Eq. (1.B.7) and the result is

$$\begin{aligned} T_{sr} &= T_s + \frac{2m_s m_r}{(m_s+m_r)^2}\left[T_r - T_s + \frac{m_s - m_r}{3k_B}(\mathbf{u}_s - \mathbf{u})\cdot(\mathbf{u}_r - \mathbf{u})\right] \\ &\quad -\frac{m_s}{3k_B}(\mathbf{u}_{sr} - \mathbf{u})^2 \\ &= \overline{T}_s + \frac{2m_s m_r}{(m_s+m_r)^2}\left[\overline{T}_r - \overline{T}_s + \frac{m_r}{6k_B}(\mathbf{u}_s - \mathbf{u}_r)^2\right]. \end{aligned} \tag{1.195}$$

In the last equality we have introduced the temperature

$$\overline{T}_s \equiv T_s - \frac{m_s}{3k_B}(\mathbf{u}_s - \mathbf{u})^2, \tag{1.196}$$

which is defined by (1.188), except that the peculiar velocity $\mathbf{V} = \mathbf{v} - \mathbf{u}$ is replaced by $\mathbf{v} - \mathbf{u}_s$. Equation (1.195) shows that T_{sr} is positive definite. Note that, in the case of self-collisions ($s = r$), $\mathbf{u}_{ss} = \mathbf{u}_s$ and $T_{ss} = \overline{T}_s$. For cross collisions, $T_{sr} \neq T_{rs}$ in general. For instance, if $\mathbf{u}_s = \mathbf{u}_r$ and $m_s \gg m_r$, one has $T_{sr} \simeq \overline{T}_s$ and $T_{rs} \simeq \overline{T}_r$.

The kinetic model defined by Eqs. (1.182), (1.184), and (1.193)–(1.195) was proposed by Gross and Krook (1956) and is usually referred to as the Gross–Krook (GK) model. For certain values of the parameters of the mixture, an H-theorem may be proved (Morse, 1964). In general, the GK model is still very complicated because f_{sr} is a highly nonlinear functional of f_s and f_r through n_s, $\mathbf{u}_s$, $\mathbf{u}_r$, $\overline{T}_s$, and $\overline{T}_r$. A simpler version of the GK model (Sirovich, 1962; Hamel, 1966) can be obtained by expanding f_{sr} around the local equilibrium distribution characterized by $\mathbf{u}_s$ and $\overline{T}_s$, retaining terms of first order in $T_{sr} - \overline{T}_s$ and second order in $\mathbf{u}_{sr} - \mathbf{u}_s$, and replacing $[(\mathbf{v}-\mathbf{u}_s)\cdot(\mathbf{u}_{sr}-\mathbf{u}_s)]^2$ by $\frac{1}{3}(\mathbf{v}-\mathbf{u}_s)^2(\mathbf{u}_{sr}-\mathbf{u}_s)^2$. In this approximation,

$$\begin{aligned} f_{sr}(\mathbf{v}) &= n_s\left(\frac{m_s}{2\pi k_B \overline{T}_s}\right)^{3/2} \exp\left[-\frac{m_s}{2k_B\overline{T}_s}(\mathbf{v}-\mathbf{u}_s)^2\right] \\ &\quad \times\left\{1 + \frac{m_s}{k_B\overline{T}_s}(\mathbf{v}-\mathbf{u}_s)\cdot(\mathbf{u}_{sr}-\mathbf{u}_s)\right. \\ &\quad + \left[\frac{m_s}{2k_B\overline{T}_s}(\mathbf{v}-\mathbf{u}_s)^2 - \frac{3}{2}\right] \\ &\quad \left.\times\left[\frac{T_{sr}-\overline{T}_s}{\overline{T}_s} + \frac{m_s}{3k_B\overline{T}_s}(\mathbf{u}_{sr}-\mathbf{u}_s)^2\right]\right\}. \end{aligned} \tag{1.197}$$

Since the right-hand sides of Eqs. (1.185) and (1.186) are of first order in T_{sr} and second order in $\mathbf{u}_{sr}$, the model given by Eq. (1.197) is still consistent with the momentum and energy transfer equations of the Boltzmann equation for Maxwell molecules.

The original GK model was proposed for Maxwell molecules, as indicated by the choice (1.194). However, for repulsive interaction potentials $\phi_{sr}(r) = K_{sr}r^{-w}$, the transfer equations (1.B.1) and (1.B.7) are still valid in the first Sonine approximation, except that the parameters λ_{sr} depend on the partial temperatures of each species (Goldman and Sirovich, 1967). As a consequence, the relations (1.193) and (1.195) also apply, but the collision frequencies can be identified as (Garzó and López de Haro, 1995)

$$\nu_{sr} = \mathrm{A}(w)n_r \left(K_{sr}\frac{m_s + m_r}{m_s m_r} \right)^{2/w} \left(\frac{2k_B \overline{T}_s}{m_s} + \frac{2k_B \overline{T}_r}{m_r} \right)^q, \tag{1.198}$$

where $q = 1/2 - 2/w$ and $\mathrm{A}(w)$ is a numerical constant to be adjusted to optimize the agreement with the Boltzmann equation.

The GK model, as well as its simplified version (1.197), have an important drawback. Let us consider the special case of mechanically identical particles ($m_s = m$ and $K_{sr} = K$), so particles of each species are only distinguished by a tag. This situation does not imply that all the species have the same mean velocity and/or the same temperature. We can imagine, for instance, that particles with a given tag are initially concentrated in a small region and they subsequently spread out (self-diffusion problem); or they can initially have a smaller (or larger) mean kinetic energy than the rest of the system and evolve in time until equipartition of energy is reached (thermalization problem). Since the tag of the particles is not a mechanical property, the *total* distribution function $f = \sum_s f_s$ must obey the same closed kinetic equation as a single gas.[6] This physical principle is obviously respected by the Boltzmann equation, as shown by the property (1.131). Since the relaxation model (1.182) is a generalization of the BGK model to mixtures, the relation equivalent to (1.131) is

$$\sum_{s=1}^{N}\sum_{r=1}^{N} \nu_{sr}\left(f_s - f_{sr}\right) = \nu(f - f_L), \quad m_s = m, \quad K_{sr} = K, \tag{1.199}$$

where $\nu = \nu_s = \sum_r \nu_{sr}$ is the common collision frequency of all the species and f_L is the local equilibrium distribution function of the whole system characterized by the total density n, the flow velocity $\mathbf{u}$, and the total temperature T. However, the highly nonlinear structure of the reference function f_{sr} prevents the property (1.199) to be generally verified. This limitation is not relevant in the case of disparate-mass binary mixtures or in those nonequilibrium steady

[6]In particular, if f is initially the equilibrium Maxwell–Boltzmann distribution, it will remain in that state, unless the boundary conditions impose a nonequilibrium state.

states where the existence of different temperatures is a consequence of the particles having different masses, so the temperatures become equal in the limit of identical particles.

In any case, it seems desirable to construct a model having the structure of Eq. (1.182) but complying with the property (1.199). A simple possibility consists of carrying out steps similar to those made in the derivation of Eq. (1.197) from (1.184), except that the expansion is made about the local equilibrium distribution

$$f_s^L(\mathbf{v}) = n_s \left(\frac{m_s}{2\pi k_B T}\right)^{3/2} e^{-m_s V^2/2k_B T} \tag{1.200}$$

characterized by the number density n_s, and the temperature T and the flow velocity $\mathbf{u}$ of the *whole* system. More explicitly, the reference distribution function f_{sr} in this alternative model is

$$\begin{aligned} f_{sr}(\mathbf{v}) &= f_s^L(\mathbf{v})\left\{1 + \frac{m_s}{k_B T}\mathbf{V}\cdot(\mathbf{u}_{sr} - \mathbf{u}) \right. \\ &\left. + \left(\frac{m_s V^2}{2k_B T} - \frac{3}{2}\right)\left[\frac{T_{sr} - T}{T} + \frac{m_s}{3k_B T}(\mathbf{u}_{sr} - \mathbf{u})^2\right]\right\}. \end{aligned} \tag{1.201}$$

This model was proposed by Garzó et al. (1989) and will be referred to as the GSB model. The physical idea behind this model is that the main global effect of collisions on particles of species s is to produce a tendency towards the local equilibrium state f_s^L. The peculiarities of collisions with particles of species r are taken into account as a correction factor through a quadratic function of $\mathbf{V}$ with coefficients depending on the properties of species s and r. The reference function (1.201) does not intend to capture the specific details of the collisions of type s-r and so it depends on the local state of the system as a whole. This is close to the spirit of the BGK model for a single gas.

In the case of mechanically equivalent Maxwell molecules, Eq. (1.194) yields $\nu_{sr} = x_r \nu$. This property is still valid for other potentials if the temperature-dependence of ν_{sr} occurs through the global temperature T instead of the partial temperatures. For equal masses, Eqs. (1.193) and (1.195) become

$$\mathbf{u}_{sr} = \frac{1}{2}(\mathbf{u}_s + \mathbf{u}_r), \tag{1.202}$$

$$T_{sr} + \frac{m}{3k_B}(\mathbf{u}_{sr} - \mathbf{u})^2 = \frac{1}{2}(T_s + T_r), \tag{1.203}$$

respectively. Using all of this in Eq. (1.201), one gets

$$\sum_r \nu_{sr} f_{sr} = \nu f_s^L \left[1 + \frac{m}{2k_B T}\mathbf{V}\cdot(\mathbf{u}_s - \mathbf{u}) + \left(\frac{mV^2}{2k_B T} - \frac{3}{2}\right)\frac{T_s - T}{2T}\right]. \tag{1.204}$$

Taking into account that $f_s^L = x_s f_L$, then $\sum_{s,r} \nu_{sr} f_{sr} = \nu f_L$, what proves that the GSB model verifies the property (1.199).

The GSB model has been successfully applied to several nonequilibrium problems, such as the vapor flow condensing onto a plane condensed phase in the presence of a non-condensable gas (Sone et al., 1992; Aoki and Doi, 1994), the color conductivity problem (Marín et al., 1996a), and the diffusion of particles in shear flows (Garzó and López de Haro, 1992; Santos and Garzó, 1992). In general, the results obtained from the GSB model agree well with exact and numerical results obtained from the Boltzmann equation (Marín et al., 1994; Marín et al., 1996b; Aoki et al., 1998). However, the price to be paid by the GSB model to be consistent with the BGK model in the limit of identical particles, while keeping the structure of Eq. (1.182), is that the reference distribution (1.201) is not positive definite. As a consequence, there might be solutions of the kinetic model where the positiveness of the distribution function is lost for large velocities. Although this shortcoming may not be relevant to the low-degree moments related to the hydrodynamic fields and their fluxes, it can be significant for physical effects associated with the high-energy population.

In the GK and GSB models each one of the Boltzmann collision terms $J_{sr}[f_s, f_r]$ is modeled as a single relaxation-time term. This approach captures essentially the binary character of the collisions. However, the time evolution of the distribution function f_s is governed by the sum $\sum_r J_{sr}[f_s, f_r]$, which represents the net effect of collisions on particles of species s. It is then natural to construct a kinetic model which approximates the above sum, rather than the individual terms, by a unique relaxation-time term. In this approach,

$$\sum_{r=1}^{N} J_{sr}[f_s, f_r] \to -\widetilde{\nu}_s \left(f_s - f_s^R\right), \tag{1.205}$$

where $\widetilde{\nu}_s$ is an effective collision frequency (whose choice is crucial) and f_s^R is a reference distribution function given by

$$f_s^R(\mathbf{v}) = n_s \left(\frac{m_s}{2\pi k_B \widetilde{T}_s}\right)^{3/2} \exp\left[-\frac{m_s}{2k_B \widetilde{T}_s} \left(\mathbf{v} - \widetilde{\mathbf{u}}_s\right)^2\right]. \tag{1.206}$$

Here the effective velocity $\widetilde{\mathbf{u}}_s$ and the effective temperature $\widetilde{T}_s$ are to be determined by imposing certain consistency conditions. The conservation of the number of particles of species s is trivially satisfied by Eq. (1.205). The conservation of the total momentum and energy implies that

$$\sum_{s=1}^{N} \widetilde{\nu}_s m_s \int d\mathbf{v} \,\{\mathbf{V}, V^2\} \left(f_s - f_s^R\right) = \{\mathbf{0}, 0\}, \tag{1.207}$$

leading to the requirements

$$\sum_{s=1}^{N} \widetilde{\nu}_s \rho_s \left(\widetilde{\mathbf{u}}_s - \mathbf{u}_s\right) = \mathbf{0}, \tag{1.208}$$

$$\sum_{s=1}^{N} \widetilde{\nu}_s \left[3 n_s k_B \left(\widetilde{T}_s - T_s\right) + \rho_s \left(\widetilde{\mathbf{u}}_s - \mathbf{u}\right)^2\right] = 0. \tag{1.209}$$

This provides 4 constraints which are not sufficient to determine the $4N$ unknowns. As done before, we require the model to give the same transfer equations as the Boltzmann equation for Maxwell molecules, namely

$$-\widetilde{\nu}_s \int d\mathbf{v} \, \{\mathbf{V}, V^2\} \left(f_s - f_s^R\right) = \sum_{r=1}^{N} \int d\mathbf{v} \, \{\mathbf{V}, V^2\} J_{sr}[f_s, f_r]. \tag{1.210}$$

From here one gets

$$\widetilde{\mathbf{u}}_s = \mathbf{u}_s + \sum_{r=1}^{N} \frac{\nu_{sr}}{\widetilde{\nu}_s} \frac{m_r}{m_s + m_r} \left(\mathbf{u}_r - \mathbf{u}_s\right), \tag{1.211}$$

$$\begin{aligned} \widetilde{T}_s &= T_s - \frac{m_s}{3k_B} \left(\widetilde{\mathbf{u}}_s - \mathbf{u}\right)^2 + 2 \sum_{r=1}^{N} \frac{\nu_{sr}}{\widetilde{\nu}_s} \frac{m_s m_r}{(m_s + m_r)^2} \\ &\quad \times \left[T_r - T_s + \frac{m_s - m_r}{3k_B} (\mathbf{u}_r - \mathbf{u}) \cdot (\mathbf{u}_s - \mathbf{u})\right], \end{aligned} \tag{1.212}$$

where ν_{sr} is given by Eq. (1.194). The quantities $\widetilde{\mathbf{u}}_s$ and $\widetilde{T}_s$ depend, not only on $\mathbf{u}_s$ and T_s, but also on the flow velocities and the partial temperatures of the remaining species. It can be from Eq. (1.212) that $\widetilde{T}_s > 0$ if $\widetilde{\nu}_s \geq \frac{1}{2} \sum_r \nu_{sr}$ (Andries et al., 2002). The kinetic model consisting of Eqs. (1.205), (1.206), (1.211), and (1.212) was proposed by Andries et al. (2002) and will be referred to as the AAP model. This model has the nice property of verifying an H-theorem (Andries et al., 2002). In the limit of identical particles, ν_{sr} is simply proportional to the mole fraction x_r, i.e. $\nu_{sr} = \nu x_r$, and $\widetilde{\nu}_s$ is a constant, i.e. $\widetilde{\nu}_s = \widetilde{\nu}$. The choice $\widetilde{\nu} = \frac{1}{2}\nu$ leads to $\widetilde{\mathbf{u}}_s = \mathbf{u}$ and $\widetilde{T}_s = T$ and, therefore, the AAP model is consistent with the BGK equation, i.e.

$$\sum_{s=1}^{N} \widetilde{\nu}_s \left(f_s - f_s^R\right) = \widetilde{\nu}(f - f_L), \quad m_s = m, \quad K_{sr} = K. \tag{1.213}$$

The natural choice of $\widetilde{\nu}_s$ in the general case is $\widetilde{\nu}_s = \frac{1}{2} \sum_r \nu_{sr} = \frac{1}{2}\nu_s$. The fact that the effective collision frequency $\widetilde{\nu}_s$ appearing in Eq. (1.205) differs from the net collision frequency ν_s is perhaps the main limitation of the AAP model.

It is straightforward to apply the Chapman–Enskog method to the above kinetic models. To first order in the gradients, the GK and GSB models become identical since $\mathbf{u}_{sr} - \mathbf{u}$ is of first order, while $T_{sr} - T$ is at least of second order (Garzó et al., 1989). Therefore, from Eqs. (1.184) or (1.201) one gets

$$\sum_{r=1}^{N} \nu_{sr} f_{sr} \to \nu_s f_s^L \left[1 + \frac{m_s}{k_B T} \mathbf{V} \cdot \sum_{r=1}^{N} \frac{\nu_{sr}}{\nu_s} (\mathbf{u}_{sr} - \mathbf{u}) \right]. \tag{1.214}$$

The expressions for the shear viscosity and the thermal conductivity appearing in Eqs. (1.144) and (1.145) are (Garzó et al., 1989)

$$\eta_0 = p \sum_{s=1}^{N} \frac{x_s}{\nu_s}, \tag{1.215}$$

$$\kappa_0 = \frac{5}{2} p k_B \sum_{s=1}^{N} \frac{x_s}{m_s \nu_s}. \tag{1.216}$$

The $N \times N$ matrix of the phenomenological transport coefficients L_{sr} appearing in Eq. (1.143) is

$$\mathsf{L} = -T \mathsf{A}^{-1} \cdot \mathsf{B}, \tag{1.217}$$

where $A_{ss} = 0$,

$$A_{sr} = \frac{\nu_{sr}}{\rho_r} \frac{m_r}{m_s + m_r} + \sum_{r' \neq s} \frac{\nu_{sr'}}{\rho_s} \frac{m_{r'}}{m_s + m_{r'}}, \quad s \neq r, \tag{1.218}$$

and

$$B_{sr} = \delta_{sr} - \frac{\rho_r}{\rho}. \tag{1.219}$$

In the binary case ($N = 2$), L_{11} is given by Eq. (1.149) with the mutual diffusion coefficient

$$D_0 = \frac{m_1 + m_2}{m_1 m_2} \frac{n_2 k_B T}{n \nu_{12}}. \tag{1.220}$$

This expression coincides with the one obtained from the Boltzmann equation for Maxwell molecules, Eq. (1.150), when Eq. (1.194) is used. On the other hand, the expressions for the shear viscosity and the thermal conductivity coefficients do not coincide with those derived from the Boltzmann equation. In particular, in the limit $x_1 \to 1$, Eq. (1.215) yields $\eta_0 = n_1 k_B T / \nu_{11} = m_1 k_B T / 2\lambda_{11}$, while the exact result from the Boltzmann equation is $\eta_0 = m_1 k_B T / \lambda'_{11}$. Therefore, the use of Eq. (1.194) makes the GK and GSB models underestimate the shear viscosity in the one-component limit by a factor $\lambda'_{11} / 2\lambda_{11} \simeq 0.775$. The same happens in the limit of identical particles. Nevertheless, given the flexibility of kinetic models to accommodate the collision frequencies by optimizing the agreement with the Boltzmann equation, we can choose the parameters ν_{sr} to be given by

$$\nu_{sr} = \frac{m_s + m_r}{2m_s m_r} \lambda'_{sr} n_r \tag{1.221}$$

instead of Eq. (1.194). While the choice (1.194) is more convenient in diffusion problems, the choice (1.221) is more adequate for shearing problems. To focus on the influence of the mole fraction and mass and force constant ratios on the shear viscosity, it is convenient to define the dimensionless shear viscosity coefficient

$$\eta_0^* = \frac{\lambda'_{11}}{m_1 k_B T} \eta_0, \tag{1.222}$$

so η_0^* is normalized to 1 in the limit $x_1 \to 1$. Figure 1.3 shows η_0^* versus the mole fraction x_1 for equal force constants ($K_{11} = K_{12} = K_{22}$) and two mass ratios. The kinetic models succeed in capturing the main qualitative features of η_0^*: it increases as the concentration of species 1 decreases if $m_1 < m_2$, while the opposite happens if $m_1 > m_2$.

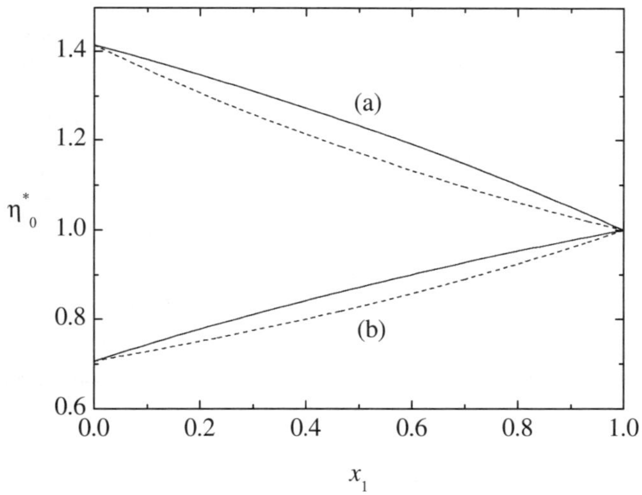

Figure 1.3. Plot of the dimensionless shear viscosity η_0^* as a function of the mole fraction x_1 for a binary mixture of Maxwell molecules with equal force constants ($K_{11} = K_{12} = K_{22}$) and mass ratios (a) $m_1/m_2 = \frac{1}{2}$ and (b) $m_1/m_2 = 2$. The solid lines are the predictions of the Boltzmann equation, while the dashed lines are the predictions of the GK, GSB, and AAP kinetic models.

As for the AAP model, to first order in the gradients one has

$$\widetilde{\nu}_s f_s^R \to \widetilde{\nu}_s f_s^L \left[1 + \frac{m_s}{k_B T} \mathbf{V} \cdot (\widetilde{\mathbf{u}}_s - \mathbf{u}) \right]. \tag{1.223}$$

By following standard steps of the Chapman–Enskog method one can easily get the transport coefficients (Andries et al., 2002). The expressions for the shear viscosity and the thermal conductivity coefficients are identical to Eqs. (1.215) and (1.216), except that ν_s must be replaced by $\widetilde{\nu}_s$. In the evaluation of the mass fluxes, the dependence on the collision frequency $\widetilde{\nu}_s$ cancels and so the diffusion coefficients L_{sr} coincide with those obtained from the GK and GSB models, Eqs. (1.217)–(1.219).

In conclusion, the different kinetic models for mixtures based on the BGK equation have advantages as well as disadvantages. Depending on the problem of interest a given model can be preferable over other ones. The choice of the model is then a compromise between simplicity, accuracy, and consistency with exact properties.

APPENDIX 1.A: Collisional moments for a single Maxwell gas

In this Appendix some collisional moments used in this Chapter are listed. They are restricted to Maxwell molecules, in which case the collision rate $gB(g,\chi)$ is independent of the relative velocity g.

Of course, the first non-trivial collisional moments are those of second degree. They are given by (Truesdell and Muncaster, 1980)

$$\int d\mathbf{v} m\mathbf{V}\mathbf{V} J[f,f] = -\nu\left(\mathsf{P} - p\mathsf{I}\right), \tag{1.A.1}$$

where $\mathbf{V} = \mathbf{v} - \mathbf{u}$ is the peculiar velocity, P is the pressure tensor, $p = \frac{1}{3}\mathrm{tr}\,\mathsf{P}$ is the hydrostatic pressure, and ν is an effective collision frequency given by

$$\nu = 3n\mathrm{A}_2, \tag{1.A.2}$$

with

$$\mathrm{A}_r \equiv \int d\Omega \sin^r\left(\frac{\chi}{2}\right)\cos^r\left(\frac{\chi}{2}\right) gB(g,\chi). \tag{1.A.3}$$

The numerical value of A_2 is $\mathrm{A}_2 = 1.3703\sqrt{2K/m}$.

Next, we consider the third-degree moments. It is convenient to introduce the homogeneous polynomials

$$Y_{2|i}(\mathbf{V}) = V^2 V_i, \tag{1.A.4}$$

$$Y_{0|ijk}(\mathbf{V}) = V_i V_j V_k - \frac{1}{5}V^2(V_i\delta_{jk} + V_j\delta_{ik} + V_k\delta_{ij}), \tag{1.A.5}$$

ten of which are linearly independent. The corresponding collisional moments are (Truesdell and Muncaster, 1980)

$$\frac{1}{n}\int d\mathbf{v} Y_{2|i}(\mathbf{V}) J[f,f] = -\frac{2}{3}\nu M_{2|i}, \tag{1.A.6}$$

$$\frac{1}{n}\int d\mathbf{v} Y_{0|ijk}(\mathbf{V}) J[f,f] = -\frac{3}{2}\nu M_{0|ijk}, \tag{1.A.7}$$

where

$$M_{2|i} = \frac{1}{n}\int d\mathbf{v} Y_{2|i}(\mathbf{V}) f(\mathbf{v}), \tag{1.A.8}$$

$$M_{0|ijk} = \frac{1}{n}\int d\mathbf{v} Y_{0|ijk}(\mathbf{V}) f(\mathbf{v}) \tag{1.A.9}$$

are the associated moments of the distribution function.

Finally, we consider the following set of fourth-degree polynomials (Truesdell and Muncaster, 1980):

$$Y_{4|0}(\mathbf{V}) = V^4, \tag{1.A.10}$$

$$Y_{2|ij}(\mathbf{V}) = V^2(V_iV_j - \frac{1}{3}V^2\delta_{ij}), \tag{1.A.11}$$

$$\begin{aligned} Y_{0|ijkl}(\mathbf{V}) &= V_iV_jV_kV_l - \frac{1}{7}V^2(V_iV_j\delta_{kl} + V_iV_k\delta_{jl} \\ &+V_iV_l\delta_{kj} + V_jV_k\delta_{il} + V_jV_l\delta_{ik} + V_kV_l\delta_{ij}) \\ &+\frac{1}{35}V^4(\delta_{ij}\delta_{kl} + \delta_{ik}\delta_{jl} + \delta_{il}\delta_{jk}). \end{aligned} \tag{1.A.12}$$

There are 15 independent polynomials, say $Y_{4|0}$, $Y_{2|xx}$, $Y_{2|xy}$, $Y_{2|xz}$, $Y_{2|yy}$, $Y_{2|yz}$, $Y_{0|xxxx}$, $Y_{0|xxxy}$, $Y_{0|xxxz}$, $Y_{0|xyyy}$, $Y_{0|xzzz}$, $Y_{0|yyyy}$, $Y_{0|yyyz}$, $Y_{0|yzzz}$, and $Y_{0|zzzz}$. If we denote by $M_{2r|ij...}$ the moments of f corresponding to the polynomials $Y_{2r|ij...}(\mathbf{V})$, then the corresponding collisional moments are (Truesdell and Muncaster, 1980):

$$\frac{1}{n}\int d\mathbf{v}Y_{4|0}(\mathbf{V})J[f,f] = -\frac{2}{3}\nu[M_{4|0} + \frac{1}{\rho^2}(P_{ij}P_{ij} - 18p^2)], \tag{1.A.13}$$

$$\begin{aligned} \frac{1}{n}\int d\mathbf{v}Y_{2|ij}(\mathbf{V})J[f,f] &= -\frac{7}{6}\nu[M_{2|ij} + \frac{1}{7\rho^2}(4P_{ik}P_{kj} - 15pP_{ij} \\ &-\frac{4}{3}P_{kl}P_{kl}\delta_{ij} + 15p^2\delta_{ij})], \end{aligned} \tag{1.A.14}$$

$$\begin{aligned} \frac{1}{n}\int d\mathbf{v}Y_{0|xxxx}(\mathbf{V})J[f,f] &= -\nu_1M_{0|xxxx} + \frac{3}{35}(2\nu - \nu_1)\frac{1}{\rho^2} \\ &\times[8P_{xx}^2 + 3(P_{yy}^2 + P_{zz}^2) + 2P_{yy}P_{zz} \\ &-8P_{xx}(P_{yy} + P_{zz}) \\ &-16(P_{xy}^2 + P_{xz}^2) + 4P_{yz}^2], \end{aligned} \tag{1.A.15}$$

$$\begin{aligned} \frac{1}{n}\int d\mathbf{v}Y_{0|xxxy}(\mathbf{V})J[f,f] &= -\nu_1M_{0|xxxy} - \frac{3}{7}(2\nu - \nu_1)\frac{1}{\rho^2}[2P_{yz}P_{xz} \\ &+P_{xy}(3P_{yy} + P_{zz} - 4P_{xx})]. \end{aligned} \tag{1.A.16}$$

In Eqs. (1.A.14) and (1.A.15), $\rho = mn$ is the mass density and the collision frequency ν_1 is given by

$$\nu_1 = \frac{7}{12}\left(4 - 5\frac{\mathrm{A}_4}{\mathrm{A}_2}\right)\nu = 1.8731\nu, \tag{1.A.17}$$

where in the last equality we have taken into account that $\mathrm{A}_4/\mathrm{A}_2 = 0.15779$. The collision frequencies ν and ν_1 determine the eigenvalues of the linearized collision operator associated with the eigenfunctions of second-, third-, and fourth-degree: $\lambda_{02}^0 = \nu$, $\lambda_{03}^0 = \frac{3}{2}\nu$, $\lambda_{11}^0 = \frac{2}{3}\nu$, $\lambda_{04}^0 = \nu_1$, $\lambda_{12}^0 = \frac{7}{6}\nu$, and $\lambda_{20}^0 = \frac{2}{3}\nu$.

The explicit expressions for the collisional moments of fifth-degree are given by Truesdell and Muncaster (1980). Those of sixth-, seventh-, and eighth-degree can be found in (Sabbane and Tij, 2002).

APPENDIX 1.B: Collisional moments for a multi-component Maxwell gas

In the case of a mixture, the first-degree moments are related to the momentum transfer due to collisions between particles of different species. They are given by (Holway, 1966; Goldman and Sirovich, 1967; Andries et al., 2002)

$$m_s \int d\mathbf{v}\, \mathbf{V} J_{sr}[f_s, f_r] = -\frac{\lambda_{sr}}{m_s m_r} \left(\rho_r \mathbf{j}_s - \rho_s \mathbf{j}_r\right), \tag{1.B.1}$$

where

$$\mathbf{j}_s = m_s \int d\mathbf{v}\, \mathbf{V} f_s(\mathbf{v}) \tag{1.B.2}$$

is the mass flux of species s relative to the flow velocity $\mathbf{u}$ and

$$\lambda_{sr} = 4\pi \times 0.422 \left(K_{sr} \frac{m_s m_r}{m_s + m_r}\right)^{1/2}, \tag{1.B.3}$$

K_{sr} being the proportionality constant in the potential function $\phi_{sr}(r) = K_{sr} r^{-4}$.

The second-degree collisional moments are (Holway, 1966; Goldman and Sirovich, 1967)

$$\begin{aligned} m_s \int d\mathbf{v}\mathbf{V}\mathbf{V} J_{sr}[f_s, f_r] &= \frac{\lambda'_{sr}}{(m_s + m_r) m_s} \left[\left(\rho_s p_r + \rho_r p_s - \frac{2}{3}\mathbf{j}_s \cdot \mathbf{j}_r\right) \mathsf{I} \right. \\ &\quad \left. - \rho_s \mathsf{P}_r - \rho_r \mathsf{P}_s + \mathbf{j}_s \mathbf{j}_r + \mathbf{j}_r \mathbf{j}_s \right] \\ &\quad - \frac{\lambda_{sr}}{(m_s + m_r) m_s} \left[2\left(\frac{m_s}{m_r} \rho_r \mathsf{P}_s - \rho_s \mathsf{P}_r\right) \right. \\ &\quad \left. + \left(1 - \frac{m_s}{m_r}\right) (\mathbf{j}_s \mathbf{j}_r + \mathbf{j}_r \mathbf{j}_s)\right], \end{aligned} \tag{1.B.4}$$

where

$$\lambda'_{sr} = 6 \times 1.3703 \left(K_{sr} \frac{m_s m_r}{m_s + m_r}\right)^{1/2}, \tag{1.B.5}$$

$\rho_s = m_s n_s$ is the mass density of species s,

$$\mathsf{P}_s = m_s \int d\mathbf{v}\, \mathbf{V}\mathbf{V} f_s \tag{1.B.6}$$

is the partial pressure tensor of species s, and $p_s = \frac{1}{3} \operatorname{tr} \mathsf{P}_s$ is the partial hydrostatic pressure. Note that in the case of particles mechanically equivalent ($K_{sr} = K$, $m_s = m$), one has $\lambda'_{sr} = (m/n)\nu$, where ν is defined by Eq. (1.A.2).

Taking the trace in Eq. (1.B.4), one gets

$$\frac{m_s}{2} \int d\mathbf{v}\, V^2 J_{sr}[f_s, f_r] = -\frac{\lambda_{sr} n_s n_r}{m_s + m_r} \left[3k_B (T_s - T_r) + \frac{m_r - m_s}{\rho_s \rho_r} \mathbf{j}_s \cdot \mathbf{j}_r\right], \tag{1.B.7}$$

where $T_s = p_s / n_s k_B$ is the partial temperature of species s.

APPENDIX 1.C: H-theorem for the ellipsoidal statistical (ES) model

In the ES model, the entropy production due to collisions is

$$\sigma_s^{(c)} = k_B\zeta \int d\mathbf{v}\, (f - f_R) \ln f. \tag{1.C.1}$$

Since $h(x) \equiv x \ln x$ is a convex function, it follows that $h'(x)(y - x) \leq h(y) - h(x)$ for any pair of positive numbers x and y. Consequently, $(x - y)\ln x \geq x \ln x - y \ln y + y - x$. Applying this convexity relation in Eq. (1.C.1), one gets

$$\frac{1}{k_B\zeta}\sigma_s^{(c)} \geq \int d\mathbf{v}\, f \ln f - \int d\mathbf{v}\, f_R \ln f_R. \tag{1.C.2}$$

It is now convenient to introduce the *information theory* distribution f_{IT} as the one that minimizes the integral $\int d\mathbf{v}\, f \ln f$ subjected to the constraints of reproducing the actual density, flow velocity, and pressure tensor of f. By standard use of Lagrange multipliers, one can easily get

$$f_{\text{IT}}(\mathbf{v}) = n\left(\frac{mn}{2\pi}\right)^{3/2} (\det \mathsf{P})^{-1/2} \exp\left(-\frac{mn}{2}\mathsf{P}^{-1} : \mathbf{V}\mathbf{V}\right). \tag{1.C.3}$$

Note that the information theory distribution f_{IT} is formally equivalent to the reference distribution f_R, Eq. (1.173), except that in the latter the tensor P is replaced by R. By definition,

$$\int d\mathbf{v}\, f \ln f \geq \int d\mathbf{v}\, f_{\text{IT}} \ln f_{\text{IT}}, \tag{1.C.4}$$

and so the inequality (1.C.2) yields

$$\begin{aligned} \frac{1}{k_B\zeta}\sigma_s^{(c)} &\geq \int d\mathbf{v}\, f_{\text{IT}} \ln f_{\text{IT}} - \int d\mathbf{v}\, f_R \ln f_R \\ &= \frac{n}{2} \ln \frac{\det \mathsf{R}}{\det \mathsf{P}}. \end{aligned} \tag{1.C.5}$$

To proceed, let us consider the diagonal representation of the pressure tensor P:

$$\mathsf{P} = \begin{pmatrix} p_1 & 0 & 0 \\ 0 & p_2 & 0 \\ 0 & 0 & p_3 \end{pmatrix}, \tag{1.C.6}$$

where p_i are the eigenvalues of P. We also define the auxiliary tensors

$$\mathsf{P}' = \begin{pmatrix} p_2 & 0 & 0 \\ 0 & p_3 & 0 \\ 0 & 0 & p_1 \end{pmatrix}, \quad \mathsf{P}'' = \begin{pmatrix} p_3 & 0 & 0 \\ 0 & p_1 & 0 \\ 0 & 0 & p_2 \end{pmatrix}. \tag{1.C.7}$$

Note that the identity tensor can be written as $\mathsf{I} = (\mathsf{P} + \mathsf{P}' + \mathsf{P}'')/3p$, as a consequence of the identity $p_1 + p_2 + p_3 = 3p$. In terms of the above tensors, the tensor R can be recast into the form

$$\mathsf{R} = \frac{1}{\text{Pr}}\left[\left(\text{Pr} - \frac{2}{3}\right)\mathsf{P} + \frac{1}{3}\mathsf{P}' + \frac{1}{3}\mathsf{P}''\right]. \tag{1.C.8}$$

According to the Brunn–Minkowsky inequality (Andries et al., 2000),

$$\det \sum_i a_i \mathsf{A}_i \geq \prod_i (\det \mathsf{A}_i)^{a_i}, \tag{1.C.9}$$

where A_i are positive symmetric matrices and a_i are positive coefficients verifying $\sum_i a_i = 1$. Application of the inequality (1.C.9) to Eq. (1.C.8) leads to

$$\det \mathsf{R} \geq \det \mathsf{P}. \tag{1.C.10}$$

Using (1.C.10) in Eqs. (1.C.5) one finally has

$$\sigma_s^{(c)} \geq 0. \tag{1.C.11}$$

The equality in (1.C.11) only holds if $f = f_R$, which implies that $\mathsf{R} = \mathsf{P}$. Equation (1.174) shows that this is only possible if $\mathsf{P} = p\mathsf{I}$, i.e. $f = f_L$.

Chapter 2

SOLUTION OF THE BOLTZMANN EQUATION FOR UNIFORM SHEAR FLOW

1. Introduction

In many textbooks (Reif, 1965), the derivation of the shear viscosity of a gas by using elementary kinetic theory assumes the existence of a simple shear flow characterized by a linear velocity profile of a component of the flow velocity, say u_x, along a normal direction, say the y-direction, while the density and the temperature remain spatially uniform (See Fig. 2.1). Implicit in those arguments and conditions is the assumption that the shear rate $\partial u_x/\partial y$ is very small, and so the xy-element of the pressure tensor is just proportional to this shear rate.

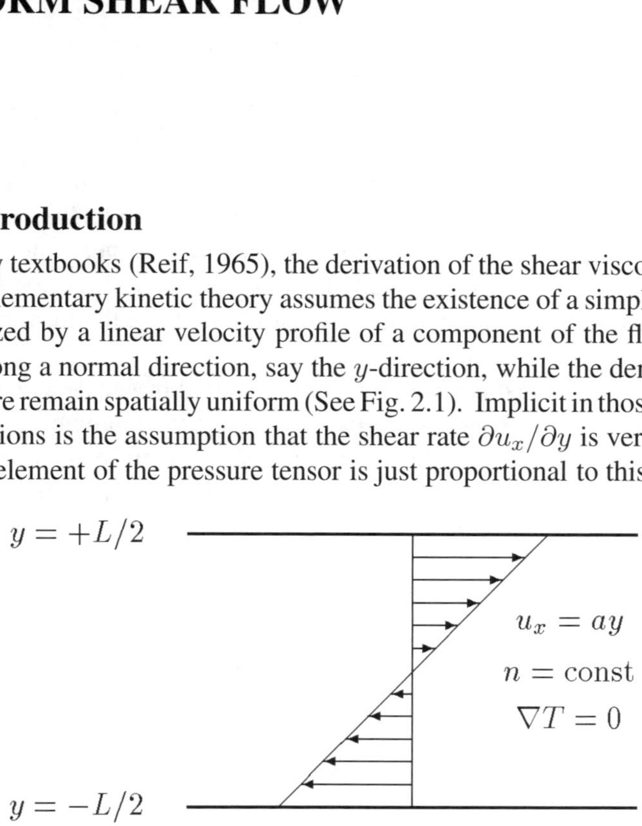

Figure 2.1. Sketch of the uniform shear flow.

What is surprising is that the equations of statistical mechanics, in general, and the Boltzmann equation, in particular, admit a solution compatible with the above conditions, namely $\nabla n = \nabla T = 0$ and $\mathbf{u}(\mathbf{r}) = u_x(y)\widehat{\mathbf{x}}$ (where $\widehat{\mathbf{x}}$ is a unit vector along the x-direction) with

$$\frac{\partial u_x}{\partial y} \equiv a = \mathrm{const}, \tag{2.1}$$

regardless of the strength of the constant shear rate a (Dufty et al., 1986a).[1] This is perhaps the simplest nonequilibrium state one can think of since fourteen of the possible hydrodynamic gradients are zero and the only nonzero gradient is a constant. As a counterpart, due to the shearing motion and the associated viscous heating, the total energy, and hence the temperature, monotonically increases in time. Nevertheless, this effect can be controlled by introducing an (artificial) external force that removes the heat produced in the system, so that a steady state is achieved (Evans and Morriss, 1990).

Henceforth we will refer to this state as *uniform shear flow* (USF).[2] Despite its apparent simplicity, this problem has proven to be useful to shed light on the intricacies and complexities associated with the *nonlinear* response of the system to the action of strong shearing (Kawasaki and Gunton, 1973; Hoover, 1983; Hanley, 1983; Evans and Morriss, 1984). By strong shearing we mean values of the shear rate comparable to or greater than the inverse of a characteristic microscopic time; for dilute gases the characteristic time is of the order of the mean free time τ. In such situation Newton's law (1.106) breaks down, so that the element P_{xy} is no longer proportional to the shear rate and normal stresses are present ($P_{xx} \neq P_{yy} \neq P_{zz}$). More generally, the velocity distribution function is expected to exhibit high distortions from the local equilibrium distribution.

While for a system constituted by atoms or small molecules the shear rates required to observe non-Newtonian effects are not easily attainable in laboratory conditions, this is not the case when one considers systems made of *mesoscopic* particles such as polymeric solutions, colloidal suspensions, or granular matter. In the latter systems the characteristic relaxation times are much longer than in atomic fluids and thus a non-Newtonian behavior can be observed in real experiments. The main motivation along this book is to investigate the transport properties in far from equilibrium shear flow states of systems structurally simple, in the sense that they are constituted by particles (not necessarily at atomic scale) interacting via pair-wise central forces. The importance of the USF lies in the fact that it can be considered as the prototype state for analyzing rheological properties under extreme conditions.

In this chapter we obtain some exact results derived from the nonlinear Boltzmann equation for a gas of Maxwell molecules under USF. This is one of the rare exact solutions of the Boltzmann equation in non-homogeneous situations, the corresponding rheological properties having been first obtained a long time ago by Ikenberry and Truesdell (1956). This exact solution is of great significance

[1]The notation $\dot{\gamma}$ to refer to the shear rate is widely extended in the literature. However, across this book we will denote the shear rate as a.

[2]It is also known as homo-energetic simple shear flow (Truesdell and Muncaster, 1980) to distinguish it from the planar Couette flow that will be studied in Chapters 5 and 6.

in providing insight for the type of phenomena that can occur in conditions far away from equilibrium. We will take advantage of the fact that in this state all the space dependence of the distribution function occurs through its dependence on the *peculiar* velocity $\mathbf{V} = \mathbf{v} - \mathbf{u}(\mathbf{r})$. In other words, when adopting a *local* Lagrangian frame of reference moving with the flow velocity, the distribution function becomes uniform. This feature, along with the properties of the collision operator for Maxwell molecules, allows one to get a solvable hierarchy of equations for the velocity moments. The most remarkable outcome of the analysis is the fact that, for any value of the shear rate a, all the (scaled) velocity moments of degree equal to or larger than a certain critical value $k_c(a)$ are divergent. As a increases, the value of k_c decreases, but it is always larger than $k = 2$, so that the (scaled) second-degree moments (which are related to the elements of the pressure tensor) reach well-defined values. This singular behavior of the moments reflects the existence of an algebraic high-velocity tail in the distribution function, which is confirmed by Monte Carlo simulations. Note that the high-velocity tail has much practical interest since the population of particles with high energy has a special influence on problems such as chemical reactions or other processes with a high activation energy.

A physically important quantity that cannot be expressed in terms of velocity moments is the nonequilibrium Boltzmann entropy. This precludes the possibility of exploiting the knowledge of the velocity moments for arbitrary shear rates to determine the entropy. Nevertheless, the velocity distribution function can be obtained in the form of a perturbation expansion in powers of the shear rate. This allows one to get the first few terms in the power series expansion of the nonequilibrium entropy and construct nonlinear approximants from the expansion.

2. The Boltzmann equation for uniform shear flow

As said above, the USF is characterized at a macroscopic level by

$$n = \text{const}, \tag{2.2}$$

$$u_i(\mathbf{r}) = a_{ij} r_j, \quad a_{ij} = a\delta_{ix}\delta_{jy}, \tag{2.3}$$

$$\nabla T = 0. \tag{2.4}$$

The tensor $a_{ij} = \nabla_j u_i$ is the (transpose of) the rate-of-strain tensor. Note that, without loss of generality, we have chosen $\mathbf{u} = \mathbf{0}$ at $y = 0$. At a more fundamental level, we define the USF state as the one that is spatially homogeneous when the velocities of particles are referred to a Lagrangian frame moving with the velocity field $\mathbf{u}(\mathbf{r})$. Consequently, the distribution function has the form

$$f(\mathbf{r}, \mathbf{v}, t) = f(\mathbf{V}, t), \tag{2.5}$$

where $\mathbf{V} \equiv \mathbf{v} - \mathbf{u}(\mathbf{r})$ is the peculiar velocity. This means that if two points $\mathbf{r}_1$ and $\mathbf{r}_2$ are separated by a distance $\Delta y = y_2 - y_1$ along the y-direction, then the average number of particles at $\mathbf{r}_2$ moving with a velocity (v_x, v_y, v_z) is the same as the average number of particles at $\mathbf{r}_1$ moving with a velocity $(v_x - a\Delta y, v_y, v_z)$. In this frame, the Boltzmann equation (1.25) becomes (in the absence of external forces)

$$\frac{\partial}{\partial t} f - \frac{\partial}{\partial V_i} a_{ij} V_j f = J[f, f]. \tag{2.6}$$

Since in this equation the convective operator $\mathbf{v} \cdot \nabla$ has given rise to the operator $-aV_y \partial/\partial V_x$ on velocity space, it does not require the application of boundary conditions. This point, however, does not preclude the question, what are the boundary conditions in the laboratory (or Eulerian) frame compatible with the USF? The usual boundary conditions used to generate the USF in computer simulations are the Lees–Edwards periodic boundary conditions (Lees and Edwards, 1972; Dufty et al., 1986b). In this case, the system is understood to be contained inside a unit cell located between $y = -L/2$ and $y = +L/2$. When a particle leaves the unit cell through the top surface at $y = +L/2$ with a velocity $\mathbf{v}$ it is reentered at the lower surface with a new x-component of the velocity given by $v_x - aL$. A complementary boundary condition is imposed at $y = -L/2$. More specifically, the Lees–Edwards boundary conditions for $f(\mathbf{r}, \mathbf{v}, t)$ are

$$f(y = \pm L/2, \mathbf{v}, t) = f(y = \mp L/2, \mathbf{v} \mp aL\widehat{\mathbf{x}}, t). \tag{2.7}$$

They are just a generalization of the conventional periodic boundary conditions employed in molecular dynamics of systems in equilibrium (Allen and Tildesley, 1987; Rapaport, 1995; Frenkel and Smit, 1996), so that now a linear velocity profile is attained. Alternatively, the USF state can also be reached by using the so-called "bounce-back" boundary conditions (Cercignani, 1989; Cercignani and Cortesse, 1994), according to which when a particle hits a plate it bounces back in the frame of reference moving with the plate, namely

$$f(y = \pm L/2, \mathbf{v}, t) = f(y = \pm L/2, -\mathbf{v} \pm aL\widehat{\mathbf{x}}, t). \tag{2.8}$$

Behind these bounce-back conditions is the symmetry property $f(\mathbf{V}, t) = f(-\mathbf{V}, t)$, which is consistent with Eq. (2.6). Note that neither the nonlocal conditions (2.7) nor the local conditions (2.8) are intended to represent the interaction of the particles with physical walls. However, they provide an idealized representation of the bulk properties of a fluid under shear. This allows one to clearly identify the nonlinear transport properties of the system without the inconveniencies associated with boundary layers.

Henceforth, we will take for granted that the system remains in the USF state and will focus on the solution to Eq. (2.6).[3] It is interesting to notice that Eq. (2.6) can also be interpreted as representing a *homogeneous* state in the laboratory frame under the action of the non-conservative external force $F_i = -ma_{ij}V_j$. Actually, the latter is an *inertial* force due to the change from the Eulerian frame to the Lagrangian frame. In addition, Eq. (2.6) is invariant under the transformations

$$V_z \to -V_z, \quad (V_x, V_y) \to (-V_x, -V_y), \quad (V_x, a) \to (-V_x, -a). \tag{2.9}$$

The price to be paid for maintaining a uniform temperature is the absence of heat flux compensating for the viscous heating produced by the shear field. Consequently, no stationary state is possible and the temperature increases with time. This is quite evident from the energy balance equation (1.44), which in the conditions of USF reads

$$\frac{dT(t)}{dt} = -\frac{2}{3nk_B} a P_{xy}. \tag{2.10}$$

On the other hand, one expects that for long times the influence of the initial condition $f(\mathbf{V}, 0)$ has completely disappeared and the time-dependence of the distribution function occurs only through a *functional* dependence on the temperature. This corresponds to a *normal* solution to Eq. (2.6). As we will see this requires a certain scaling.

Let us introduce the scaled quantities (Dufty et al., 1986b)

$$\overline{\mathbf{V}} = R(t)\mathbf{V}, \tag{2.11}$$

$$\overline{f}(\overline{\mathbf{V}}, t) = [R(t)]^{-3} f(\mathbf{V}, t), \tag{2.12}$$

where $R(t)$ is so far an arbitrary function of time. Then, Eq. (2.6) reduces to

$$\frac{\partial}{\partial t}\overline{f} - \frac{\partial}{\partial \overline{V}_i}[a_{ij}\overline{V}_j + \alpha(t)\overline{V}_i]\overline{f} = \overline{J}[\overline{f}, \overline{f}], \tag{2.13}$$

where

$$\alpha(t) \equiv -\frac{d}{dt} \ln R(t) \tag{2.14}$$

and

$$\overline{J}[\overline{\mathbf{V}}|\overline{f}, \overline{f}] \equiv [R(t)]^{-3} J[\mathbf{V}|f, f]. \tag{2.15}$$

Note that, in general, $\overline{J}[\overline{f}, \overline{f}]$ is a *modified* collision operator that differs from the original Boltzmann operator $J[\overline{f}, \overline{f}]$ acting on the scaled function $\overline{f}(\overline{\mathbf{V}})$.

[3]However, as we will analyze in Section 3.8, the USF becomes unstable under long wavelength perturbations. In what follows, we will assume that the size L of the system is smaller than the critical wavelength, so that the instability is suppressed.

Nevertheless, in the special case of particles interacting via a repulsive potential of the form $\phi(r) = Kr^{-w}$, the collision rate $gB(g, \chi)$ is proportional to $g^{1-(4/w)}$ [cf. Eq. (1.17)], so that

$$\overline{J}[\overline{f}, \overline{f}] = [R(t)]^{(4/w)-1} J[\overline{f}, \overline{f}]. \tag{2.16}$$

In that case, Eq. (2.13) can be written as

$$\frac{\partial}{\partial \overline{t}} \overline{f} - \frac{\partial}{\partial \overline{V}_i} [\overline{a}_{ij}(\overline{t}) \overline{V}_j + \overline{\alpha}(\overline{t}) \overline{V}_i] \overline{f} = J[\overline{f}, \overline{f}], \tag{2.17}$$

where

$$\overline{t} \equiv \int_0^t dt' [R(t')]^{(4/w)-1}, \tag{2.18}$$

$$\overline{a}_{ij}(\overline{t}) \equiv a_{ij} [R(t)]^{1-(4/w)}, \quad \overline{\alpha}(\overline{t}) \equiv \alpha(t) [R(t)]^{1-(4/w)}. \tag{2.19}$$

Thus, for r^{-w}-potentials, Eqs. (2.6) and (2.17) are completely equivalent. From a physical point of view, Eq. (2.6) represents a system subjected to USF with a *constant* shear rate a, while Eq. (2.17) represents a system with a *time-dependent* shear rate $\overline{a}$ and in presence of a non-conservative *external* force $-m\overline{\alpha}\overline{\mathbf{V}}$. This equivalence holds regardless of the choice of the scaling function $R(t)$, which controls the relationship between the actual velocity and time variables and the respective scaled variables.[4] In particular, $R(t)$ can be chosen to impose any given time dependence of the (scaled) temperature.

The most plausible physical choice is $R(t) \propto [T(t)]^{-1/2}$, so that

$$[R(t)]^{(4/w)-1} \propto \nu(t), \tag{2.20}$$

where $\nu(t)$ is an effective collision frequency. Consequently, $\overline{\mathbf{V}}$ is just (proportional to) the velocity relative to the thermal velocity, $\overline{t}$ is (proportional to) the number of collisions per particle and $\overline{a}$ is (proportional to) the shear rate relative to the collision frequency. In order to estimate $R(t)$, let us resort to the solution of Eq. (2.10) for very long times. In that limit, the shear rate relative to the collision frequency tends to zero, so that it is legitimate to assume the validity of Newton's law (1.106), which in the geometry of the USF reads $P_{xy}(t) = -a\eta_0(t)$, with $\eta_0(t) \propto T(t)/\nu(t) \propto [T(t)]^{[1+(4/w)]/2}$. As a consequence, the solution of Eq. (2.10) for long times is

$$T(t) = T_0 \left[1 + \left(1 - \frac{4}{w}\right) \alpha_0 t\right]^{2/[1-(4/w)]}, \tag{2.21}$$

[4] In the absence of shearing ($a = 0$), this equivalence implies that the long-time solution of Eq. (2.17) is just the Maxwell–Boltzmann distribution with a time-dependent temperature (Garzó et al., 1990).

where T_0 and α_0 are positive constants. By making $R(t) = [T(t)/T_0]^{-1/2}$, we have

$$\alpha(t) = \frac{\alpha_0}{1 + (1 - 4/w)\,\alpha_0 t}, \qquad \bar{t} = \frac{\ln\left[1 + (1 - 4/w)\,\alpha_0 t\right]}{(1 - 4/w)\,\alpha_0}. \tag{2.22}$$

Two interesting limiting cases correspond to Maxwell molecules ($w = 4$) and hard spheres ($w \to \infty$). In the former case one simply has

$$T(t) = T_0 e^{2\alpha_0 t}, \quad \alpha = \alpha_0, \quad \bar{t} = t, \tag{2.23}$$

while

$$T(t) = T_0(1 + \alpha_0 t)^2, \quad \alpha(t) = \frac{\alpha_0}{1 + \alpha_0 t}, \quad \bar{t} = \frac{\ln(1 + \alpha_0 t)}{\alpha_0}, \tag{2.24}$$

for hard spheres.

3. Moment equations for a gas of Maxwell molecules. Rheological properties

In the special case of Maxwell molecules, the scaling laws discussed in the previous Section happen to be quite simple since, according to Eqs. (2.18) and (2.19), $\bar{t} = t$ and $\bar{a} = a$ with independence of $R(t)$. This is basically due to the fact that the collision rate $gB(g, \chi)$ is independent of the relative velocity g and, therefore, the collision frequency is constant in time. As a consequence, although the temperature grows in time, the shear rate measured in units of the collision frequency remains constant, so that the departure of the system from equilibrium, as measured by a/ν, does not change in time. This implies that the distribution of velocities relative to the thermal velocity $(2k_B T/m)^{1/2}$ is expected to reach, after a sufficient number of collisions per particle, a stationary form that only depends on the value of the control parameter a/ν. In order to analyze this asymptotic form, as well as the transient regime, it is convenient to scale the velocities with a factor $R(t) = e^{-\alpha t}$, with $\alpha =$ const, as suggested by Eq. (2.23). In that case, Eqs. (2.13) or (2.17) are identical to

$$\frac{\partial}{\partial t}\overline{f} - \frac{\partial}{\partial \overline{V}_i}[a_{ij}\overline{V}_j + \alpha\overline{V}_i]\overline{f} = J[\overline{f}, \overline{f}]. \tag{2.25}$$

We must emphasize that Eqs. (2.6) and (2.25) are totally equivalent in the case of Maxwell molecules, with independence of the value of α. In our description, we will choose α as a function of the shear rate a by the condition that the scaled temperature $\overline{T}(t) = e^{-2\alpha t}T(t)$ reaches a constant value in the long-time limit, so that the term $-m\alpha\overline{\mathbf{V}}$ plays the role of a "thermostat" force. This kind of thermostat forces is usually employed in nonequilibrium molecular dynamics simulations (Hoover, 1983; Evans and Morriss, 1984; Evans and

Hoover, 1986; Evans and Morriss, 1990). Therefore, in the case of Maxwell molecules, the *transformed* Boltzmann equation (2.25) *without* a thermostat is formally equivalent to the *original* Boltzmann equation (2.6) *with* a thermostat force added. As we will discuss in Chapter 3, this equivalence breaks down for non-Maxwell molecules. Henceforth, we will adopt the point of view behind Eq. (2.25) and, for the sake of simplicity, will drop the bars.

Although the explicit solution to Eq. (2.25) is not known, it can be exactly solved by the moment method. This is due to the fact that, in the case of Maxwell molecules, a collisional moment of degree k is a bilinear combination of moments of f of degrees k' and k'', such that $k' + k'' = k$ (Truesdell and Muncaster, 1980; McLennan, 1989). This allows one, in principle, to solve recursively the hierarchy of moment equations. We will call *asymmetric* moments those that vanish for solutions exhibiting the same invariance properties (2.9) as Eq. (2.25). The remaining moments will be referred to as *symmetric*. To clarify this point, let us introduce the moments

$$\langle V_x^{k_1} V_y^{k_2} V_z^{k_3} \rangle = \frac{1}{n} \int d\mathbf{V} V_x^{k_1} V_y^{k_2} V_z^{k_3} f. \tag{2.26}$$

Then, $\langle V_x^{k_1} V_y^{k_2} V_z^{k_3} \rangle$ is a symmetric moment if k_1+k_2 and k_3 are even numbers. The total number of independent moments of degree $k = k_1 + k_2 + k_3$ is $\frac{1}{2}(k+1)(k+2)$; if $k =$ odd, all the moments are asymmetric, while if $k =$ even, there are $(\frac{k}{2} + 1)^2$ symmetric moments and $\frac{k}{4}(k + 2)$ asymmetric moments.

The second-degree moments of the distribution function are related to the pressure tensor (1.40). The evolution equations for the elements of the pressure tensor are especially interesting because they describe the so-called viscoelastic effects (Jou et al., 2001), which play an important role in the analysis of high-frequency perturbations, as for instance in light-scattering experiments.

By multiplying both sides of Eq. (2.25) by $V_i V_j$ and integrating over $\mathbf{V}$, one gets

$$\frac{\partial}{\partial t} P_{ij} + a_{ik} P_{jk} + a_{jk} P_{ik} + 2\alpha P_{ij} = -\nu (P_{ij} - p\delta_{ij}), \tag{2.27}$$

where we have made use of the second-degree collisional moments (see Appendix 1.A). Here, ν is an effective collision frequency defined as

$$\nu = 3n\mathrm{A}_2, \tag{2.28}$$

where

$$\mathrm{A}_r \equiv \int d\Omega \sin^r \left(\frac{\chi}{2}\right) \cos^r \left(\frac{\chi}{2}\right) gB(g, \chi). \tag{2.29}$$

The numerical value of A_2 is $\mathrm{A}_2 \simeq 1.3703\sqrt{2K/m}$ (Truesdell and Muncaster, 1980). For Maxwell molecules, the collision frequency ν coincides with the eigenvalue λ_{02}^0 of the linearized Boltzmann collision operator [cf. Eq. (1.90)]. From Eq. (1.111), it follows that

$$\nu = \frac{p}{\eta_0}, \tag{2.30}$$

where η_0 is the Navier–Stokes shear viscosity. It is convenient to choose ν^{-1} as the time unit. This means that we use dimensionless quantities $t^* = \nu t$, $a^* = a/\nu$, and $\alpha^* = \alpha/\nu$. The physical meaning of the reduced shear rate a^* is clear. It represents a *uniformity* parameter, namely $a^* \sim \ell/h$, where $\ell \sim \sqrt{2k_BT/m}/\nu$ is the mean free path and $h \sim \sqrt{2k_BT/m}\,(\partial u_x/\partial y)^{-1}$ is the characteristic hydrodynamic length of the problem. The quantity a^* is in fact the only relevant parameter controlling the departure from equilibrium.

The two asymmetric moments P_{xz} and P_{yz} are simply given by

$$P_{yz}(t^*) = P_{yz}(0)e^{-(1+2\alpha^*)t^*}, \tag{2.31}$$

$$P_{xz}(t^*) = [P_{xz}(0) - a^*t^*P_{yz}(0)]e^{-(1+2\alpha^*)t^*}. \tag{2.32}$$

The following combination of symmetric moments has a similar time behavior:

$$P_{yy}(t^*) - P_{zz}(t^*) = [P_{yy}(0) - P_{zz}(0)]e^{-(1+2\alpha^*)t^*}. \tag{2.33}$$

The remaining elements of the pressure tensor are P_{xx}, P_{yy}, and P_{xy}, since P_{zz} is obtained from P_{yy} by Eq. (2.33). If we eliminate P_{xx} in favor of the hydrostatic pressure $p = \frac{1}{3}\mathrm{tr}\,\mathsf{P}$, we get the following set of coupled equations:

$$\frac{\partial}{\partial t^*}p + \frac{2a^*}{3}P_{xy} + 2\alpha^* p = 0, \tag{2.34}$$

$$\frac{\partial}{\partial t^*}P_{yy} + 2\alpha^* P_{yy} = p - P_{yy}, \tag{2.35}$$

$$\frac{\partial}{\partial t^*}P_{xy} + a^* P_{yy} + 2\alpha^* P_{xy} = -P_{xy}. \tag{2.36}$$

This set of first-order linear differential equations leads to a closed third-order linear differential equation for $p(t^*)$,

$$\left(\frac{\partial}{\partial t^*} + 2\alpha^*\right)\left(\frac{\partial}{\partial t^*} + 2\alpha^* + 1\right)^2 p = \frac{2}{3}a^{*2}p. \tag{2.37}$$

Once solved, one gets P_{xy} and P_{yy} as

$$P_{xy} = -\frac{3}{2a^*}\left(2\alpha^* p + \frac{\partial p}{\partial t^*}\right), \tag{2.38}$$

$$P_{yy} = \frac{3}{2a^{*2}} \left[2\alpha^*(1+2\alpha^*)p + (1+4\alpha^*)\frac{\partial p}{\partial t^*} + \frac{\partial^2 p}{\partial t^{*2}} \right]. \tag{2.39}$$

The three roots of the characteristic equation corresponding to Eq. (2.37) are

$$\ell_1 = 2\alpha^* - \lambda, \tag{2.40}$$

$$\ell_{2,3} = \left(2\alpha^* + 1 + \frac{1}{2}\lambda \right) \pm \mathrm{i}\omega, \tag{2.41}$$

where

$$\lambda = \frac{4}{3} \sinh^2 \left[\frac{1}{6} \cosh^{-1}(1+9a^{*2}) \right], \tag{2.42}$$

$$\omega = \left[\lambda \left(\frac{3}{4}\lambda + 1 \right) \right]^{1/2}, \tag{2.43}$$

and i is the imaginary unit. Note that λ and $-(\frac{1}{2}\lambda+1) \pm \mathrm{i}\omega$ are the roots of the cubic equation

$$a^{*2} = \frac{3}{2}\lambda(1+\lambda)^2. \tag{2.44}$$

The general solution of Eq. (2.37) is then

$$p(t^*) = e^{(\lambda - 2\alpha^*)t^*} \left[A + e^{-(1+3\lambda/2)t^*} (B\cos\omega t^* + C\sin\omega t^*) \right], \tag{2.45}$$

where A, B, and C are constants given by the initial conditions.

So far, α^* is arbitrary. If one takes $\alpha^* = 0$, one recovers the results corresponding to the solution of the USF problem without scaling (Truesdell and Muncaster, 1980). In that case, the temperature grows exponentially in time when $t^* \gg 1$, in agreement with the simple analysis made at the end of Section 2.2 [cf. Eq. (2.23)]. On the other hand, as said before, here we choose α^* under the condition that the pressure reaches a stationary value in the long-time limit. Consequently, we take $\alpha^* = \lambda/2$, so that α^* is a nonlinear function of the shear rate a^* given by

$$\alpha^*(a^*) = \frac{2}{3} \sinh^2 \left[\frac{1}{6} \cosh^{-1}(1+9a^{*2}) \right], \tag{2.46}$$

which is the real root of the cubic equation

$$a^{*2} = 3\alpha^*(1+2\alpha^*)^2. \tag{2.47}$$

The heating rate $\alpha^*(a^*)$ exhibits a nonlinear dependence on the shear rate, as illustrated in Fig. 2.2. The time dependence of p, P_{yy}, and P_{xy} is then

$$p(t^*) = A + e^{-(1+3\alpha^*)t^*} (B\cos\omega t^* + C\sin\omega t^*), \tag{2.48}$$

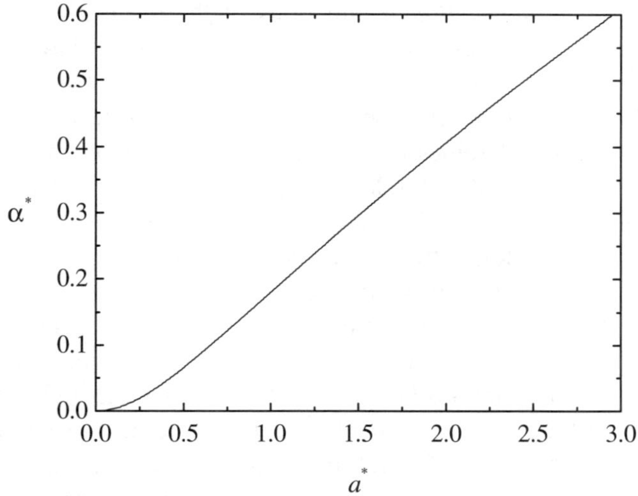

Figure 2.2. Heating rate α^* as a function of the shear rate a^*.

$$P_{yy}(t^*) = \frac{1}{1+2\alpha^*}\left\{A - \frac{1}{2}e^{-(1+3\alpha^*)t^*}\left[\left(B + \frac{\omega}{\alpha^*}C\right)\cos\omega t^* + \left(C - \frac{\omega}{\alpha^*}B\right)\sin\omega t^*\right]\right\}, \tag{2.49}$$

$$P_{xy}(t^*) = -\frac{3}{2a^*}\{2A\alpha^* - e^{-(1+3\alpha^*)t^*}[((1+\alpha^*)B - \omega C)\cos\omega t^* + ((1+\alpha^*)C + \omega B)\sin\omega t^*]\}, \tag{2.50}$$

where we have made use of Eqs. (2.38), (2.39), and (2.47). The relaxation time of p, P_{yy}, and P_{xy} towards their stationary values is $(1+3\alpha^*)^{-1}$, while the remaining elements of the pressure tensor have a longer relaxation time, namely $(1+2\alpha^*)^{-1}$, according to Eqs. (2.31)–(2.33).

From the above general analysis, it follows that the *reduced* elements $P^*_{ij}(t) = P_{ij}(t)/p(t)$ of the pressure tensor reach well defined stationary values for sufficiently long times:

$$\frac{P_{xx}}{p} = \frac{1+6\alpha^*}{1+2\alpha^*}, \tag{2.51}$$

$$\frac{P_{yy}}{p} = \frac{P_{zz}}{p} = \frac{1}{1+2\alpha^*}, \tag{2.52}$$

$$\frac{P_{xy}}{p} = -\frac{3\alpha^*}{a^*} = -\frac{a^*}{(1+2\alpha^*)^2}. \tag{2.53}$$

These values are *highly nonlinear* functions of the (arbitrary) shear rate, in contrast to the behavior described by the *linear* Newton's law, Eq. (1.106), according to which $P_{xx}/p = P_{yy}/p = P_{zz}/p = 1$ and $-P_{xy}/p = a^*$. This is an example of *non-Newtonian* behavior and it occurs because the solution is not restricted to small values of $a^* = a/\nu$. Non-Newtonian properties are usually linked to structurally complex materials, such as colloidal suspensions, polymeric liquids, gels, ... (Bird et al., 1987a; Bird et al., 1987b). In fact, there is a science –rheology– that studies the mechanical behavior which does not follow the classical models of the perfect or inviscid fluid, linear viscous or Newtonian fluid, linear elastic or Hookean solid, and plasticity (Anderson, 1989). In other words, rheology is the science of flow and deformation of matter and describes the interrelations between force, deformation, and time. What is interesting here is the existence of rheological properties in a *structurally simple* fluid, such as a dilute gas of Maxwell molecules. In this case, these properties are associated with the behavior of the gas in situations far from equilibrium.

In order to characterize the deviations from Newton's law, it is convenient to introduce the following three (dimensionless) material functions (Bird, 1983; Bird et al., 1987a; Bird et al., 1987b; McLennan, 1989; Dealy, 1995):

$$F_\eta(a^*) = -\lim_{t\to\infty} \frac{\nu}{a}\frac{P_{xy}(t)}{p(t)}, \tag{2.54}$$

$$\Psi_1(a^*) = \lim_{t\to\infty} \frac{\nu^2}{a^2}\frac{P_{yy}(t)-P_{xx}(t)}{p(t)}, \tag{2.55}$$

$$\Psi_2(a^*) = \lim_{t\to\infty} \frac{\nu^2}{a^2}\frac{P_{zz}(t)-P_{yy}(t)}{p(t)}. \tag{2.56}$$

The *viscosity function* (or nonlinear shear viscosity) $F_\eta(a^*)$ is a measure of the breakdown of the linear relationship between P_{xy} and the shear rate, while the first and second *viscometric functions* $\Psi_{1,2}(a^*)$ represent the normal stress differences. From Eqs. (2.51)–(2.53) one gets

$$F_\eta(a^*) = \frac{1}{[1+2\alpha^*(a^*)]^2}, \tag{2.57}$$

$$\Psi_1(a^*) = -\frac{2}{[1+2\alpha^*(a^*)]^3}, \tag{2.58}$$

$$\Psi_2(a^*) = 0, \tag{2.59}$$

where we have made use of Eq. (2.47). The fact that the second viscometric function is identically zero is a particular property of the Maxwell interaction. As we will see later, $\Psi_2(a^*) \neq 0$ for other potentials, although its magnitude is always much smaller than that of $\Psi_1(a^*)$.

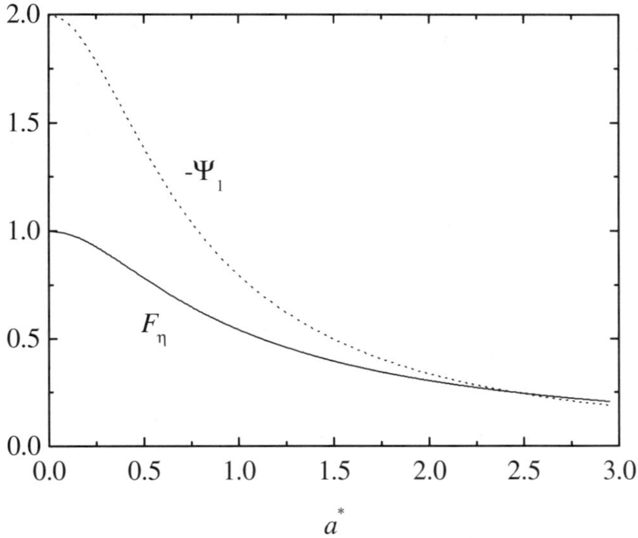

Figure 2.3. Viscosity function F_η and first viscometric function Ψ_1 as functions of the shear rate a^*.

Figure 2.3 shows the shear-rate dependence of the functions $F_\eta(a^*)$ and $-\Psi_1(a^*)$. It is apparent that both functions are monotonically decreasing functions of the shear rate. In the case of the viscosity function, this effect is usually referred to as *shear thinning* (Evans and Hoover, 1986; Bird et al., 1987a; Evans and Morriss, 1990); it implies that the shearing produces an inhibition of the momentum transport in the sense that the actual value of $|P_{xy}|$ is smaller than the one predicted by Newton's law. A fluid exhibiting shear thinning is said to be *pseudoplastic* (Bird et al., 1987a). Regarding the first viscometric function, its negative sign indicates that P_{xx} is larger than P_{yy}. This means that the anisotropy of the system induces a breakdown of the equipartition of energy. For very small shear rates ($a^* \ll 1$), the difference between these two normal stresses grows as $P_{xx} - P_{yy} \approx 2pa^{*2}$, since $\Psi_1(0) = -2$. This is consistent with the results obtained from the Chapman–Enskog expansion through Burnett order, specialized to the geometry of USF and to the Maxwell interaction (Chapman and Cowling, 1970). Since, obviously, $3p$ is an upper bound of $P_{xx} - P_{yy}$, this difference grows more slowly than proportionally to a^{*2} for finite shear rates, so that $-\Psi_1(a^*)$ decreases as a^* increases.

It is interesting to make a qualitative comparison between the exact rheological behavior obtained from the Boltzmann equation for a dilute Maxwell gas with those observed in dense gases or real complex fluids. In the case of a dense gas, there are collisional transfer contributions to the momentum flux in addition to the kinetic contributions. As a matter of fact, the former contributions can be even larger than the latter ones for a fairly dense gas. An analysis of the Enskog equation for hard spheres (Santos et al., 1998) shows that, while the kinetic part of the viscosity function exhibits shear thinning, the collisional part is responsible for a crossover of the total viscosity function from shear thinning to shear *thickening* at sufficiently large shear rates.[5] The relation $P_{xx} > P_{yy}$ is kept for a dense gas, but P_{yy} is clearly larger than P_{zz}. In fact, at a given value of the density and the shear rate, the difference $P_{yy} - P_{zz}$ can even be larger than the difference $P_{xx} - P_{yy}$. More important differences appear in the case of complex fluids. For polymeric liquids (Bird, 1983; Bird et al., 1987a), experiments show that the first viscometric function Ψ_1 is positive (i.e. $P_{yy} > P_{xx}$), while the second viscometric function Ψ_2 is negative (i.e. $P_{yy} > P_{zz}$), but the magnitude of the latter is much smaller than that of Ψ_1. In fact the relation $\Psi_2 = 0$ (Weissenberg hypothesis) has been used for a long time in the study of the dynamics of polymeric liquids (Bird et al., 1987a).

It is worthwhile taking advantage of the explicit expressions (2.57) and (2.58) to analyze the asymptotic behaviors of the rheological quantities for small and large shear rates. To do that, it is first necessary to consider the properties of the heating rate α^*, seen as a function of $z \equiv a^{*2}$. It is straightforward to check that Eq. (2.46) can be rewritten as

$$\begin{aligned} \alpha^* &= \frac{1}{6} z^{1/3} \left[\left(9 + z^{-1} + 9\sqrt{1 + \tfrac{2}{9} z^{-1}} \right)^{1/3} \right. \\ &\quad \left. + \left(9 + z^{-1} - 9\sqrt{1 + \tfrac{2}{9} z^{-1}} \right)^{1/3} - 2z^{-1/3} \right]. \end{aligned} \tag{2.60}$$

We see that the only singularity for finite z is a branch point at $z = -\frac{2}{9}$. In particular, the origin $z = 0$ is a *regular* point, what implies that the expansion of α^* in powers of z converges for $|z| < \frac{2}{9}$.[6] As a consequence, the expansion of the rheological quantities in powers of the shear rate (which is nothing but the application to USF of the Chapman–Enskog expansion for those quantities) is a *convergent* one. On the other hand, the expansion converges only for values of the shear rate a^* less than $\sqrt{2}/3 \simeq 0.47$. This is quite a large value of the radius of convergence, as shown by the fact that $F_\eta(\sqrt{2}/3) \simeq 0.80$ and $\Psi_1(\sqrt{2}/3) \simeq -1.43$. The first few terms in the expansion of α^*, F_η, and Ψ_1 in powers of $z \equiv a^{*2}$ are

$$\alpha^* = \frac{1}{3} z \left(1 - \frac{4}{3} z + \frac{28}{9} z^2 + \cdots \right), \tag{2.61}$$

[5] A fluid whose viscosity function increases with shear rate is sometimes called *dilatant* (Bird et al., 1987a).
[6] This value does not have a special significance beyond being the radius of convergence of the series expansion of α^* in powers of $z = a^{*2}$. In fact, Eq. (2.60) is well defined for any *real* value of the shear rate.

$$F_\eta = 1 - \frac{4}{3}z + \frac{28}{9}z^2 + \cdots, \tag{2.62}$$

$$\Psi_1 = -2\left(1 - 2z + \frac{16}{3}z^2 + \cdots\right). \tag{2.63}$$

It must be pointed out that the fact that the origin $z = 0$ is a regular point of the rheological functions cannot be extrapolated to other situations. As we will see in Chapter 3, the results obtained from the BGK model for hard spheres show that the expansion of the viscosity function in powers of the shear rate is only asymptotic (Santos et al., 1986a).

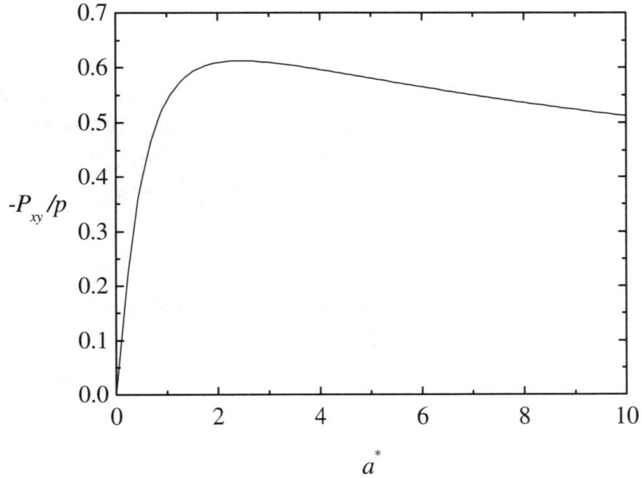

Figure 2.4. Relative shear stress $-P_{xy}/p$ as a function of the shear rate a^*.

Now we consider the behavior for large shear rates. From Eq. (2.60), one sees that the natural variable to expand about the point at infinity is $z^{-1/3}$. The results are

$$a^* = \tfrac{1}{2}\left(\tfrac{2}{3}\right)^{1/3} z^{1/3}\left[1 - \left(\tfrac{2}{3}\right)^{2/3} z^{-1/3} + \tfrac{1}{9}\left(\tfrac{2}{3}\right)^{-2/3} z^{-2/3} + \cdots\right], \tag{2.64}$$

$$F_\eta = \left(\tfrac{2}{3}\right)^{-2/3} z^{-2/3}\left[1 - \left(\tfrac{2}{3}\right)^{2/3} z^{-1/3} + \tfrac{1}{9}\left(\tfrac{2}{3}\right)^{-2/3} z^{-2/3} + \cdots\right], \tag{2.65}$$

$$\Psi_1 = -3z^{-1}\left[1 - \left(\tfrac{2}{3}\right)^{-1/3} z^{-1/3} + \tfrac{1}{3}\left(\tfrac{2}{3}\right)^{-2/3} z^{-2/3} + \cdots\right]. \tag{2.66}$$

Since $\Psi_1 = (P_{yy} - P_{xx})/pa^{*2}$ and $p = (P_{xx} + 2P_{yy})/3$, the asymptotic behavior $\Psi_1 \approx -3/a^{*2}$ implies that $P_{yy}/p \to 0$ and $P_{xx}/p \to 3$ when $a^* \to \infty$. It means that in this limit practically all the particles move along the flow direction. As a consequence, the (reduced) viscous pressure or shear stress $-P^*_{xy} = -P_{xy}/p$ must vanish in the limit of infinitely large shear rates. This is confirmed by Eq. (2.65), according to which $-P_{xy}^* \approx \left(\frac{2}{3}\right)^{-2/3} a^{*-1/3}$. Thus, the shear stress has a non-monotonic dependence on the shear rate, so that it reaches a maximum value at a certain shear rate. From Eqs. (2.47) and (2.53), it follows that the maximum value is $-P_{xy}^* = \frac{\sqrt{6}}{4} \simeq 0.61$ and takes place at $a^* = \sqrt{6} \simeq 2.45$, which corresponds to $\alpha^* = \frac{1}{2}$ and $F_\eta = \frac{1}{4}$. The shear-rate dependence of $-P_{xy}^*$ is displayed in Fig. 2.4. The non-monotonic behavior of the shear stress is a dramatic consequence of the shear-thinning effect and the tendency of the particles to move along the x-direction for large shear rates. One might argue that the decrease of the shear stress for shear rates larger than $a^* = \sqrt{6}$ is inconsistent with a steady USF since the "resistance" of the fluid to flow would be smaller and smaller as the shearing increases. However, this paradoxical situation arises from an identification of the ratio $-P_{xy}/p$ with the "friction coefficient" $-P_{xy}/P_{yy}$, which measures the tangential force relative to the normal force exerted on a shearing plane. While for small shear rates both ratios are comparable, they become quite different for large shear rates. In fact, in the limit $a^* \to \infty$, Eqs. (2.52), (2.53), and (2.64) imply that $-P_{xy}/P_{yy} \approx \frac{3}{2}\left(-P_{xy}/p\right)^{-1} \approx \left(\frac{3}{2}\right)^{1/3} a^{*1/3}$. As a consequence, the friction coefficient $-P_{xy}/P_{yy}$ monotonically increases with the shear rate, as will be illustrated in Section 5.4 when comparing with the results of the Couette flow. Moreover, the USF at $a^* = \sqrt{6}$ is linearly stable if the separation L between the boundaries is small enough, as will be discussed in Section 3.8.

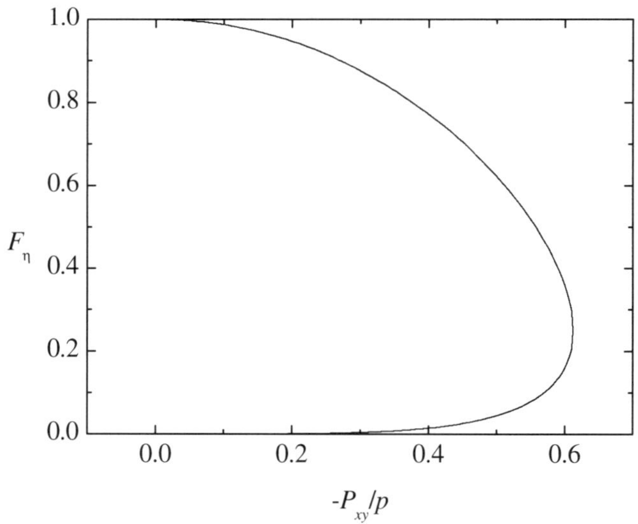

Figure 2.5. Viscosity function F_η as a function of the relative shear stress $-P_{xy}/p$.

In the context of non-Newtonian fluids (Bird et al., 1987a; Bird et al., 1987b; Anderson, 1989), it is also usual to express the viscosity function F_η in terms of the shear stress P_{xy}. Elimination of the shear rate a^* between Eqs. (2.47) and (2.53) yields a quadratic equation for α^* with coefficients depending on $(P_{xy}/p)^2$. Once the quadratic equation is solved, Eq. (2.57) gives

$$F_\eta = \frac{1}{4}\left(1 \pm \sqrt{1 - \frac{8}{3}\left(\frac{P_{xy}}{p}\right)^2}\right)^2 . \tag{2.67}$$

Here, the positive sign corresponds to $a^* \leq \sqrt{6}$ and the negative sign corresponds to $a^* \geq \sqrt{6}$. Both branches are needed to give the full dependence of F_η on $-P_{xy}/p$, which is plotted in Fig. 2.5. The upper branch represents the shear-thinning effect in the range where the shear stress increases with the shear rate, while the lower branch corresponds to the range of shear rates where the magnitude of the shear stress decreases. In the limits of small values of the shear stress, the viscosity function behaves as

$$F_\eta \approx \begin{cases} 1 - \frac{4}{3}\left(P_{xy}/p\right)^2, & a^* \to 0, \\ \frac{4}{9}\left(P_{xy}/p\right)^4, & a^* \to \infty. \end{cases} \tag{2.68}$$

4. Third- and fourth-degree velocity moments

The most relevant information in the understanding of physical phenomena in nonequilibrium fluids is provided by the low-degree velocity moments (i.e. the pressure tensor and the heat flux), which are related to the population of molecules with energies of the order of or less than the mean kinetic energy. On the other hand, the high-energy population plays a crucial role in processes such as chemical reactions with a high activation energy or the controlled thermonuclear fusion of a confined hydrogen plasma. An indirect information on the high-energy population is obtained from the high-degree moments. Thus, the knowledge of those moments is useful to complement the general picture and gain insight into the main features of the velocity distribution. In addition, high-degree moments are also important in theories such as Grad's method (Grad, 1949), generalized moment methods (Weinert, 1982), or extended irreversible thermodynamics (Jou et al., 2000). In the case of the USF, the transport properties are obtained from the second-degree velocity moments, since the third-degree moments are expected to vanish for long times because of symmetry reasons. Therefore, beyond the pressure tensor, the first relevant moments containing information about the velocity distribution function are of fourth degree.

For the sake of completeness, let us start by analyzing the time behavior of the third-degree moments, i.e. moments of the form (2.26) with $k_1 + k_2 + k_3 = 3$. This includes the three components of the heat flux vector [cf. Eq. (1.45)]. The total number of independent third-degree moments is 10. Since all of them are asymmetric, one expects that they go to zero in the long-time limit. For computational purposes, it is convenient to work with the following moments:

$$M_{2|i}(t) = \frac{1}{n} \int d\mathbf{V}\, Y_{2|i}(\mathbf{V}) f(\mathbf{V}, t), \tag{2.69}$$

$$M_{0|ijk}(t) = \frac{1}{n} \int d\mathbf{V}\, Y_{0|ijk}(\mathbf{V}) f(\mathbf{V}, t), \tag{2.70}$$

where (Truesdell and Muncaster, 1980)

$$Y_{2|i}(\mathbf{V}) = V^2 V_i, \tag{2.71}$$

$$Y_{0|ijk}(\mathbf{V}) = V_i V_j V_k - \frac{1}{5} V^2 (V_i \delta_{jk} + V_j \delta_{ik} + V_k \delta_{ij}) \tag{2.72}$$

are homogeneous third-degree polynomials. As the set of 10 independent third-degree moments, we take

$$\{M_{2|x}, M_{2|y}, M_{2|z}, M_{0|xxy}, M_{0|xxz}, M_{0|xyy}, M_{0|yyz}, M_{0|xzz}, M_{0|yzz}, M_{0|xyz}\}. \tag{2.73}$$

From Eq. (2.25) and making use of the third-degree collisional moments (cf. Appendix 1.A), one gets the following set of equations:

$$\left(\partial_{t^*} + \frac{3}{2} + 3\alpha^*\right)\left(\frac{1}{4} M_{0|xxz} + M_{0|yyz}\right) = 0, \tag{2.74}$$

$$\left(\partial_{t^*} + \frac{3}{2} + 3\alpha^*\right)\left(\frac{1}{4} M_{0|xxy} + M_{0|yzz}\right) = 0, \tag{2.75}$$

$$\begin{pmatrix} \partial_{t^*} + \frac{2}{3} + 3\alpha^* & 0 & 2a^* \\ 0 & \partial_{t^*} + \frac{3}{2} + 3\alpha^* & -\frac{2}{5} a^* \\ \frac{1}{5} a^* & a^* & \partial_{t^*} + \frac{3}{2} + 3\alpha^* \end{pmatrix} \cdot \begin{pmatrix} M_{2|z} \\ M_{0|yyz} \\ M_{0|xyz} \end{pmatrix} = 0, \tag{2.76}$$

$$\begin{pmatrix} \partial_{t^*} + \frac{2}{3} + 3\alpha^* & \frac{7}{5} a^* & 2a^* & 0 \\ \frac{2}{5} a^* & \partial_{t^*} + \frac{2}{3} + 3\alpha^* & 0 & 2a^* \\ \frac{8}{25} a^* & 0 & \partial_{t^*} + \frac{3}{2} + 3\alpha^* & \frac{8}{5} a^* \\ 0 & \frac{8}{25} a^* & -\frac{23}{20} a^* & \partial_{t^*} + \frac{3}{2} + 3\alpha^* \end{pmatrix}$$

$$\cdot \begin{pmatrix} M_{2|x} \\ M_{2|y} \\ M_{0|xxy} \\ M_{0|xyy} \end{pmatrix} = a^* \left(\frac{1}{4} M_{0|xxy} + M_{0|yzz}\right) \begin{pmatrix} 0 \\ 0 \\ 0 \\ 1 \end{pmatrix}, \tag{2.77}$$

$$\left(\partial_{t^*} + \frac{3}{2} + 3\alpha^*\right) M_{0|xzz} = a^* \left(\frac{2}{5} M_{2|y} + \frac{2}{5} M_{0|xxy} - M_{0|yzz}\right). \tag{2.78}$$

In these equations, $\partial_{t^*} \equiv d/dt^*$ and we have again nondimensionalized quantities with the collision frequency ν.

The characteristic relaxation times are ℓ_σ^{-1}, where ℓ_σ ($\sigma = 1, \ldots, 10$) are the roots of the characteristic equations. More specifically, $\ell_1 = \ell_2 = \ell_3 = \frac{3}{2} + 3\alpha^*$, ℓ_{4-6} are the roots of the cubic equation

$$\left(\frac{3}{2} + 3\alpha^* - \ell\right)^2 \left(\frac{2}{3} + 3\alpha^* - \ell\right) = \frac{1}{3} a^{*2}, \tag{2.79}$$

and ℓ_{7-10} are the roots of the quartic equation

$$\left(\frac{3}{2} + 3\alpha^* - \ell\right)^2 \left(\frac{2}{3} + 3\alpha^* - \ell\right)^2 = 2a^{*2} \left(\frac{31}{36} + 3\alpha^* - \ell\right). \tag{2.80}$$

Figure 2.6 shows the shear-rate dependence of the real parts of the eigenvalues ℓ_σ. Except ℓ_7, all the eigenvalues increase monotonically with the shear rate. As a matter of fact, ℓ_7 reaches a minimum value $\ell_7|_{\text{min}} \simeq 0.32$ at $a^* \simeq 2.26$. Since ℓ_7 is the smallest eigenvalue, it dominates the long-time behavior of the third-degree moments $\langle V^2 V_x \rangle$, $\langle V^2 V_y \rangle$, $\langle V_x^2 V_y \rangle$, and $\langle V_y^2 V_x \rangle$, which then decay as $e^{-\ell_7 t^*}$. For shear rates smaller than $a^* \simeq 31$, ℓ_7 remains smaller than its equilibrium value $\frac{2}{3}$. This means that, if $a^* < 31$, the characteristic relaxation time for the above third-degree moments is longer than that of equilibrium.

It must be recalled that we are still working with the scaled velocities $\overline{\mathbf{V}} = e^{-\alpha^* t}\mathbf{V}$, although we had dropped the bars. In terms of the *unscaled* velocities, we have, for instance, $\langle V^2 V_y \rangle \sim e^{-(\ell_7 - 3\alpha^*)t^*}$. For shear rates larger than $a^* = \sqrt{18/31} \simeq 0.76$, it turns out that ℓ_7 is smaller than $3\alpha^*$. This implies that these (unscaled) moments increase in time. Truesdell and Muncaster (1980) refer to this behavior as an *instability* in the heat flux solution. However, this is simply a consequence of viscous heating and after scaling the third-degree moments with the thermal velocity, all of them decay to zero.

Next, we are going to analyze the time evolution of the fourth-degree moments and the shear-rate dependence of their stationary values (Santos et al., 1993; Santos and Garzó, 1995a). As said at the beginning of this Section, these are the first non-trivial moments not related to transport properties. Therefore, they can provide information about the population of particles with velocities larger than the thermal velocity. We will see that the fourth-degree moments diverge in time beyond a certain critical value of the shear rate, thus suggesting an algebraic high-velocity tail. Let us consider the following set of fourth-degree polynomials (Truesdell and Muncaster, 1980):

$$Y_{4|0}(\mathbf{V}) = V^4, \tag{2.81}$$

$$Y_{2|ij}(\mathbf{V}) = V^2 \left(V_i V_j - \frac{1}{3} V^2 \delta_{ij}\right), \tag{2.82}$$

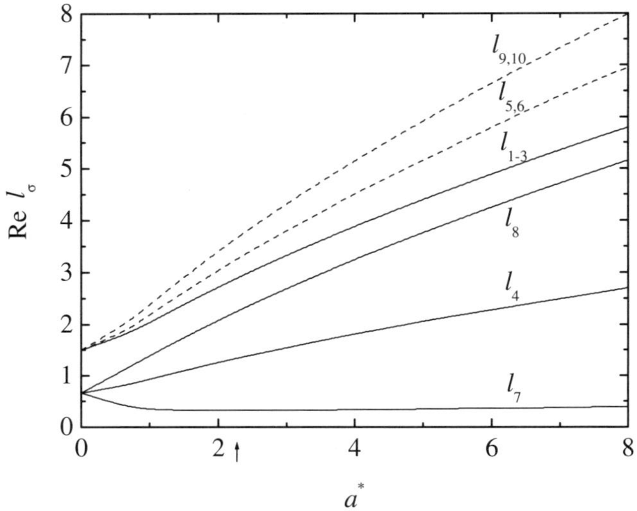

Figure 2.6. Shear-rate dependence of the real parts of the eigenvalues corresponding to the evolution of the third-degree moments. The solid (dashed) lines refer to the real (complex) eigenvalues. The arrow indicates the location of the minimum value of ℓ_7.

$$
\begin{aligned}
Y_{0|ijkl}(\mathbf{V}) &= V_iV_jV_kV_l - \frac{1}{7}V^2\left(V_iV_j\delta_{kl} + V_iV_k\delta_{jl} + V_iV_l\delta_{kj}\right. \\
&\quad \left. + V_jV_k\delta_{il} + V_jV_l\delta_{ik} + V_kV_l\delta_{ij}\right) \\
&\quad + \frac{1}{35}V^4(\delta_{ij}\delta_{kl} + \delta_{ik}\delta_{jl} + \delta_{il}\delta_{jk}). \qquad (2.83)
\end{aligned}
$$

We denote by $M_{2r|ij\ldots}$ the moments of f corresponding to the polynomials $Y_{2r|ij\ldots}(\mathbf{V})$. Our choice for the 15 independent fourth-degree moments is

$$
\begin{aligned}
&\{M_{4|0}, M_{2|xx}, M_{2|yy}, M_{2|xy}, M_{2|xz}, M_{2|yz}, M_{0|xxxx}, M_{0|yyyy}, M_{0|zzzz}, \\
&M_{0|xxxy}, M_{0|xxxz}, M_{0|xyyy}, M_{0|xzzz}, M_{0|yyyz}, M_{0|yzzz}\}. \qquad (2.84)
\end{aligned}
$$

Taking velocity moments of degree four in Eq. (2.13) and making use of their corresponding collisional moments (cf. Appendix 1.A), one gets a closed set of equations, involving moments of degrees two and four only. For simplicity, we will assume that the second-degree moments P_{ij} take their stationary values (2.51)–(2.53). Because of the symmetries of Eq. (2.25), the symmetric and asymmetric moments are uncoupled. Although the relevant moments are the symmetric ones, we first analyze the time evolution of the asymmetric moments, for the sake of completeness. There are 6 asymmetric moments, which verify the following set of equations:

$$
(\partial_{t^*} + 4\alpha^* + \nu_1^*)(M_{0|yyyz} - M_{0|yzzz}) = 0, \qquad (2.85)
$$

$$\begin{pmatrix} \partial_{t^*}+4\alpha^*+\frac{7}{6} & \frac{9}{7}a^* & 0 & -2a^* \\ \frac{2}{7}a^* & \partial_{t^*}+4\alpha^*+\frac{7}{6} & -2a^* & 0 \\ 0 & -\frac{12}{49}a^* & \partial_{t^*}+4\alpha^*+\nu_1^* & -\frac{11}{14}a^* \\ -\frac{12}{49}a^* & 0 & \frac{12}{7}a^* & \partial_{t^*}+4\alpha^*+\nu_1^* \end{pmatrix}$$
$$\cdot\begin{pmatrix} M_{2|xz} \\ M_{2|yz} \\ M_{0|xxxz}+M_{0|xzzz} \\ M_{0|yyyz}+M_{0|yzzz} \end{pmatrix} = \frac{1}{2}a^*(M_{0|yyyz}-M_{0|yzzz})\begin{pmatrix} 0 \\ 0 \\ 1 \\ 0 \end{pmatrix}, \tag{2.86}$$

$$\begin{aligned}(\partial_{t^*}+4\alpha^*+\nu_1^*)(M_{0|xxxz}-M_{0|xzzz}) &= \frac{7}{2}a^*(M_{0|yyyz}+M_{0|yzzz}) \\ &\quad -\frac{1}{2}a^*(M_{0|yyyz}-M_{0|yzzz}).\end{aligned} \tag{2.87}$$

In these equations, $\nu_1^* = \nu_1/\nu$, where ν_1 is given by

$$\nu_1 = \frac{7}{12}\left(4-5\frac{\mathrm{A}_4}{\mathrm{A}_2}\right)\nu, \tag{2.88}$$

with the coefficients A_r being defined in Eq. (2.29). For the true Maxwell interaction, one gets $\nu_1^* \simeq 1.8731$ (Alterman et al., 1962), while $\nu_1^* = \frac{7}{4}$ in the case of the isotropic scattering model defined on p. 23. The eigenvalues associated with the asymmetric fourth-degree moments are $4\alpha^* + \nu_1^*$ (double) and the roots of the characteristic quartic equation corresponding to Eq. (2.86), all of them having a positive real part. Consequently, all the asymmetric moments decay to zero in the long-time limit.

The set of equations for the 9 symmetric moments are given by

$$(\partial_{t^*}+4\alpha^*+\nu_1^*)\mathcal{M}_9 = 0, \tag{2.89}$$

where

$$\mathcal{M}_9 = 3M_{0|xxxx} - 4M_{0|yyyy} - 4M_{0|zzzz}, \tag{2.90}$$

and

$$(\delta_{\sigma\sigma'}\partial_{t^*} + \mathcal{L}_{\sigma\sigma'})\mathcal{M}_{\sigma'} = \mathcal{C}_\sigma, \quad \sigma = 1,\ldots,8. \tag{2.91}$$

Here, $\boldsymbol{\mathcal{M}}$ is a vector made of the following 8 moments

$$\boldsymbol{\mathcal{M}} = \begin{pmatrix} M_{4|0} \\ M_{2|xx} \\ M_{2|yy} \\ M_{0|yyyy} \\ M_{0|zzzz} \\ M_{2|xy} \\ M_{0|xxxy} \\ M_{0|xyyy} \end{pmatrix}, \tag{2.92}$$

$\boldsymbol{\mathcal{C}}$ is a vector of elements

$$\boldsymbol{\mathcal{C}} = a^* \mathcal{M}_9 \begin{pmatrix} 0 \\ 0 \\ 0 \\ 0 \\ 0 \\ \frac{1}{3} \\ \frac{5}{14} \\ -\frac{1}{7} \end{pmatrix} + \frac{(2k_BT/m)^2}{(1+2\alpha^*)^2} \begin{pmatrix} \frac{1}{2}(5+18\alpha^*+12\alpha^{*2}) \\ \alpha^*(1+\alpha^*) \\ -\frac{1}{4}\alpha^*(3+2\alpha^*) \\ -\frac{9}{35}\alpha^*(4-9\alpha^*)(2-\nu_1^*) \\ \frac{9}{35}\alpha^*(1+9\alpha^*)(2-\nu_1^*) \\ -\frac{1}{24}a^*\frac{7+6\alpha^*}{1+2\alpha^*} \\ -\frac{18}{7}a^*\frac{\alpha^*}{1+2\alpha^*}(2-\nu_1^*) \\ \frac{27}{14}a^*\frac{\alpha^*}{1+2\alpha^*}(2-\nu_1^*) \end{pmatrix}, \tag{2.93}$$

and the square matrix $\boldsymbol{\mathcal{L}}$ is

$$\boldsymbol{\mathcal{L}} = 4\alpha^* \boldsymbol{\mathcal{I}} + \boldsymbol{\mathcal{L}}', \tag{2.94}$$

where $\boldsymbol{\mathcal{I}}$ is the 8×8 identity matrix and

$$\boldsymbol{\mathcal{L}}' = \begin{pmatrix} \frac{2}{3} & 0 & 0 & 0 & 0 & 4a^* & 0 & 0 \\ 0 & \frac{7}{6} & 0 & 0 & 0 & \frac{32a^*}{21} & 2a^* & 0 \\ 0 & 0 & \frac{7}{6} & 0 & 0 & -\frac{10a^*}{21} & 0 & 2a^* \\ 0 & 0 & 0 & \nu_1^* & 0 & -\frac{96a^*}{245} & 0 & -\frac{12a^*}{7} \\ 0 & 0 & 0 & 0 & \nu_1^* & \frac{24a^*}{245} & \frac{12a^*}{7} & \frac{12a^*}{7} \\ \frac{7a^*}{15} & \frac{2a^*}{7} & \frac{9a^*}{7} & -\frac{7a^*}{3} & -\frac{a^*}{3} & \frac{7}{6} & 0 & 0 \\ 0 & \frac{15a^*}{49} & -\frac{6a^*}{49} & -\frac{5a^*}{2} & -\frac{5a^*}{14} & 0 & \nu_1^* & 0 \\ 0 & -\frac{6a^*}{49} & \frac{15a^*}{49} & 2a^* & \frac{a^*}{7} & 0 & 0 & \nu_1^* \end{pmatrix}. \tag{2.95}$$

According to Eq. (2.91), $\boldsymbol{\mathcal{L}}$ represents the evolution operator for the moments $\boldsymbol{\mathcal{M}}$. In the limit of zero shear rate, $\boldsymbol{\mathcal{L}}$ reduces to the linearized Boltzmann collision operator projected onto the subspace spanned by the polynomials (2.81)–(2.83).

Equation (2.89) shows that the combination $\mathcal{M}_9 = 6\langle V_y^2 V_z^2 \rangle - \langle V_y^4 \rangle - \langle V_z^4 \rangle$ decays to zero with a characteristic time $\ell_9^{-1} = (4\alpha^* + \nu_1^*)^{-1}$. This exact behavior is similar to that of $\langle V_y^2 \rangle - \langle V_z^2 \rangle$, Eq. (2.33). More in general, it can be easily shown that those eigenfunctions of the linearized Boltzmann collision operator that do not depend on V_x give rise to moments decaying to zero. The only symmetric moment of degree $2k$ belonging to the above class is

$$\frac{1}{2}\langle (V_y + \mathrm{i}\, V_z)^{2k} + (V_y - \mathrm{i}\, V_z)^{2k} \rangle = \sum_{q=0}^{k} (-1)^q \begin{pmatrix} 2k \\ 2q \end{pmatrix} \langle V_y^{2q} V_z^{2k-2q} \rangle. \quad (2.96)$$

The solution of Eq. (2.91) can be written as

$$\boldsymbol{\mathcal{M}}(t^*) = e^{-\boldsymbol{\mathcal{L}}t^*} \cdot [\boldsymbol{\mathcal{M}}(0) - \boldsymbol{\mathcal{M}}^{\mathrm{ss}}] + \boldsymbol{\mathcal{M}}^{\mathrm{ss}}, \quad (2.97)$$

where

$$\mathcal{M}_\sigma^{\mathrm{ss}} = (\boldsymbol{\mathcal{L}}^{-1})_{\sigma\sigma'} \mathcal{C}_{\sigma'}. \quad (2.98)$$

Upon writing Eq. (2.97), we have assumed that $\mathcal{M}_9 = 0$, so that $\mathcal{C}_\sigma$ is constant.

The time evolution of $\mathcal{M}_\sigma$, $\sigma = 1, \ldots, 8$, is governed by the eigenvalues ℓ_σ of the matrix $\mathcal{L}_{\sigma\sigma'}$. Figure 2.7 shows the shear-rate dependence of $\ell_9 = 4\alpha^* + \nu_1^*$ and of the real parts of the eigenvalues ℓ_σ ($\sigma = 1, \ldots, 8$) obtained by numerically solving the corresponding eighth-degree characteristic equation with $\nu_1^* = 1.8731$. At equilibrium ($a^* = 0$), their values are $\ell_1 = \frac{2}{3}$, $\ell_2 = \ell_3 = \ell_4 = \frac{7}{6}$, and $\ell_5 = \cdots = \ell_9 = \nu_1^*$. For nonzero shear rates, the degeneracy of some of these eigenvalues is broken. The most remarkable feature of the figure is that one of the eigenvalues (here denoted as ℓ_1) monotonically decreases when the shear rate increases and eventually vanishes at a certain critical shear rate $a_c^* \simeq 6.846$. This implies that for shear rates larger than a_c^*, the fourth-degree moments do not relax towards stationary values but grow exponentially with time.

Although Eq. (2.97) gives the full time behavior of $\boldsymbol{\mathcal{M}}$ (see Appendix 2.A), the asymptotic behavior for times much larger than $\max\{(\mathrm{Re}\,\ell_\sigma)^{-1}, \sigma = 2\ldots, 9\} \leq \frac{6}{7}$ is

$$\mathcal{M}_\sigma(t^*) \approx \mathcal{M}_\sigma^{\mathrm{ss}} + \mathcal{M}_\sigma^0 e^{-\ell_1 t^*}, \quad (2.99)$$

where $\mathcal{M}_\sigma^0$ is a constant depending on the initial conditions. As an illustration, we have plotted in Fig. 2.8 the time evolution of $\langle V^4 \rangle$ relative to its local equilibrium value $\langle V^4 \rangle_L = \frac{15}{4}(2k_B T/m)^2$ for two values of the shear rate: $a^* \simeq 6.45$ ($\alpha^* = \frac{6}{5}$) and $a^* \simeq 7.33$ ($\alpha^* = \frac{4}{3}$) (cf. Appendix 2.A). While in the first case ($a^* < a_c^*$) $\langle V^4 \rangle$ reaches a stationary value for times t^* larger than $\ell_1^{-1} \simeq 46.5$, it grows exponentially with a characteristic time $-\ell_1^{-1} \simeq 37.1$ in the second case ($a^* > a_c^*$). For shear rates smaller than a_c^*, the steady-state values of the moments are given by Eq. (2.98). Their explicit forms as functions of the shear rate are listed in Appendix 2.B.

Before closing this Section, it is worthwhile mentioning that the exact knowledge of the fourth-degree moments in the USF has been exploited (Brey et al.,

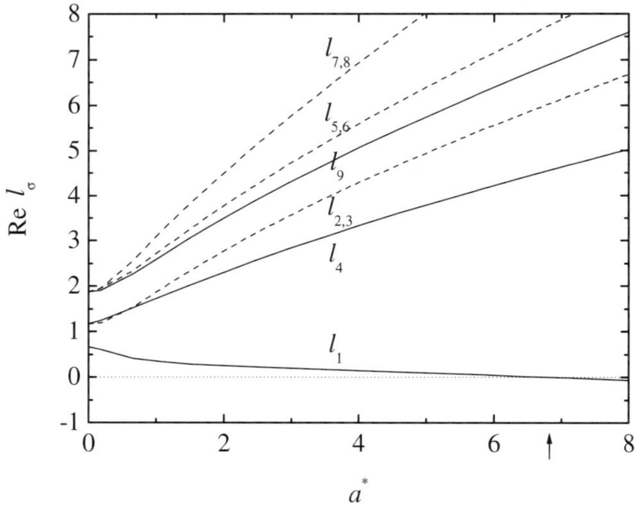

Figure 2.7. Shear-rate dependence of the real parts of the eigenvalues corresponding to the evolution of the symmetric fourth-degree moments. The solid (dashed) lines refer to the real (complex) eigenvalues. The arrow indicates the location of the critical shear rate $a_c^* \simeq 6.846$.

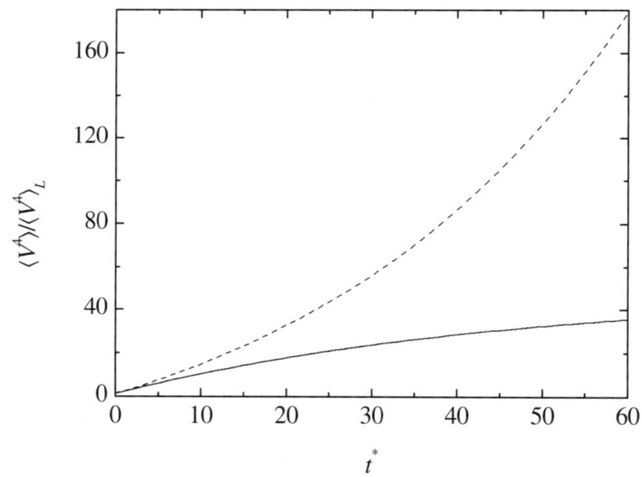

Figure 2.8. Time evolution of $\langle V^4 \rangle$ relative to its local equilibrium value $\langle V^4 \rangle_L = \frac{15}{4}(2k_BT/m)^2$ for two values of the shear rate: $a^* \simeq 6.45$ (solid line) and $a^* \simeq 7.33$ (dashed line). In both cases the initial condition is an anisotropic Gaussian distribution consistent with the stationary second-degree moments.

1993; Santos et al., 1995) to show that a variational principle proposed to characterize thermostatted nonequilibrium states (Evans and Baranyai, 1991; Baranyai, 1996) is not generally valid.

5. Singular behavior of the velocity moments

The divergence of the fourth-degree moments for $a^* > a_c^* \simeq 6.846$ stimulates the analysis of higher-degree moments to investigate whether they diverge for smaller values of the above critical shear rate and whether there exists a finite range of shear rates for which *all* the moments adopt a stationary form. This study will provide information about the high-energy population in the USF state.

In order to consider moments of an arbitrary degree k, the use of velocity polynomials of the form (2.71), (2.72), (2.81)–(2.83) is rather cumbersome. Thus we use here the polynomials defined by Eqs. (1.86) and (1.87), which are eigenfunctions of the linearized Boltzmann equation for Maxwell molecules (McLennan, 1989). The polynomials $\{\Psi_{\mathbf{k}}(\boldsymbol{\xi}),\ \mathbf{k} \equiv (r, \ell, m)\}$ constitute a complete set of orthonormal functions with the inner product (1.83). Due to the invariance properties of USF, we can restrict ourselves to the subset $\{\Psi_{\mathbf{k}}\}$ with $k \equiv 2r + \ell =$ even and m =even. In this subset, there exist $(\frac{k}{2} + 1)^2$ independent polynomials of degree k. This choice implies that we will focus on what we called *symmetric* moments.

Let us assume that the velocity distribution function can be represented in terms of the set $\{\Psi_{\mathbf{k}}\}$, i.e.

$$f(\mathbf{V}, t) = f_L(\mathbf{V}) \sum_{\mathbf{k}} M_{\mathbf{k}}(t^*)\Psi_{\mathbf{k}}(\boldsymbol{\xi}), \tag{2.100}$$

where the dimensionless moments $M_{\mathbf{k}}(t^*)$ are defined as

$$M_{\mathbf{k}}(t^*) = \frac{1}{n} \int d\mathbf{V}\ \Psi_{\mathbf{k}}^{\dagger}(\boldsymbol{\xi})\ f(\mathbf{V}, t). \tag{2.101}$$

In Eq. (2.100), f_L is the local equilibrium distribution defined by Eq. (1.59). Inserting the expansion (2.100) into Eq. (2.25) and performing the inner product with $\Psi_{\mathbf{k}}$, one arrives at the following hierarchy

$$\begin{aligned} &\frac{\partial}{\partial t^*} M_{\mathbf{k}} + a^* \sum_{\mathbf{k}'} \langle \Psi_{\mathbf{k}'} | \xi_y \frac{\partial}{\partial \xi_x} | \Psi_{\mathbf{k}} \rangle^{\dagger} M_{\mathbf{k}'} + \alpha^* \sum_{\mathbf{k}'} \langle \Psi_{\mathbf{k}'} | \boldsymbol{\xi} \cdot \frac{\partial}{\partial \boldsymbol{\xi}} | \Psi_{\mathbf{k}} \rangle^{\dagger} M_{\mathbf{k}'} \\ &= -\lambda_{\mathbf{k}}^0 M_{\mathbf{k}} + {\sum_{\mathbf{k}', \mathbf{k}''}}' J_{\mathbf{k}\mathbf{k}'\mathbf{k}''}\ M_{\mathbf{k}'}\ M_{\mathbf{k}''}. \end{aligned} \tag{2.102}$$

Here,

$$\begin{aligned} J_{\mathbf{k}\mathbf{k}'\mathbf{k}''} &= \frac{n}{\nu} \pi^{-3} \int d\boldsymbol{\xi} \int d\boldsymbol{\xi}_1 \int d\Omega\, g B(g, \chi) e^{-(\xi^2 + \xi_1^2)} \\ &\quad \times \Psi_{\mathbf{k}}^{\dagger}(\boldsymbol{\xi}) \left[\Psi_{\mathbf{k}'}(\boldsymbol{\xi}') \Psi_{\mathbf{k}''}(\boldsymbol{\xi}_1') - \Psi_{\mathbf{k}'}(\boldsymbol{\xi}) \Psi_{\mathbf{k}''}(\boldsymbol{\xi}_1) \right], \end{aligned} \tag{2.103}$$

and $\lambda_{\mathbf{k}}^0$ are the eigenvalues (1.90) of the linearized collision operator, except that here they are nondimensionalized with respect to ν, i.e.

$$\lambda_{\mathbf{k}}^0 \equiv \lambda_{r\ell}^0 = \frac{n}{\nu}\int d\Omega\, gB(g,\chi)\left[1+\delta_{r0}\delta_{\ell 0}-\cos^{2r+\ell}\left(\frac{\chi}{2}\right)P_\ell\left(\cos\frac{\chi}{2}\right)\right. \\ \left. -\sin^{2r+\ell}\left(\frac{\chi}{2}\right)P_\ell\left(\sin\frac{\chi}{2}\right)\right] \tag{2.104}$$

The prime in the last summation of Eq. (2.102) denotes the constraints $k'+k''=k$, and $k', k''>0$. The explicit expression for the coefficients $J_{\mathbf{k}\mathbf{k}'\mathbf{k}''}$ is known (McLennan, 1989) but it will not be needed in the subsequent discussion.

The brackets appearing in Eq. (2.102) are zero if $k'>k$. Consequently, the hierarchy (2.102) can be recast into the form

$$\frac{\partial}{\partial t^*}M_{\mathbf{k}} + \sum_{\substack{\mathbf{k}' \\ (k'=k)}} \mathcal{L}_{\mathbf{k}\mathbf{k}'}M_{\mathbf{k}'} = B_{\mathbf{k}}, \tag{2.105}$$

where, for a given degree k, $\mathcal{L}_{\mathbf{k}\mathbf{k}'}$ is a $(\frac{k}{2}+1)^2 \times (\frac{k}{2}+1)^2$ square matrix given by

$$\mathcal{L}_{\mathbf{k}\mathbf{k}'} = (\lambda_{\mathbf{k}}^0 + ka^*)\delta_{\mathbf{k}\mathbf{k}'} + a^*\langle\Psi_{\mathbf{k}'}|\xi_y\frac{\partial}{\partial\xi_x}|\Psi_{\mathbf{k}}\rangle^\dagger \tag{2.106}$$

and $B_{\mathbf{k}}$ is a bilinear combination of moments of order *less* than k. As in the case of the fourth-degree moments, $\boldsymbol{\mathcal{L}}$ represents the evolution operator of the moments $M_{\mathbf{k}}$. In the special case of $k=4$, Eq. (2.105) is equivalent to Eqs. (2.89) and (2.91). A tedious but standard calculation yields, for $2r'+\ell'=2r+\ell$, (Montanero et al., 1996b)

$$\begin{aligned}\langle\Psi_{r'\ell'm'}|\xi_y\frac{\partial}{\partial\xi_x}|\Psi_{r\ell m}\rangle &= R_{r,\ell,m}\delta_{\ell',\ell}\delta_{m,m'-2} + R^\dagger_{r',\ell',m'}\delta_{\ell',\ell}\delta_{m',m-2} \\ &\quad + S_{r,\ell,m}\delta_{\ell,\ell'-2}\delta_{m,m'-2} + S^\dagger_{r',\ell',m'}\delta_{\ell',\ell-2}\delta_{m',m-2} \\ &\quad + S_{r,\ell,-m}\delta_{\ell,\ell'-2}\delta_{m,m'+2} \\ &\quad + S^\dagger_{r',\ell',-m'}\delta_{\ell',\ell-2}\delta_{m',m+2} + \mathrm{i}\frac{m}{2}\delta_{\ell',\ell}\delta_{m',m},\end{aligned} \tag{2.107}$$

where

$$R_{r,\ell,m} = \frac{\mathrm{i}}{2}\frac{2r+\ell+\frac{3}{2}}{(2\ell+3)(2\ell-1)}\left[\frac{(\ell+m+2)!(\ell-m)!}{(\ell+m)!(\ell-m-2)!}\right]^{1/2}, \tag{2.108}$$

$$S_{r,\ell,m} = \frac{\mathrm{i}}{2}\frac{1}{2\ell+3}\left[\frac{r(r+\ell+\frac{3}{2})(\ell+m+4)!}{(2\ell+1)(2\ell+5)(\ell+m)!}\right]^{1/2}. \tag{2.109}$$

This completes the explicit knowledge of the matrix $\mathcal{L}_{\mathbf{kk'}}$.

The time evolution of all the moments $M_{\mathbf{k}}$ of the same degree k is described by the corresponding eigenvalues $\lambda_{\mathbf{k}}(a^*)$ of the matrix $\mathcal{L}_{\mathbf{kk'}}$, which are the roots of an algebraic equation of degree $(\frac{k}{2}+1)^2$. Obviously, $\lambda_{\mathbf{k}}(0) = \lambda^0_{\mathbf{k}}$. In the particular case of $k = 4$ those eigenvalues are the ℓ_σ plotted in Fig. 2.7. The long-time behavior of the moments $M_{\mathbf{k}}$ of the same degree k is governed by the eigenvalue with the smallest real part. We will denote such an eigenvalue by $\widetilde{\lambda}_k(a^*)$. Consequently, the moments of degree k diverge to infinity when the real part of $\widetilde{\lambda}_k(a^*)$ becomes negative. From the numerical solution of the characteristic equation, it turns out that $\widetilde{\lambda}_k(a^*)$ corresponds to $\lambda_{\frac{k}{2}0}(a^*)$ and is a real number (Montanero et al., 1996b).

The eigenvalues $\widetilde{\lambda}_k(a^*)$ for $k = 4, 6, \ldots, 36$ have been evaluated by Montanero et al. (1996b). Figure 2.9 shows $\widetilde{\lambda}_k$ as a function of the shear rate for $k = 4, 6, 8$, and 20. For a given degree k, the eigenvalue $\widetilde{\lambda}_k(a^*)$ monotonically

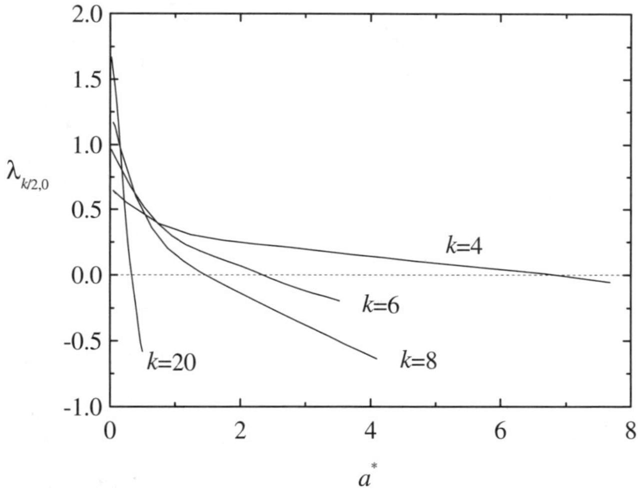

Figure 2.9. Shear-rate dependence of the smallest eigenvalue for $k = 4, 6, 8$, and 20.

decreases as the shear rate increases and eventually changes sign at a certain critical value $a_c^{(k)}$. This extends the results already discussed in Section 2.4 for $k = 4$. Table 2.1 gives the numerical values of $a_c^{(k)}$ for $k = 4, 6, \ldots, 36$. In addition to the actual collision rate for Maxwell molecules, we have also considered the isotropic Maxwell model (Hendriks and Nieuwenhuizen, 1982), for which the collision rate $gB(g, \chi)$ is assumed to be independent of the scattering angle χ. It can be observed that the value of $a_c^{(k)}$ decreases as the degree k increases. This means that if $a^* \geq a_c^{(k)}$, all the moments of degree equal to or greater than k diverge as $t^* \to \infty$. On the other hand, if $a^* < a_c^{(k)}$ all the moments of degree less than k reach stationary values in the long-time limit.

Table 2.1. Values of the critical shear rate $a_c^{(k)}$ associated with moments of degree $k = 4, 6, \ldots, 36$.

k	$a_c^{(k)}$	
	Actual scattering	*Isotropic scattering*
4	6.846	7.746
6	2.346	3.344
8	1.450	1.667
10	0.817	0.940
12	0.618	0.633
14	0.502	0.479
16	0.427	0.388
18	0.373	0.327
20	0.333	0.283
22	0.301	0.250
24	0.276	0.224
26	0.255	0.202
28	0.238	0.185
30	0.223	0.171
32	0.210	0.158
34	0.199	0.147
36	0.189	0.138

The relevant question now is whether there exists a nonzero lower bound for the set $\{a_c^{(k)}, k \geq 4\}$. In other words, is $\lim_{k\to\infty} a_c^{(k)} \equiv a_c^{(\infty)}$ different from zero? If that were the case, *all* the moments would reach stationary values for shear rates smaller than $a_c^{(\infty)}$. In order to clarify this question, a log-log plot of $a_c^{(k)}$ versus k is shown in Fig. 2.10. We see that the points corresponding to the largest values of k tend to align. This suggests that the asymptotic behavior of $a_c^{(k)}$ for large k is of the form

$$a_c^{(k)} \sim k^{-\mu}. \tag{2.110}$$

A linear fit gives $\mu \simeq 0.91$ for actual Maxwell molecules and $\mu \simeq 1.14$ for the isotropic model. Equation (2.110) implies that $a_c^{(\infty)} = 0$. While no rigorous proof of the law (2.110) has so far been given, it is strongly supported by the numerical results presented in Table 2.1. This implies that for *any* value of the shear rate, there exists a critical value $k_c(a^*)$ such that *all* the moments of degree equal to or larger than k_c diverge as $t^* \to \infty$. The apparent influence of

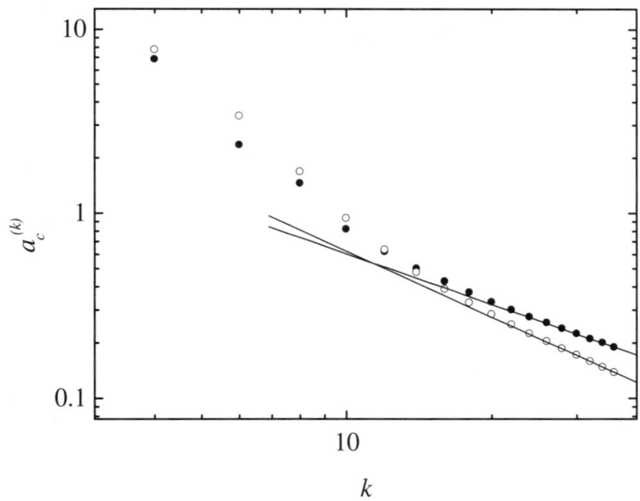

Figure 2.10. Log-log plot of $a_c^{(k)}$ versus k for the actual scattering model (filled circles) and the isotropic scattering model (open circles). The straight lines are linear fits of the last 4 points.

the scattering model on the value of μ might be due to the fact that the values of k considered in Table 2.1 are not sufficiently large. From that point of view, it is tempting to conjecture that $\mu = 1$ for Maxwell molecules, both for the actual and the isotropic scattering models.

So far, the analysis has been restricted to the hierarchy of moments, rather than dealing with the Boltzmann equation itself. In this sense, we only have an *indirect* information about the behavior of the distribution function. On the other hand, the divergence of the moments suggests that the high-velocity population decays more slowly than a certain power, even for small shear rates. The simplest scenario corresponds to a stationary distribution function with an asymptotic behavior of the form

$$f(\mathbf{V}) \sim V^{-5-\sigma(a^*)} \tag{2.111}$$

for large V, where the exponent $\sigma(a^*)$ is a certain shear-rate dependent function to be determined. For the sake of simplicity let us consider $\langle V^k \rangle$ as a representative moment of degree k. In that case,

$$\langle V^k \rangle = \int_0^\infty dV\, V^k F(V), \tag{2.112}$$

where

$$F(V) = \frac{1}{n} V^2 \int d\widehat{\mathbf{V}}\, f(\mathbf{V}) \tag{2.113}$$

is the probability distribution for the magnitude of the velocity. The asymptotic behavior (2.111) yields a high-velocity tail

$$F(V) \sim V^{-3-\sigma(a^*)}. \tag{2.114}$$

As a consequence, the moment $\langle V^k \rangle$ diverges if $k-3-\sigma(a^*) \geq -1$, i.e. $k \geq 2+$

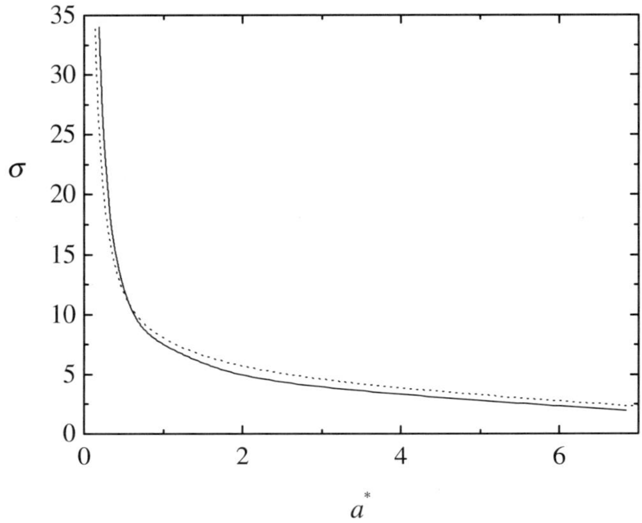

Figure 2.11. Exponent σ as a function of the shear rate for the actual scattering model (solid line) and the isotropic scattering model (dotted line).

$\sigma(a^*)$; the critical shear rate $a^* = a_c^{(k)}$ would be then the solution of $\sigma(a_c^{(k)}) = k-2$. This allows one to interpret the critical shear rates $a_c^{(4)}, a_c^{(6)}, a_c^{(8)}, \ldots$ as those values of a^* at which the function $\sigma(a^*)$ takes even integer values. Since the second-degree moments are finite, the function $\sigma(a^*)$ is positive definite for any shear rate, but it decreases as the shear rate increases. It seems natural to conjecture that $\lim_{a^*\to\infty} \sigma(a^*) = 0$. In the opposite limit, Eq. (2.110) implies that $\sigma(a^*) \sim a^{*-1/\mu}$ for small shear rates. Figure 2.11 shows the shear-rate dependence of σ using the values of Table 2.1 supplemented by an adequate interpolation.

The rigorous proof of the asymptotic behavior (2.114) is still lacking.[7] An alternative approach to overcome this absence is to resort to a numerical solution of the Boltzmann equation by means of the Direct Simulation Monte Carlo (DSMC) method (Bird, 1994; Alexander and Garcia, 1997; Oran et al., 1998).

[7]However, the algebraic high-velocity tail of the velocity distribution function has been derived from the two-dimensional Boltzmann–Fokker–Planck equation of a Maxwell gas under USF (Acedo et al., 2002).

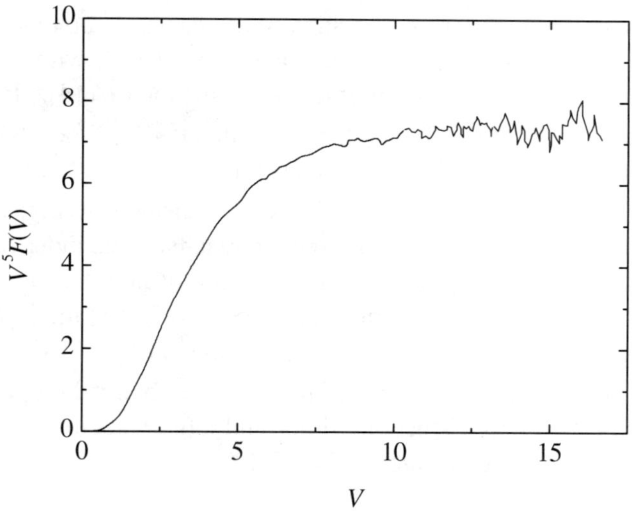

Figure 2.12. Plot of $V^5F(V)$ as a function of the velocity V at the shear rate $a^* = a_c^{(4)} \simeq 5.847$ for a two-dimensional gas. The velocity is measured in units of the thermal velocity $(2k_BT/m)^{1/2}$.

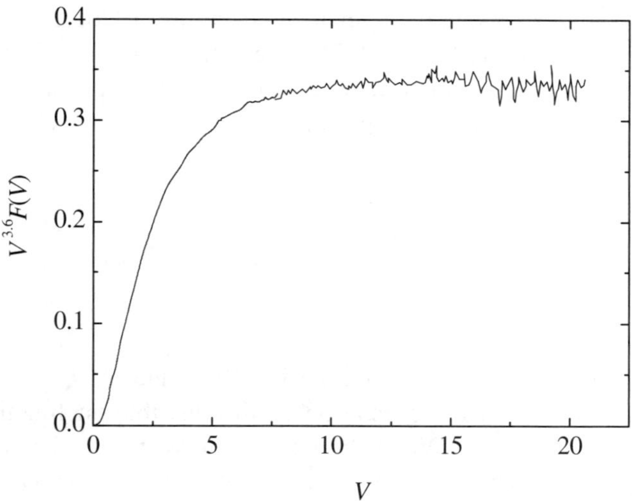

Figure 2.13. Plot of $V^{3.6}F(V)$ as a function of the velocity V at the shear rate $a^* = 12$ for a two-dimensional gas. The velocity is measured in units of the thermal velocity $(2k_BT/m)^{1/2}$.

This investigation has been carried out by Montanero et al. (1997b), who analyzed in detail the USF state of a *two-dimensional* gas of Maxwell molecules. First, by following similar steps as those described in this Section for the three-dimensional case, they obtained the critical shear rates $a_c^{(k)}$ up to $k = 240$. The results strongly supported the asymptotic behavior (2.110) with $\mu = 1$. Next, the DSMC results were used to confirm the behavior (2.114). In particular, the critical shear rate for the fourth-degree moments is $a_c^{(4)} \simeq 5.847$. Thus, at $a^* = 5.847$ the function $F(V)$ must behave as $F(V) \sim V^{-5}$, in accordance with Eq. (2.114), since $\sigma(a_c^{(4)}) = 2$. This is confirmed by Fig. 2.12, where the simulation results for $V^5 F(V)$ are seen to reach a value independent of the velocity for $V \gtrsim 10(2k_BT/m)^{1/2}$. As a measure of how far the high-energy population is from that of equilibrium at $a^* \simeq 5.847$, it is interesting to note that, at equilibrium $V^5 F(V) \sim 10^{-37}(2k_BT/m)^2$ at $V = 10(2k_BT/m)^{1/2}$. For shear rates $a^* > a_c^{(4)}$ the values of $\sigma(a^*)$ were obtained empirically from log-log plots of F versus V for long times and large velocities. For instance, $\sigma \simeq 0.6$ for $a^* = 12$, as confirmed in Fig. 2.13.

6. Perturbation expansion of the distribution function

While all the previous analytical results are exact consequences of the Boltzmann equation (2.25), it must be acknowledged that the corresponding velocity distribution function is not known. As an alternative, we can look for a perturbation solution in powers of the shear rate a^*. This is nothing else but the Chapman–Enskog expansion (Chapman and Cowling, 1970; Dorfman and van Beijeren, 1977), specialized to the USF state. Of course, the practical usefulness of such an expansion is restricted to small values of the shear rate, i.e. to near equilibrium states. Let us represent the *stationary* velocity distribution function as the expansion

$$f(\mathbf{V}; a^*) = n\left(\frac{m}{2k_BT}\right)^{3/2} \varphi_0(\boldsymbol{\xi})\left[1 + \sum_{q=1}^{\infty} \Phi^{(q)}(\boldsymbol{\xi})\, a^{*q}\right], \tag{2.115}$$

where the reduced velocity $\boldsymbol{\xi}$ is defined in (1.88) and $\varphi_0(\boldsymbol{\xi}) = \pi^{-3/2}e^{-\xi^2}$. Equation (2.115) gives a representation of the distribution function alternative to that of Eq. (2.100). While the latter expresses f/f_L as an expansion in the orthogonal set $\{\Psi_{\mathbf{k}}(\boldsymbol{\xi})\}$, the functions $\Phi^{(q)}(\boldsymbol{\xi})$ correspond to the successive Chapman–Enskog approximations to the distribution function. Since the moments $M_{\mathbf{k}}$ defined by Eq. (2.101) cannot converge for $a^* > a_c^{(k)}$ and $a_c^{(k)} \stackrel{k\to\infty}{\longrightarrow} 0$, the expansion (2.115) might be only asymptotic. The heating rate $\alpha^*(a^*)$, given explicitly by Eq. (2.46), can also be expanded as

$$\alpha^* = \sum_{q=0}^{\infty} \alpha^{(q)} a^{*q}. \tag{2.116}$$

According to Eq. (2.61), $\alpha^{(q)} = 0$ if $q =$ odd, $\alpha^{(0)} = 0$, $\alpha^{(2)} = \frac{1}{3}$, $\alpha^{(4)} = -\frac{4}{9}$, $\alpha^{(6)} = \frac{28}{27}, \ldots$. Substitution of Eqs. (2.115) and (2.116) into Eq. (2.25) gives

$$\Phi^{(q-1)}\xi_y \frac{\partial}{\partial \xi_x} \ln \varphi_0 + \xi_y \frac{\partial}{\partial \xi_x} \Phi^{(q-1)} + \sum_{q'=2}^{q} \alpha^{(q')} \left[\Phi^{(q-q')} \boldsymbol{\xi} \cdot \frac{\partial}{\partial \boldsymbol{\xi}} \ln \varphi_0 \right.$$
$$\left. + \frac{\partial}{\partial \boldsymbol{\xi}} \cdot \boldsymbol{\xi} \Phi^{(n-n')} \right] = \mathcal{L}\Phi^{(q)} + \sum_{q'=1}^{q-1} \mathcal{K}\left[\Phi^{(q')}, \Phi^{(q-q')} \right]. \tag{2.117}$$

Here $\mathcal{K}$ is the bilinear operator

$$\begin{aligned} \mathcal{K}\left[X_1, X_2\right] &\equiv -\frac{n}{\nu\varphi_0} J\left[\varphi_0 X_1, \varphi_0 X_2\right] \\ &= \frac{n}{\nu} \int d\boldsymbol{\xi}_1 \int d\Omega \, gB(g,\chi)\varphi_0(\boldsymbol{\xi}_1) \\ &\quad \times \left[X_1(\boldsymbol{\xi})X_2(\boldsymbol{\xi}_1) - X_1(\boldsymbol{\xi}')X_2(\boldsymbol{\xi}_1') \right], \end{aligned} \tag{2.118}$$

while $\mathcal{L}$ is the (dimensionless) linearized Boltzmann collision operator defined by Eq. (1.78), i.e.

$$\mathcal{L}X = \mathcal{K}[X,1] + \mathcal{K}[1,X]. \tag{2.119}$$

As said above, the basis functions $\{\Psi_{\mathbf{k}}(\boldsymbol{\xi})\}$ defined in Eq. (1.87) are eigenfunctions of $\mathcal{L}$, so that

$$\mathcal{L}\Psi_{r\ell m}(\boldsymbol{\xi}) = \lambda_{r\ell}^0 \Psi_{r\ell m}(\boldsymbol{\xi}), \tag{2.120}$$

where the (dimensionless) eigenvalues $\lambda_{r\ell}^0$ are given by Eq. (2.104).

It can be checked by induction that $\Phi^{(q)}(\boldsymbol{\xi})$ is a polynomial in $\boldsymbol{\xi}$ of degree $2q$. Consequently, $\Phi^{(q)}(\boldsymbol{\xi})$ can be expressed as a linear combination of a *finite* number of eigenfunctions:

$$\Phi^{(q)}(\boldsymbol{\xi}) = \sum_{\substack{\mathbf{k} \\ 2\leq k\leq 2q}} M_{\mathbf{k}}^{(q)} \Psi_{\mathbf{k}}(\boldsymbol{\xi}). \tag{2.121}$$

This exact property is peculiar of the Maxwell interaction. From Eqs. (2.100), (2.115), and (2.121) one gets

$$M_{\mathbf{k}}(a^*) = \sum_{q=k/2}^{\infty} M_{\mathbf{k}}^{(q)} a^{*q}. \tag{2.122}$$

On account of the invariance properties (2.9), it follows the symmetry property $M_{r\ell m}^{(q)} = (-1)^q M_{r,\ell,-m}^{(q)}$. As a consequence,

$$\langle \Phi^{(q)} | \Phi^{(q')} \rangle = \sum_{\substack{\mathbf{k} \\ 2\leq k\leq \min(2q,2q')}} M_{\mathbf{k}}^{(q)\dagger} M_{\mathbf{k}}^{(q')}, \tag{2.123}$$

which vanishes if $q+q'=$ odd. Thus, in order to know $f(\mathbf{V};a^*)$ through order q in a^* we need to determine the coefficients $\{M_{\mathbf{k}}^{(q')}, k \leq 2q', q' \leq q\}$.

The hierarchy (2.117) can be solved recursively. Provided that $\Phi^{(q')}$ with $q'<q$ are known, $\Phi^{(q)}$ can be written as

$$\begin{aligned}\Phi^{(q)} &= \mathcal{L}^{-1}\Bigg\{-2\xi_x\xi_y\Phi^{(q-1)}+\xi_y\frac{\partial}{\partial\xi_x}\Phi^{(q-1)}+\sum_{q'=2}^{q}\alpha^{(q')}\Bigg[-2\xi^2\Phi^{(q-q')}\\ &\quad+\frac{\partial}{\partial\boldsymbol{\xi}}\cdot\boldsymbol{\xi}\Phi^{(q-q')}\Bigg]-\sum_{q'=1}^{q-1}\mathcal{K}[\Phi^{(q')},\Phi^{(q-q')}]\Bigg\}. \qquad (2.124)\end{aligned}$$

The expression enclosed between braces is a polynomial of degree $2q$. Once it is expressed as a linear combination of $\{\Psi_{\mathbf{k}}, k \leq 2q\}$, the action of the operator $\mathcal{L}^{-1}$, on account of Eq. (2.120), is straightforward and one can get $\Phi^{(q)}$ explicitly. Following this process, the velocity distribution function through third order in the shear rate (super-Burnett order) has been obtained (Montanero and Santos, 1996). More specifically,

$$\Phi^{(1)}(\boldsymbol{\xi})=-2\xi_x\xi_y=-\Psi_{022}^{(-)}(\boldsymbol{\xi}), \qquad (2.125)$$

$$\begin{aligned}\Phi^{(2)}(\boldsymbol{\xi}) &= \Psi_{022}^{(+)}(\boldsymbol{\xi})-\tfrac{\sqrt{3}}{3}\Psi_{020}(\boldsymbol{\xi})+\tfrac{\sqrt{105}}{70}\left(\tfrac{4}{\lambda_{04}^0}-1\right)\Psi_{040}(\boldsymbol{\xi})\\ &\quad-\tfrac{\sqrt{3}}{2}\left(\tfrac{4}{\lambda_{04}^0}-1\right)\Psi_{044}^{(+)}(\boldsymbol{\xi})-\tfrac{10\sqrt{42}}{147}\Psi_{120}(\boldsymbol{\xi})+\tfrac{\sqrt{30}}{6}\Psi_{200}(\boldsymbol{\xi})\\ &= \left(\tfrac{8}{35\lambda_{04}^0}+\tfrac{1}{245}\right)\xi^4-\left(\tfrac{8}{7\lambda_{04}^0}-\tfrac{34}{49}\right)\xi^2(\xi_x^2+\xi_y^2)\\ &\quad+\left(\tfrac{8}{\lambda_{04}^0}-2\right)\xi_x^2\xi_y^2-\tfrac{29}{21}\xi^2+\tfrac{4}{7}\xi_x^2-\tfrac{10}{7}\xi_y^2+\tfrac{5}{4}, \qquad (2.126)\end{aligned}$$

$$\begin{aligned}\Phi^{(3)}(\boldsymbol{\xi}) &= \tfrac{4}{3}\Psi_{022}^{(-)}(\boldsymbol{\xi})+\tfrac{\sqrt{21}}{7}\left(\tfrac{20}{7{\lambda_{04}^0}^2}+\tfrac{132}{49\lambda_{04}^0}-1\right)\Psi_{042}^{(-)}(\boldsymbol{\xi})\\ &\quad-\sqrt{3}\left(\tfrac{4}{{\lambda_{04}^0}^2}+\tfrac{2}{\lambda_{04}^0}-1\right)\Psi_{044}^{(-)}(\boldsymbol{\xi})+\tfrac{8\sqrt{14}}{36\,015}\left(\tfrac{3402}{\lambda_{04}^0}-193\right)\Psi_{122}^{(-)}(\boldsymbol{\xi})\\ &\quad+\tfrac{3}{4}\tfrac{(4-\lambda_{04}^0)(4-4\lambda_{04}^0+\lambda_{06}^0)}{\lambda_{04}^0\lambda_{06}^0}\left[\tfrac{\sqrt{22}}{11}\Psi_{062}^{(-)}(\boldsymbol{\xi})-\sqrt{10}\Psi_{066}^{(-)}(\boldsymbol{\xi})\right]\\ &\quad+\tfrac{10\sqrt{462}}{539}\tfrac{33{\lambda_{04}^0}^2-97\lambda_{04}^0-140}{\lambda_{04}^0(69\lambda_{04}^0-49)}\Psi_{142}^{(-)}(\boldsymbol{\xi})\\ &\quad+\tfrac{\sqrt{14}}{490}\tfrac{4872{\lambda_{04}^0}^2-20\,761\lambda_{04}^0-26\,208}{\lambda_{04}^0(24\lambda_{04}^0+49)}\Psi_{222}^{(-)}(\boldsymbol{\xi}), \qquad (2.127)\end{aligned}$$

where we have called

$$\Psi_{r\ell m}^{(\pm)}(\boldsymbol{\xi})\equiv\frac{e^{\mathrm{i}\frac{\pi}{4}(-1\pm1)}}{\sqrt{2}}\left[\Psi_{r\ell m}(\boldsymbol{\xi})\pm\Psi_{r\ell -m}(\boldsymbol{\xi})\right]. \qquad (2.128)$$

The numerical values of the eigenvalues appearing in the above equations are $\lambda_{04}^0 \simeq 1.87313$ and $\lambda_{06}^0 \simeq 2.45323$ (Alterman et al., 1962). The results (2.125)–(2.127) have been used (Montanero et al., 1995b) to compute averages of the forms $\langle \ln |V_i| \rangle$ and $\langle V_j V_k \ln |V_i| \rangle$ through order a^{*3} and refute an extremum property of the Gaussian thermostat (Evans and Baranyai, 1992; Sarman et al., 1994).

7. Nonequilibrium entropy

An important application of the above exact expansion of the distribution function is the evaluation of the corresponding expansion in powers of the shear rate of the nonequilibrium entropy in the USF (Montanero and Santos, 1996). The concept of entropy is well established for systems in equilibrium. Let us denote by $s_0(n, e)$ the equilibrium entropy per unit volume of a fluid with a number density n and an internal energy per unit volume e. For a monatomic ideal gas (Reif, 1965), s_0 is given by Eq. (1.49). For states close to equilibrium, a natural extension of the entropy is to take the *local equilibrium* version of Eq. (1.49), where n and e are replaced by their local values $n(\mathbf{r}, t)$ and $e(\mathbf{r}, t)$ (de Groot and Mazur, 1984). Beyond the hypothesis of local equilibrium, a more general approach is provided by the so-called extended irreversible thermodynamics (Jou et al., 1992; Jou et al., 1999; Jou et al., 2000; Jou et al., 2001). This theory assumes that the nonequilibrium entropy depends on the dissipative fluxes, in addition to the hydrodynamic fields, and leads to relaxational transport equations compatible with the positiveness of the entropy production.

Far from equilibrium, Boltzmann's H-theorem (Dorfman and van Beijeren, 1977; McLennan, 1989; Cercignani, 1990) provides a fundamental basis to extend the concept of entropy $s(\mathbf{r}, t)$ of a low density gas to nonequilibrium states. From Eq. (1.47), we can decompose the entropy as

$$s(\mathbf{r}, t) = s_0(n(\mathbf{r}, t), e(\mathbf{r}, t)) + s_{\text{ex}}(\mathbf{r}, t), \tag{2.129}$$

where the *excess* entropy per unit volume is a functional of the velocity distribution function given by

$$\begin{aligned} s_{\text{ex}}(\mathbf{r}, t) &= -k_B \int d\mathbf{v}\, f(\mathbf{r}, \mathbf{v}, t) \ln \frac{f(\mathbf{r}, \mathbf{v}, t)}{f_L(\mathbf{r}, \mathbf{v}, t)} \\ &= \langle f/f_L | \ln(f/f_L) \rangle , \end{aligned} \tag{2.130}$$

where the local equilibrium velocity distribution function is given by Eq. (1.59) and the inner product is defined by Eq. (1.83).

Of course the excess entropy must be negative, in agreement with the second principle of thermodynamics and the H-theorem (1.57). In fact, the local equilibrium distribution is the one that maximizes the entropy (1.47) subjected to given fields $n(\mathbf{r}, t)$ and $T(\mathbf{r}, t)$. A more stringent upper bound is obtained by information-theory arguments (Mead and Papanicolau, 1984; Buck and Macauly, 1991; Luzzi et al., 2002) from the knowledge of the relevant fluxes. For simplicity, here we assume that only the pressure tensor $\mathsf{P}(\mathbf{r},t)$ is known. In fact, this is the most important flux in shear flows and, in addition, the implementation of the information theory method by including the heat flux requires numerical procedures. The problem consists of obtaining the distribution function $f_{\rm IT}$ that maximizes $s_{\rm ex}$ with the constraints

$$\{n,\mathbf{0},\mathsf{P}\}=\int d\mathbf{v}\{1,\mathbf{V},m\mathbf{V}\mathbf{V}\}f. \tag{2.131}$$

By standard use of Lagrange multipliers, one can easily get (Garzó and Santos, 1995a; Jou and Criado-Sancho, 2001)

$$f_{\rm IT}(\mathbf{v})=n\left(\frac{mn}{2\pi}\right)^{3/2}(\det\mathsf{P})^{-1/2}\exp\left(-\frac{mn}{2}\mathsf{P}^{-1}:\mathbf{V}\mathbf{V}\right). \tag{2.132}$$

The associated excess entropy $s_{\rm ex,IT}$ is an upper bound of the true excess entropy and is given by

$$s_{\rm ex,IT}=-\frac{nk_B}{2}\ln\left[p^3(\det\mathsf{P})^{-1}\right]. \tag{2.133}$$

Therefore, $s_{\rm ex}\leq s_{\rm ex,IT}\leq 0$. In general, this upper bound $s_{\rm ex,IT}$ has a complex dependence on the hydrodynamic gradients. When the pressure tensor in Eq. (2.133) is replaced by its Navier–Stokes approximation (1.106) and one expands in powers of the velocity gradients, the result coincides with the first term on the right-hand side of Eq. (1.124). Therefore, in the absence of thermal gradients, the information-theory and the exact entropies tend to coincide in the limit of small shear.

In the special case of the USF, where the distribution function can be represented by Eq. (2.115), Equation (2.130) allows one to get the power expansion of $s_{\rm ex}^*(a^*)\equiv s_{\rm ex}/nk_B$ in terms of inner products involving the functions $\Phi^{(q)}(\boldsymbol{\xi})$ (Montanero and Santos, 1996). In particular, to evaluate $s_{\rm ex}^*(a^*)$ to order a^{*6} we need $\Phi^{(1)}$, $\Phi^{(2)}$, $\Phi^{(3)}$, the projection of $\Phi^{(4)}$ onto the subspace spanned by $\{\Psi_{\mathbf{k}}(\boldsymbol{\xi}),k=2,4\}$ and the projection of $\Phi^{(5)}$ onto the subspace spanned by $\{\Psi_{\mathbf{k}}(\boldsymbol{\xi}),k=2\}$. The result is

$$s_{\rm ex}^*(a^*)=-\frac{1}{2}a^{*2}+0.58417a^{*4}-1.36502a^{*6}+\mathcal{O}(a^{*8}). \tag{2.134}$$

The first term, $s_{\rm ex}^*(a^*)\approx-\frac{1}{2}a^{*2}$, corresponds to the Navier–Stokes description, in agreement with Eq. (1.124). The fact that the second term is positive indicates that the decrease of $s_{\rm ex}^*(a^*)$ as a^* increases is *smaller* than the one predicted by the extrapolation of the Navier–Stokes value, so we can expect $s_{\rm ex}^*(a^*)>-\frac{1}{2}a^{*2}$. The upper bound provided by information theory, Eq. (2.133), can be explicitly evaluated by using Eqs. (2.51)–(2.53) and the result is

$$s_{\rm ex,IT}^*(a^*)=-\frac{1}{2}\ln\frac{\left[1+2\alpha^*(a^*)\right]^3}{1+3\alpha^*(a^*)}, \tag{2.135}$$

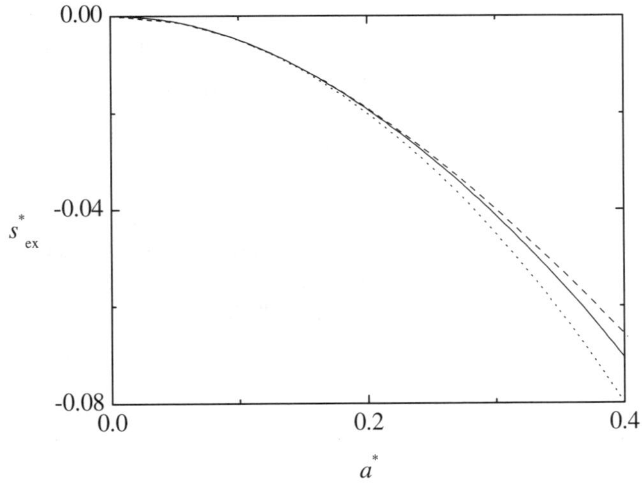

Figure 2.14. Shear-rate dependence of the reduced excess entropy, according to the Navier–Stokes approximation $s^*_{\text{ex,NS}} = -\frac{1}{2}a^{*2}$ (dotted line), the truncated expansion (2.134) (solid line), and information theory (dashed line), Eq. (2.135).

where $\alpha^*(a^*)$ is given by Eq. (2.46). The first few terms in the power expansion of $s^*_{\text{ex,IT}}(a^*)$ are

$$s^*_{\text{ex,IT}}(a^*) = -\frac{1}{2}a^{*2} + \frac{3}{4}a^{*4} - \frac{95}{54}a^{*6} + \mathcal{O}(a^{*8}). \tag{2.136}$$

Figure 2.14 shows the shear-rate dependence of the (reduced) excess entropy, as given by the Navier–Stokes approximation $s^*_{\text{ex,NS}} = -\frac{1}{2}a^{*2}$, the exact expansion (2.134) truncated after the term of order a^{*6}, and information theory, Eq. (2.135). Over the range of shear rates considered in the figure, the truncated series (2.134) practically coincides with the Padé approximant

$$s^*_{\text{ex}} \simeq -a^{*2}\frac{1 + 2 \times 0.58417a^{*2}}{2(1 + 4 \times 0.58417a^{*2})}, \tag{2.137}$$

so we can expect that it is close to the exact form within that range. We observe that the Navier–Stokes prediction is accurate for shear rates smaller than $a^* = 0.2$, even though the Navier–Stokes approximation overestimates the shear stress by about 5% at $a^* = 0.2$ [cf. Eq. (2.57) and Fig. 2.3]. The information-theory upper bound is actually a very good estimate of the true entropy up to $a^* = 0.3$.

APPENDIX 2.A: Time evolution of the fourth-degree moments

In order to get the explicit time-dependence of the fourth-degree moments from Eq. (2.97), it is convenient to diagonalize the matrix $\mathcal{L}$. Let us denote by $\{x_\sigma, \sigma = 1, \ldots, 8\}$ the eigenvectors of $\mathcal{L}$. Since the matrix $\mathcal{L}$ is not Hermitian, their eigenvectors do not form an orthogonal set. The matrix of change of basis is $\mathcal{U}$, where $\mathcal{U}_{\sigma'\sigma}$ is the σ'-component of $\boldsymbol{x}_\sigma$. Consequently,

$$e^{-\mathcal{L}t^*} = \mathcal{U} \cdot \mathcal{D}(t^*) \cdot \mathcal{U}^{-1}, \tag{2.A.1}$$

where $\mathcal{D}(t^*)$ is a diagonal matrix whose σ-element is $e^{-\ell_\sigma t^*}$, ℓ_σ ($\sigma = 1, \ldots, 8$) being the eigenvalues of $\mathcal{L}$.

As an application, we consider the cases $\alpha^* = \frac{6}{5}$ and $\alpha^* = \frac{4}{3}$, which, according to Eq. (2.47), correspond to shear rates $a^* = \frac{51}{5}\sqrt{\frac{2}{5}} \simeq 6.45$ and $a^* = \frac{22}{3} \simeq 7.33$, respectively. After calculating the corresponding eigenvalues and eigenvectors, one gets the following expressions for the time evolution of $\langle V^4 \rangle$ (Santos and Garzó, 1995a):

$$\begin{aligned}(2k_BT/m)^{-2}\langle V^4\rangle \;=\;& 182.7 - 178.2e^{-0.0215t^*} - 0.014e^{-4.40t^*}\\ &+e^{-5.82t^*}[0.46\cos(2.83t^*) + 3.03\sin(2.83t^*)]\\ &-e^{-7.48t^*}[0.012\cos(1.75t^*) + 0.011\sin(1.75t^*)]\\ &+e^{-9.52t^*}[0.64\cos(4.65t^*) - 0.42\sin(4.65t^*)],\end{aligned} \tag{2.A.2}$$

$$\begin{aligned}(2k_BT/m)^{-2}\langle V^4\rangle \;=\;& -160.4 + 164.8e^{0.0270t^*} - 0.013e^{-4.76t^*}\\ &+e^{-6.31t^*}[0.51\cos(3.21t^*) + 3.24\sin(3.21t^*)]\\ &-e^{-8.10t^*}[0.010\cos(1.91t^*) + 0.011\sin(1.91t^*)]\\ &+e^{-10.38t^*}[0.73\cos(5.05t^*) - 0.46\sin(5.05t^*)].\end{aligned} \tag{2.A.3}$$

Here, Eq. (2.A.2) refers to $\alpha^* = \frac{6}{5}$ and Eq. (2.A.3) to $\alpha^* = \frac{4}{3}$. In both cases, we have taken the anisotropic Gaussian distribution consistent with the stationary pressure tensor as the initial condition (Garzó and Santos, 1995a). In the case of Eq. (2.A.2), the shear rate $a^* \simeq 6.45$ is smaller than the critical value $a_c^* \simeq 6.846$, and so $\langle V^4\rangle/(2k_BT/m)^2$ relaxes to the steady-state value 182.7 with a characteristic time $\ell_1^{-1} = 1/0.0215 \simeq 46.5$. On the other hand, in the case of Eq. (2.A.3), the shear rate $a^* \simeq 7.33$ is larger than a_c^* and, consequently, $\langle V^4\rangle/(2k_BT/m)^2$ grows monotonically with a characteristic time $-\ell_1^{-1} = 1/0.0270 \simeq 37.1$.

The exact expression (2.A.2) corresponding to $a^* \simeq 6.45$ has been used as a benchmark to assess the reliability of the DSMC method (Montanero and Santos, 1995).

APPENDIX 2.B: Steady-state values of the fourth-degree moments

In this Appendix we provide the explicit shear-rate dependence of the stationary fourth-degree moments given by Eq. (2.98) They are given by (Santos and Garzó, 1995a)

$$M_{4|0}^{\rm ss} = \left(\frac{2k_BT}{m}\right)^2 \frac{15}{4}\frac{1}{(1+2\alpha^*)^2}\frac{N_1(\alpha^*)}{\Delta(\alpha^*)}, \tag{2.B.1}$$

$$M_{2|xx}^{\rm ss} = \left(\frac{2k_BT}{m}\right)^2 \frac{54}{7}\frac{\alpha^*}{(1+2\alpha^*)^2}\frac{N_2(\alpha^*)}{\Delta(\alpha^*)}, \tag{2.B.2}$$

$$M_{2|yy}^{\rm ss} = -\left(\frac{2k_BT}{m}\right)^2 \frac{39}{14}\frac{\alpha^*}{(1+2\alpha^*)^2}\frac{N_3(\alpha^*)}{\Delta(\alpha^*)}, \tag{2.B.3}$$

$$M_{0|yyyy}^{\rm ss} = -\left(\frac{2k_BT}{m}\right)^2 \frac{\alpha^*}{(1+2\alpha^*)^2} \frac{N_4(\alpha^*)}{\Delta(\alpha^*)}, \tag{2.B.4}$$

$$M_{0|zzzz}^{\rm ss} = \left(\frac{2k_BT}{m}\right)^2 \frac{1}{4} \frac{\alpha^*}{(1+2\alpha^*)^2} \frac{N_5(\alpha^*)}{\Delta(\alpha^*)}, \tag{2.B.5}$$

$$M_{2|xy}^{\rm ss} = -\left(\frac{2k_BT}{m}\right)^2 \frac{7}{4} \frac{a^*}{(1+2\alpha^*)^3} \frac{N_6(\alpha^*)}{\Delta(\alpha^*)}, \tag{2.B.6}$$

$$M_{0|xxxy}^{\rm ss} = -\left(\frac{2k_BT}{m}\right)^2 \frac{a^*\alpha^*}{(1+2\alpha^*)^3} \frac{N_7(\alpha^*)}{\Delta(\alpha^*)}, \tag{2.B.7}$$

$$M_{0|xyyy}^{\rm ss} = \left(\frac{2k_BT}{m}\right)^2 \frac{a^*\alpha^*}{(1+2\alpha^*)^3} \frac{N_8(\alpha^*)}{\Delta(\alpha^*)}, \tag{2.B.8}$$

where

$$\begin{aligned}\Delta(\alpha^*) &= 1 + 18.37\alpha^* + 142.8\alpha^{*2} + 608.7\alpha^{*3} + 1524\alpha^{*4} + 2167\alpha^{*5} \\ &\quad +1310\alpha^{*6} - 625.1\alpha^{*7} - 1433\alpha^{*8} - 644.3\alpha^{*9},\end{aligned} \tag{2.B.9}$$

$$\begin{aligned}N_1(\alpha^*) &= 1 + 24.37\alpha^* + 271.2\alpha^{*2} + 1796\alpha^{*3} + 7777\alpha^{*4} \\ &\quad +22968\alpha^{*5} + 47063\alpha^{*6} + 67138\alpha^{*7} + 66381\alpha^{*8} \\ &\quad +44959\alpha^{*9} + 19949\alpha^{*10} + 4639\alpha^{*11},\end{aligned} \tag{2.B.10}$$

$$\begin{aligned}N_2(\alpha^*) &= 1 + 22.61\alpha^* + 222.8\alpha^{*2} + 1258\alpha^{*3} + 4496\alpha^{*4} \\ &\quad +10636\alpha^{*5} + 16933\alpha^{*6} + 18181\alpha^{*7} + 13078\alpha^{*8} \\ &\quad +6097\alpha^{*9} + 1503\alpha^{*10},\end{aligned} \tag{2.B.11}$$

$$\begin{aligned}N_3(\alpha^*) &= 1 + 24.11\alpha^* + 249.6\alpha^{*2} + 1466\alpha^{*3} + 5423\alpha^{*4} \\ &\quad +13213\alpha^{*5} + 21581\alpha^{*6} + 23687\alpha^{*7} + 17374\alpha^{*8} \\ &\quad +8258\alpha^{*9} + 2081\alpha^{*10},\end{aligned} \tag{2.B.12}$$

$$\begin{aligned}N_4(\alpha^*) &= 1.168 + 19.22\alpha^* + 127.3\alpha^{*2} + 404.7\alpha^{*3} + 419.2\alpha^{*4} \\ &\quad -1255\alpha^{*5} - 5274\alpha^{*6} - 8776\alpha^{*7} - 8248\alpha^{*8} \\ &\quad -4755\alpha^{*9} - 1491\alpha^{*10},\end{aligned} \tag{2.B.13}$$

$$\begin{aligned}N_5(\alpha^*) &= 1.168 + 35.20\alpha^* + 426.4\alpha^{*2} + 2825\alpha^{*3} + 11503\alpha^{*4} \\ &\quad +30298\alpha^{*5} + 52674\alpha^{*6} + 60630\alpha^{*7} + 46012\alpha^{*8} \\ &\quad +22550\alpha^{*9} + 5964\alpha^{*10},\end{aligned} \tag{2.B.14}$$

$$\begin{aligned}N_6(\alpha^*) &= 1 + 22.54\alpha^* + 223.5\alpha^{*2} + 1278\alpha^{*3} + 4648\alpha^{*4} \\ &\quad +11194\alpha^{*5} + 18116\alpha^{*6} + 19691\alpha^{*7} + 14255\alpha^{*8} \\ &\quad +6664\alpha^{*9} + 1657\alpha^{*10},\end{aligned} \tag{2.B.15}$$

$$\begin{aligned}N_7(\alpha^*) &= 3.120 + 59.31\alpha^* + 488.0\alpha^{*2} + 2274\alpha^{*3} + 6592\alpha^{*4} \\ &\quad +12329\alpha^{*5} + 15019\alpha^{*6} + 11905\alpha^{*7} \\ &\quad +6041\alpha^{*8} + 1657\alpha^{*9},\end{aligned} \tag{2.B.16}$$

$$N_8(\alpha^*) = 2.315 + 44.06\alpha^* + 362.9\alpha^{*2} + 1692\alpha^{*3} + 4910\alpha^{*4} + 9188\alpha^{*5} + 11200\alpha^{*6} + 8884\alpha^{*7} + 4515\alpha^{*8} + 1242\alpha^{*9}. \tag{2.B.17}$$

The critical shear rate corresponds to the only positive real root of $\Delta(\alpha^*)$, i.e. $\alpha_c^* \simeq 1.2605$, which yields $a_c^* \simeq 6.846$. Of course, Equations (2.B.1)–(2.B.8) are only meaningful for $a^* < a_c^*$.

Chapter 3

KINETIC MODEL FOR UNIFORM SHEAR FLOW

1. Introduction

The results obtained in Chapter 2 hold for the Boltzmann equation in the case of Maxwell molecules. While the description generally applies to arbitrary values of the shear rate a, it has two limitations. On the one hand, no explicit form of the velocity distribution function $f(\mathbf{V})$ is known for finite a. On the other hand, the analysis cannot be extended to interactions different from the Maxwell one, unless additional approximations are introduced (Loose and Hess, 1988; Loose, 1988). These two limitations can be overcome by employing numerical simulations and/or kinetic models. This chapter is mainly concerned with the latter route. Specifically, we will focus on the well-known Bhatnagar–Gross–Krook (BGK) model kinetic equation (Bhatnagar et al., 1954; Welander, 1954; Cercignani, 1990) introduced in Section 1.7.

First we explicitly get the velocity distribution function as obtained from the BGK equation for Maxwell molecules. The second-degree moments are the same as those given by the Boltzmann equation, but higher-degree moments are different. In particular, the BGK moments do not diverge for any value of the shear rate, in contrast to the exact results found from the Boltzmann equation in Chapter 2.

The usefulness of the BGK model is especially relevant in the case of interactions harder than Maxwell's. In that case, the shear rate relative to the collision frequency monotonically decreases in time due to the viscous heating, so the system asymptotically tends to the local equilibrium state. The interesting point, however, is that after a few collision times the system reaches a time-dependent hydrodynamic regime in which one can identify a viscosity function that depends nonlinearly on the reduced shear rate.

From a practical point of view, it is desirable to analyze the nonequilibrium properties of the system in a steady state. This requires the introduction of a *thermostat* external force that removes the heat produced by the shearing motion. In principle, one might expect that the thermostat plays a passive role, not affecting the rheological properties. However, while it is true that in the case of Maxwell molecules there is an exact equivalence between the thermostatted and the unthermostatted states, a certain (but small) influence of the thermostat on the transport properties exists in general (Dufty et al., 1986a; Dufty et al., 1986b). According to the BGK model, the properties of the thermostatted state are independent of the interaction potential. Comparison with computer simulations shows that the rheological properties are indeed practically insensitive to the interaction law. On the other hand, the higher-degree velocity moments are clearly influenced by the potential. As a matter of fact, the simulations indicate that the harder the interaction the weaker the divergence of the moments. In the case of hard spheres, all the moments seem to be convergent. In contrast to what happens with the Boltzmann equation. the explicit knowledge of the distribution function can be exploited to get the nonequilibrium entropy by means of a numerical integration. Furthermore, one can define a nonequilibrium "thermodynamic" temperature by extending the equilibrium relationship between entropy and temperature. Here, a comparison is made between the kinetic and the thermodynamic temperatures.

This chapter ends with a study of small perturbations from the thermostatted USF state. The idea is to assume the existence of small deviations from a constant density and temperature and a linear velocity profile. By performing a Chapman–Enskog-like expansion around the nonequilibrium USF state, we get generalized shear-rate dependent transport coefficients characterizing the heat and momentum fluxes due to the hydrodynamic gradients. A related problem is the analysis of the heat transport induced by a thermal gradient in a steady state with a constant pressure, which allows one to identify a thermal conductivity tensor. Finally, the stability of the USF is considered. It is seen that the USF state becomes unstable under long-wavelength perturbations. Simulation results show that the asymptotic state presents a regular periodic behavior.

2. The BGK equation for uniform shear flow. Maxwell molecules

As described in Chapter 1, the BGK model is constructed by replacing the true Boltzmann collision operator by a single relaxation-time term:

$$J[f,f] \to -\nu(f - f_L), \tag{3.1}$$

where ν is an effective collision frequency independent of velocity and

$$f_L(\mathbf{r},\mathbf{v},t) = n\left(\frac{m}{2\pi k_B T}\right)^{3/2} \exp\left[-\frac{m(\mathbf{v}-\mathbf{u})^2}{2k_B T}\right] \tag{3.2}$$

is the local equilibrium distribution function. In the case of repulsive potentials of the form $\phi(r) = Kr^{-w}$, the collision frequency is proportional to $nT^{1/2-2/w}$. As a consequence, all the formal and scaling properties discussed for the Boltzmann equation in Section 2.2 also apply to the BGK equation.

Let us consider the case of Maxwell molecules ($w = 4$) under USF. Thus the BGK version of the Boltzmann equation (2.25) is

$$\frac{\partial}{\partial t}\overline{f} - \frac{\partial}{\partial \overline{V}_i}\left(a_{ij}\overline{V}_j + \alpha \overline{V}_i\right)\overline{f} = -\nu(\overline{f} - \overline{f}_L), \tag{3.3}$$

where the scaled quantities $\overline{\mathbf{V}}$ and $\overline{f}$ are defined by Eqs. (2.11) and (2.12), respectively, with $R(t) = e^{-\alpha t}$, α being so far an arbitrary constant. As done in Chapter 2, henceforth the bars will be dropped. The evolution equation for the pressure tensor can be easily obtained from Eq. (3.3). It turns out that this equation is exactly the same as the one obtained from the Boltzmann equation [cf. Eq. (2.27)], provided that the BGK collision frequency ν is chosen as given by Eq. (2.28). This implies that all the results derived in Section 2.3 referring to the time evolution of the pressure tensor and rheological properties are still valid in the context of the BGK model. In particular, the choice (2.46) leads to the stationary values of the (scaled) pressure tensor (2.51)–(2.53), so the rheological properties are given by Eqs. (2.57)–(2.59).

The main advantage of using the BGK equation is that it lends itself to an exact solution (Zwanzig, 1979). To that end, it is convenient to rewrite Eq. (3.3) as

$$\left(\frac{\partial}{\partial t} + \nu - 3\alpha + \Lambda\right) f = \nu f_L, \tag{3.4}$$

where

$$\Lambda \equiv -a_{ij}V_j\frac{\partial}{\partial V_i} - \alpha \mathbf{V}\cdot\frac{\partial}{\partial \mathbf{V}} \tag{3.5}$$

is an operator acting on velocity space. Equation (3.4) has the form of an inhomogeneous, linear, first-order differential equation with respect to the time variable. Its general solution is

$$\begin{aligned} f(\mathbf{V},t) &= e^{-(\nu-3\alpha)t}e^{-\Lambda t}f(\mathbf{V},0) \\ &\quad +\nu\int_0^t dt'\, e^{-(\nu-3\alpha)t'}e^{-\Lambda t'}f_L(\mathbf{V},t-t'). \end{aligned} \tag{3.6}$$

Equation (3.6) has a formal structure similar to that of Eq. (1.166). The operator Λ is the sum of a shift generator and a scaling generator. More specifically,

$$e^{-\Lambda t}X(\mathbf{V}) = X\left(e^{\alpha t}(\mathbf{V} + t\mathsf{a}\cdot\mathbf{V})\right), \tag{3.7}$$

where $X(\mathbf{V})$ is an arbitrary function and a is the (transpose of the) rate-of-strain tensor with elements $a_{ij} = a\delta_{ix}\delta_{jy}$. Since the (scaled) temperature reaches a stationary value for long times, so does the distribution f_L. As a consequence, in the limit $t \to \infty$, the indexInitial condition initial-value contribution in Eq. (3.6) vanishes (Santos and Brey, 1991a) and the distribution function becomes

$$f^*(\boldsymbol{\xi}) = \pi^{-3/2} \int_0^\infty ds\, e^{-(1-3\alpha^*)s} \exp\left[-e^{2\alpha^* s}(\boldsymbol{\xi} + s\mathsf{a}^* \cdot \boldsymbol{\xi})^2\right], \tag{3.8}$$

where

$$f^* = n^{-1} \left(\frac{2k_B T}{m}\right)^{3/2} f \tag{3.9}$$

is the reduced velocity distribution function, $\boldsymbol{\xi}$ is the reduced peculiar velocity defined by Eq. (1.88), $\mathsf{a}^* = \mathsf{a}/\nu$ is the reduced rate-of-strain tensor, and $\alpha^* = \alpha/\nu$. Equation (3.8) represents the *normal* solution of the BGK model for Maxwell molecules under USF, where the whole spatial and temporal dependence occurs through the hydrodynamic quantities. The dependence of f^* on the shear rate $a^* = a/\nu$ appears explicitly and also through the heating rate $\alpha^*(a^*)$ given by Eq. (2.46). Following the notation of Eq. (2.115), we write the Chapman–Enskog expansion of f^* as

$$f^*(\boldsymbol{\xi}) = \varphi_0(\boldsymbol{\xi}) \left[1 + \sum_{k=1}^{\infty} \Phi^{(k)}(\boldsymbol{\xi})\, a^{*k}\right]. \tag{3.10}$$

The coefficients $\Phi^{(k)}(\boldsymbol{\xi})$ can be obtained from Eq. (3.8) or, alternatively, by inserting Eq. (3.10) into the BGK kinetic equation (3.4). The first few coefficients are (Santos and Brey, 1991a)

$$\Phi^{(1)}(\boldsymbol{\xi}) = -2\xi_x\xi_y, \tag{3.11}$$

$$\Phi^{(2)}(\boldsymbol{\xi}) = 1 - \frac{2}{3}\xi^2 - 2\xi_y^2\left(1 - 2\xi_x^2\right), \tag{3.12}$$

$$\Phi^{(3)}(\boldsymbol{\xi}) = -4\xi_x\xi_y\left[\frac{5}{3} - \frac{2}{3}\xi^2 - \xi_y^2\left(3 - 2\xi_x^2\right)\right]. \tag{3.13}$$

Since α^* is regular at $a^* = 0$, so does f^*. Consequently, the Chapman–Enskog expansion (3.10) is convergent [with the same radius of convergence as $\alpha^*(a^*)$, namely $a^* = \sqrt{2}/3$].

Let us define now the *marginal* distributions

$$f_{xy}^*(\xi_x, \xi_y) = \int_{-\infty}^{\infty} d\xi_z\, f^*(\boldsymbol{\xi}), \tag{3.14}$$

$$f_x^{(+)}(\xi_x) = \int_0^\infty d\xi_y \int_{-\infty}^{\infty} d\xi_z\, f^*(\boldsymbol{\xi}), \tag{3.15}$$

$$f_y^{(+)}(\xi_y) = \int_0^\infty d\xi_x \int_{-\infty}^\infty d\xi_z\, f^*(\boldsymbol{\xi}), \tag{3.16}$$

$$f_x^*(\xi_x) = \int_{-\infty}^\infty d\xi_y \int_{-\infty}^\infty d\xi_z\, f^*(\boldsymbol{\xi}), \tag{3.17}$$

$$f_y^*(\xi_y) = \int_{-\infty}^\infty d\xi_x \int_{-\infty}^\infty d\xi_z\, f^*(\boldsymbol{\xi}). \tag{3.18}$$

Their expressions can be easily obtained from Eq. (3.8):

$$f_{xy}^*(\boldsymbol{\xi}) = \pi^{-1} \int_0^\infty ds\, e^{-(1-2\alpha^*)s} \exp\left\{-e^{2\alpha^* s}\left[(\xi_x + a^* s\xi_y)^2 + \xi_y^2\right]\right\}, \tag{3.19}$$

$$\begin{aligned} f_x^{(+)}(\xi_x) &= \frac{1}{2\sqrt{\pi}} \int_0^\infty ds\, \frac{e^{-(1-\alpha^*)s}}{(1+a^{*2}s^2)^{1/2}} \\ &\quad \times \exp\left(-e^{2\alpha^* s}\frac{\xi_x^2}{1+a^{*2}s^2}\right) \operatorname{erfc}\left[e^{\alpha^* s}\frac{a^* s\xi_x}{(1+a^{*2}s^2)^{1/2}}\right], \end{aligned} \tag{3.20}$$

$$f_y^{(+)}(\xi_y) = \frac{1}{2\sqrt{\pi}} \int_0^\infty ds\, e^{-(1-\alpha^*)s} \exp\left(-e^{2\alpha^* s}\xi_y^2\right) \operatorname{erfc}\left(e^{\alpha^* s} a^* s\xi_y\right), \tag{3.21}$$

$$\begin{aligned} f_x^*(\xi_x) &= f_x^{(+)}(\xi_x) + f_x^{(+)}(-\xi_x) \\ &= \frac{1}{\sqrt{\pi}} \int_0^\infty ds\, \frac{e^{-(1-\alpha^*)s}}{(1+a^{*2}s^2)^{1/2}} \exp\left(-e^{2\alpha^* s}\frac{\xi_x^2}{1+a^{*2}s^2}\right), \end{aligned} \tag{3.22}$$

$$\begin{aligned} f_y^*(\xi_y) &= f_y^{(+)}(\xi_y) + f_y^{(+)}(-\xi_y) \\ &= \frac{1}{\sqrt{\pi}} \int_0^\infty ds\, e^{-(1-\alpha^*)s} \exp\left(-e^{2\alpha^* s}\xi_y^2\right), \end{aligned} \tag{3.23}$$

where $\operatorname{erfc}(x)$ is the complementary error function (Abramowitz and Stegun, 1972). In Eqs. (3.22) and (3.23) we have taken into account the symmetry property $f^*(\xi_x, \xi_y, \xi_z) = f^*(-\xi_x, -\xi_y, \xi_z)$.

As said above, the BGK distribution reproduces the exact rheological properties of the Boltzmann equation. Thus, one can expect that the BGK solution (3.8) gives a good picture of the true distribution in the region of thermal velocities ($\xi \sim 1$). This is confirmed by Fig. 3.1, where we plot the ratio

$$R_y(\xi_y) = \frac{f_y^*(\xi_y)}{\pi^{-1/2} e^{-\xi_y^2}} \tag{3.24}$$

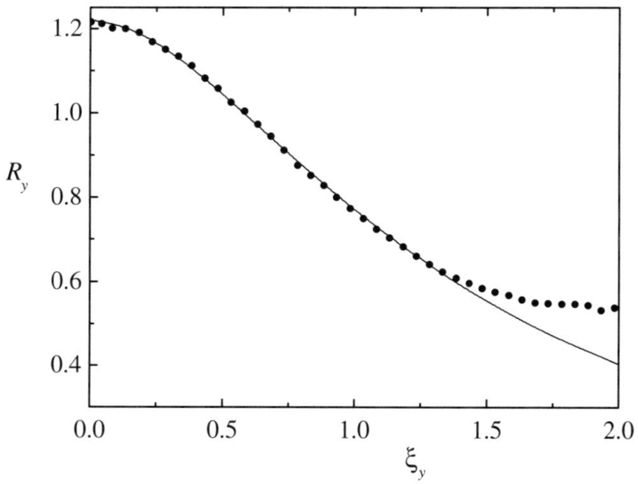

Figure 3.1. Ratio $R_y(\xi_y) = f_y^*(\xi_y)/(\pi^{-1/2}e^{-\xi_y^2})$ for $a^* = 1$. The circles are simulation data for Maxwell molecules, while the solid line is the BGK prediction.

between the marginal distribution f_y^* and its corresponding local equilibrium value, for $a^* = 1$ (Garzó, 1997). Since the shear rate is not small, the distortion from equilibrium ($R_y \neq 1$) is quite apparent. The agreement between theory and simulation is excellent up to $\xi_y \simeq 1.2$. For larger velocities, the BGK solution underestimates the simulation results. This is consistent with the high-velocity tail discussed in Section 2.5, which is absent in the BGK model.

The solution (3.8) exhibits an interesting singular behavior in the vicinity of $\boldsymbol{\xi} = \mathbf{0}$ and for sufficiently large shear rates. This happens when $1 - 3\alpha^* \leq 0$, which implies $a^* \geq 5/3$. In order to look at this point more closely, let us take $\xi_y = 0$. In that case, Eq. (3.8) becomes

$$f^*(\xi_x, 0, \xi_z) = \frac{\pi^{-3/2}}{2\alpha^*}\xi_\perp^{(1-3\alpha^*)/\alpha^*}\Gamma\left(\frac{3\alpha^* - 1}{2\alpha^*}, \xi_\perp^2\right), \tag{3.25}$$

where $\xi_\perp^2 \equiv \xi_x^2 + \xi_z^2$ and

$$\Gamma(x, \epsilon) \equiv \int_\epsilon^\infty dt\, t^{x-1}e^{-t} \tag{3.26}$$

is the incomplete gamma function (Abramowitz and Stegun, 1972). By taking into account the asymptotic behavior of $\Gamma(x, \epsilon)$ in the limit $\epsilon \to 0^+$ (Santos and

Brey, 1991a), one gets

$$f^*(\xi_x, 0, \xi_z) \approx \begin{cases} \frac{\pi^{-3/2}}{1-3\alpha^*}, & \alpha^* < \frac{1}{3} \\ -\frac{3}{2}\pi^{-3/2} \ln \xi_\perp^2, & \alpha^* = \frac{1}{3} \\ \frac{\pi^{-3/2}}{2\alpha^*}\Gamma\left(\frac{3\alpha^*-1}{2\alpha^*}\right) \xi_\perp^{(1-3\alpha^*)/\alpha^*}, & \alpha^* > \frac{1}{3} \end{cases} \tag{3.27}$$

in the limit $\xi_\perp^2 \to 0$. The above equation shows that f^* diverges when $\boldsymbol{\xi} \to \mathbf{0}$ if $\alpha^* \geq 1/3$ or, equivalently, $a^* \geq 5/3$. The origin of this divergence at vanishing velocity can be traced back to the competition between the two exponential terms e^{-s} and $e^{3\alpha^* s}$ appearing in Eq. (3.8). The first one is the survival probability (Reif, 1965), i.e. the probability that a particle has not collided after a time interval equal to s collision times. The second exponential term, on the other hand, is a reflection of the viscous heating (recall that the unscaled temperature behaves as $T(t) \sim e^{2\alpha t}$) and a consequence of the nondimensionalization of f^*. For shear rates smaller than $a^* = 5/3$ the viscous heating is not enough to overcome the effect of collisions, so that f^* is finite at $\boldsymbol{\xi} = \mathbf{0}$. However, when $a^* > 5/3$, the viscous heating effect dominates, which gives rise to a "condensation" of particles around $\boldsymbol{\xi} = \mathbf{0}$. The marginal distribution $f_{xy}^*(\xi_x, \xi_y)$ diverges at $\xi_x = \xi_y = 0$ if $\alpha^* \geq 1/2$, i.e. $a^* \geq \sqrt{6}$, whereas the distributions $f_x^{(+)}(\xi_x)$ and $f_y^{(+)}(\xi_y)$ diverge at $\xi_x = 0$ and $\xi_y = 0$, respectively, if $\alpha^* \geq 1$, i.e. $a^* \geq 3\sqrt{3}$.

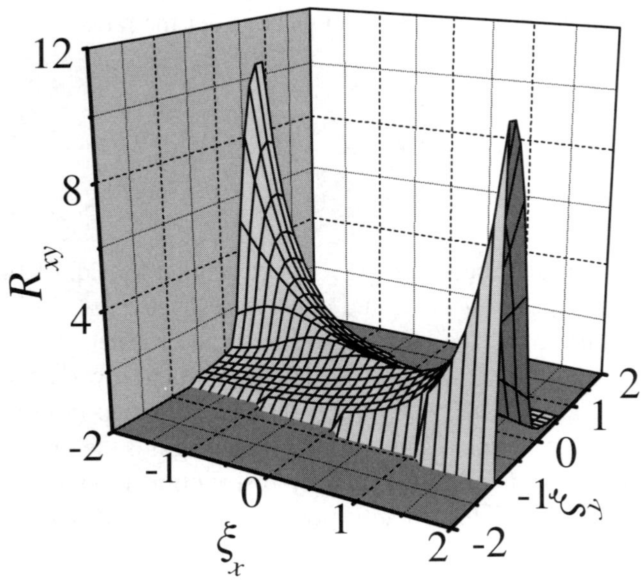

Figure 3.2. Surface plot of the ratio R_{xy} for $a^* = 1$.

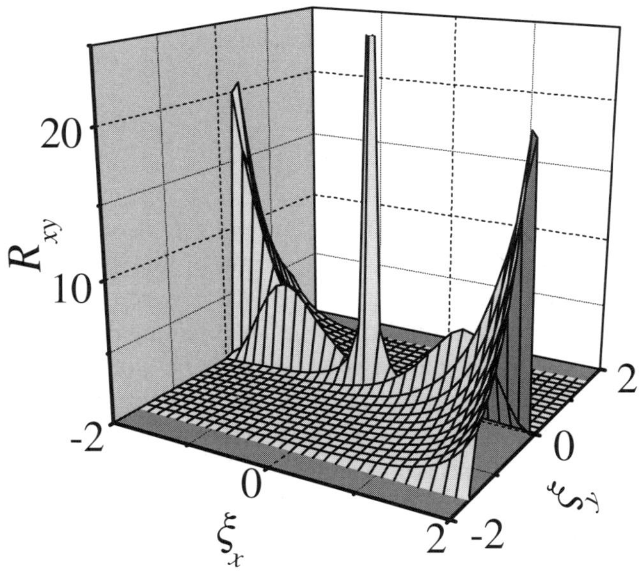

Figure 3.3. Surface plot of the ratio R_{xy} for $a^* = 4$.

In order to illustrate graphically the distortion from local equilibrium of the BGK distribution given by Eq. (3.8), let us consider the ratio

$$R_{xy}(\xi_x, \xi_y) = \frac{f^*_{xy}(\xi_x, \xi_y)}{\pi^{-1} e^{-(\xi_x^2+\xi_y^2)}}, \tag{3.28}$$

where the marginal distribution $f^*_{xy}(\xi_x, \xi_y)$ is defined by Eq. (3.14). Figures 3.2 and 3.3 show R_{xy} for $a^* = 1$ and $a^* = 4$, respectively. The distortion is much more significant in the latter case than in the former. In fact, $a^* = 4$ is beyond the shear rate $a^* = \sqrt{6}$ at which f^*_{xy} starts to diverge for vanishing velocities. The region $-0.05 \leq \xi_x, \xi_y \leq 0.05$ is magnified in Fig. 3.4, where the singular behavior in the neighborhood of the origin is clearly observed.

The explicit knowledge of the velocity distribution function of the BGK model allows one to get *all* the velocity moments. As in Eq. (2.26), we introduce the (dimensionless) moments

$$\langle \xi_x^{k_1} \xi_y^{k_2} \xi_z^{k_3} \rangle = \int d\boldsymbol{\xi}\, \xi_x^{k_1} \xi_y^{k_2} \xi_z^{k_3} f^*. \tag{3.29}$$

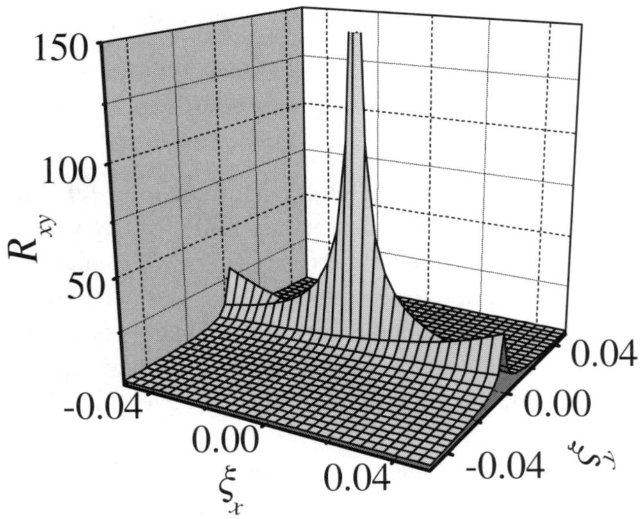

Figure 3.4. Surface plot of the ratio R_{xy} for $a^* = 4$ in the region $-0.05 \leq \xi_x, \xi_y \leq 0.05$.

Of course, the only non vanishing moments correspond to even values of k_1+k_2 and k_3. From Eq. (3.8), one has

$$
\begin{aligned}
\langle \xi_x^{k_1} \xi_y^{k_2} \xi_z^{k_3} \rangle &= \pi^{-3/2} \int_0^\infty ds\, e^{-(1-3\alpha^*)s} \int d\boldsymbol{\xi}\, \left(\xi_x - a^* s \xi_y\right)^{k_1} \xi_y^{k_2} \xi_z^{k_3} \\
&\quad \times \exp\left(-e^{2\alpha^* s} \xi^2\right) \\
&= \sum_{q=0}^{k_1} \frac{k_1!}{q!(k_1-q)!} \langle \xi_x^{k_1-q} \xi_y^{k_2+q} \xi_z^{k_3} \rangle_L \int_0^\infty ds\, \left(-a^* s\right)^q \\
&\quad \times \exp\left[-\left(1+k\alpha^*\right)s\right], \qquad (3.30)
\end{aligned}
$$

where $k = k_1 + k_2 + k_3$ and in the last step we have expanded $\left(\xi_x - a^* s \xi_y\right)^{k_1}$ and integrated over $\boldsymbol{\xi}$. In addition, $\langle \xi_x^{k_1} \xi_y^{k_2} \xi_z^{k_3} \rangle_L$ are local equilibrium moments, given by

$$
\langle \xi_x^{k_1} \xi_y^{k_2} \xi_z^{k_3} \rangle_L = \pi^{-3/2} \Gamma\left(\frac{k_1+1}{2}\right) \Gamma\left(\frac{k_2+1}{2}\right) \Gamma\left(\frac{k_3+1}{2}\right) \qquad (3.31)
$$

if k_1, k_2, and k_3 are even, being zero otherwise. Finally, after performing the s-integration in Eq. (3.30), one gets

$$
\langle \xi_x^{k_1} \xi_y^{k_2} \xi_z^{k_3} \rangle = \sum_{q=0}^{k_1} \frac{k_1!}{(k_1-q)!} (-a^*)^q \left(1+k\alpha^*\right)^{-(q+1)} \langle \xi_x^{k_1-q} \xi_y^{k_2+q} \xi_z^{k_3} \rangle_L. \qquad (3.32)
$$

It is straightforward to check that the above expression is consistent with the relationship

$$\sum_{q=0}^{k}(-1)^q \begin{pmatrix} 2k \\ 2q \end{pmatrix} \langle \xi_y^{2q}\xi_z^{2k-q}\rangle = 0, \tag{3.33}$$

which also holds in the case of the Boltzmann equation [cf. Eq. (2.96) and above]. Obviously, all the BGK moments are finite for any value of the shear rate, in contrast to what happens in the case of the Boltzmann equation. This failure of the BGK description to reproduce the singular behavior of fourth- and higher-degree moments as obtained from the Boltzmann equation for Maxwell molecules is not surprising at all. In fact, the BGK model does not intend to mimic the behavior of the true distribution function beyond the thermal velocity region.

Since the second-degree moments are identical in the Boltzmann and BGK equations, it is worthwhile comparing the shear-rate dependence of the stationary fourth-degree moments given by both kinetic equations (Garzó and Santos, 1995a). Figures 3.5–3.9 show the moments $\langle\xi_x^4\rangle$, $\langle\xi_y^4\rangle$, $\langle\xi^2\xi_y^2\rangle$, $\langle\xi^2\xi_x\xi_y\rangle$, and $\langle\xi_x^3\xi_y\rangle$, respectively, in the range $0 \leq a^* \leq 1$. Figure 3.5 shows that the

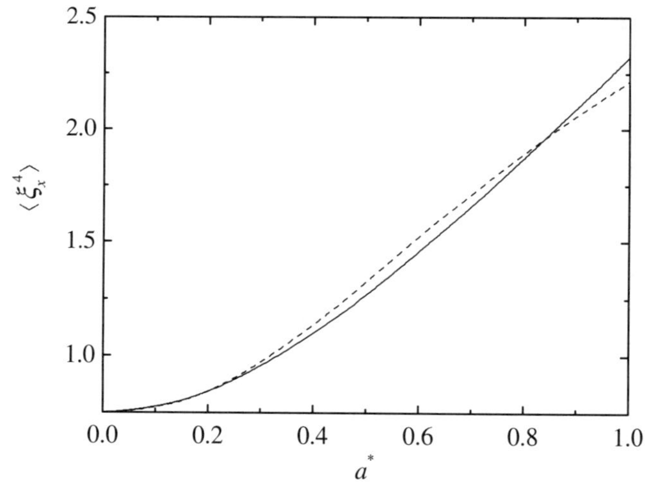

Figure 3.5. Shear-rate dependence of $\langle\xi_x^4\rangle$, according to the Boltzmann equation (solid line) and the BGK equation (dashed line).

BGK model reproduces quite well the exact behavior of $\langle\xi_x^4\rangle$ over the range of shear rates considered, the deviation from equilibrium being overestimated by the BGK model for shear rates $a^* \lesssim 0.8$. On the other hand, the agreement is only qualitative in the case of $\langle\xi_y^4\rangle$, as displayed in Fig. 3.6. An interesting prediction of the Boltzmann equation is that the moment $\langle\xi^2\xi_y^2\rangle$ is practically

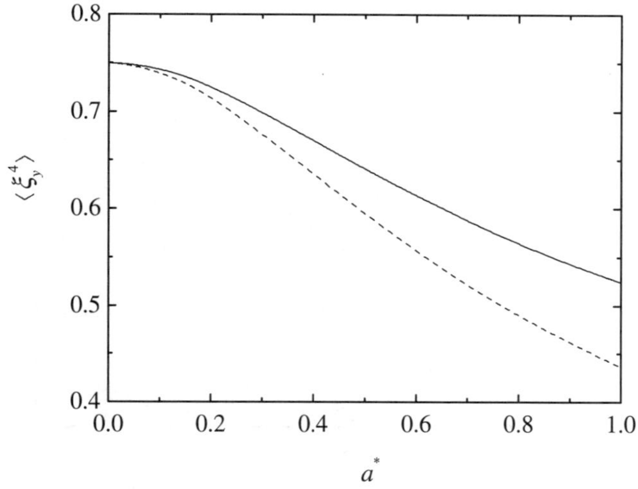

Figure 3.6. Shear-rate dependence of $\langle \xi_y^4 \rangle$, according to the Boltzmann equation (solid line) and the BGK equation (dashed line).

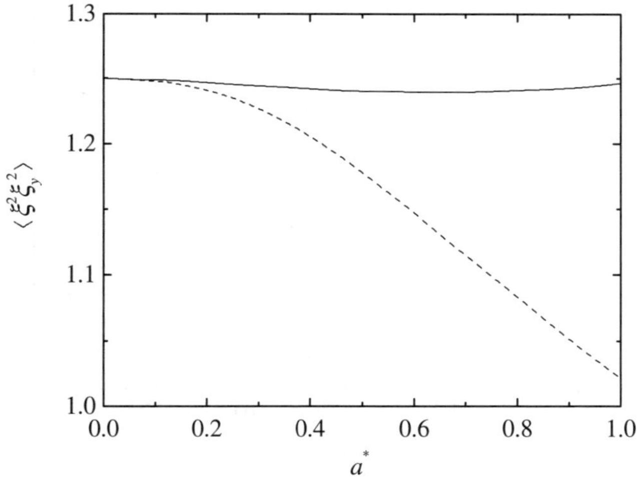

Figure 3.7. Shear-rate dependence of $\langle \xi^2 \xi_y^2 \rangle$, according to the Boltzmann equation (solid line) and the BGK equation (dashed line).

insensitive to the value of the shear rate in the range considered, as observed in Fig. 3.7. This feature, however, is not shared by the BGK model. Finally, let us consider the odd moments $\langle \xi^2 \xi_x \xi_y \rangle$ and $\langle \xi_x^3 \xi_y \rangle$. Their behavior is similar to that of $\langle \xi_x^4 \rangle$, namely the BGK equation overestimates the exact (absolute) value for

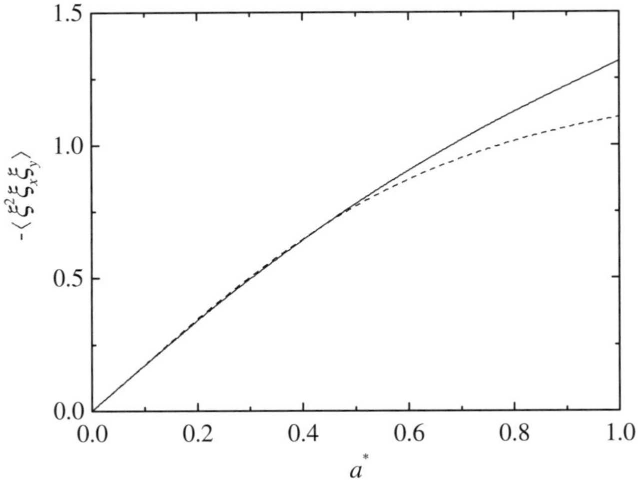

Figure 3.8. Shear-rate dependence of $-\langle \xi^2 \xi_x \xi_y \rangle$, according to the Boltzmann equation (solid line) and the BGK equation (dashed line).

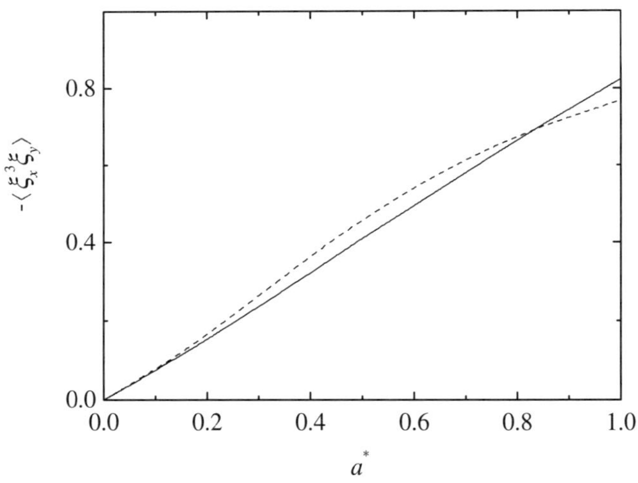

Figure 3.9. Shear-rate dependence of $-\langle \xi_x^3 \xi_y \rangle$, according to the Boltzmann equation (solid line) and the BGK equation (dashed line).

small shear rates, but it underestimates that value for large shear rates. A more exhaustive comparison can be found elsewhere (Garzó and Santos, 1995a).

In general, the BGK predictions are reasonably good for moderate values of the shear rates (say $a^* \lesssim 0.2$), especially for those fourth-degree moments in which the x-component of the velocity is the most relevant one. Note that, since the Chapman–Enskog solutions to the BGK and Boltzmann equations for Maxwell molecules coincide at the level of the Navier–Stokes order [cf. Eq. (2.125) and (3.11)], all the BGK moments (3.32) are exact to first order in a^*. However, the BGK moments differ from the exact ones to orders a^{*2} (Burnett order) and a^{*3} (super-Burnett order). For instance, the Burnett coefficients of $\langle\xi^2\xi_x^2\rangle$ and $\langle\xi^2\xi_x^2\rangle$ are $\frac{143}{42}$ and $-\frac{2}{21}$, respectively, in the Boltzmann equation, while the corresponding values in the BGK model are $\frac{10}{3}$ and $-\frac{1}{6}$.

3. Power-law repulsive potentials. Hard spheres

As mentioned in Section 3.1, one of the advantages of using a kinetic model is the possibility of obtaining explicit results for interactions other than the Maxwell one. In this Section, we will consider repulsive potentials of the form $\phi(r) = Kr^{-w}$, $w \geq 4$, for which the collision frequency is $\nu \propto nT^q$, where $q = \frac{1}{2} - \frac{2}{w}$. The most interesting case is that of hard spheres ($w \to \infty$, $q = \frac{1}{2}$), while the Maxwell case corresponds to $w = 4$ (i.e. $q = 0$). The collision frequency increases with time due to viscous heating (except for Maxwell molecules). Consequently, although the shear rate a in real units is constant, its value *relative* to the collision frequency ($a^* = a/\nu$) decreases monotonically with time ($a^* \propto T^{-q}$). Thus, as time grows, the state of the system is closer and closer to that of local equilibrium and so the existence of a *normal* solution for finite values of a^* is not evident. In this problem, normal solution means that all the time dependence of the distribution function f occurs through the dependence on the temperature. More specifically, the reduced distribution $f^*(\boldsymbol{\xi}) = n^{-1}(2k_BT/m)^{3/2}f(\mathbf{V})$ becomes a certain function of a^*, with independence of the initial preparation of the system.

3.1 Velocity distribution function

Since the relevant control parameter (a^*) is not constant in time, there is no real advantage in using scaled quantities, in contrast to what happens in the case of Maxwell molecules. Thus, we start from the BGK version of the unscaled Boltzmann equation (2.6):

$$\frac{\partial}{\partial t}f - \frac{\partial}{\partial V_i}a_{ij}V_jf = -\nu(f - f_L). \tag{3.34}$$

It is straightforward to get from the above equation the following closed set of equations for the relevant elements of the pressure tensor

$$\frac{\partial}{\partial t}p = -\frac{2}{3}aP_{xy}, \tag{3.35}$$

$$\frac{\partial}{\partial t}P_{xy} = -\nu P_{xy} - aP_{yy}, \tag{3.36}$$

$$\frac{\partial}{\partial t}P_{yy} = -\nu(P_{yy} - p), \tag{3.37}$$

$$\frac{\partial}{\partial t}P_{zz} = -\nu(P_{zz} - p). \tag{3.38}$$

Note that this is a *nonlinear* set since $\nu(t) \propto [p(t)]^q$. From Eqs. (3.35)–(3.37) one gets the following third-order differential equation (Zwanzig, 1979):

$$\left(\frac{\partial}{\partial t} + \nu\right)^2 \frac{\partial}{\partial t}p = \frac{2}{3}a^2\nu p. \tag{3.39}$$

This nonlinear equation must be solved subjected to the initial conditions at $t = t_0$, namely $p(t_0)$, $\dot{p}(t_0) = -\frac{2}{3}aP_{xy}(t_0)$, and $\ddot{p}(t_0) = \frac{2}{3}a[\nu(t_0)P_{xy}(t_0) + aP_{yy}(t_0)]$. Once Eq. (3.39) is solved, the time evolution of the collision frequency and of the elements of the pressure tensor is known, as well as that of the local equilibrium distribution f_L. This in turn allows one to write the general solution of Eq. (3.34) as

$$\begin{aligned} f(\mathbf{V},t) &= e^{-s(t)} \exp\left[(t-t_0)aV_y\frac{\partial}{\partial V_x}\right] f(\mathbf{V},t_0) \\ &\quad + \int_{t_0}^{t} dt'\, \nu(t')e^{-[s(t)-s(t')]} \exp\left[(t-t')aV_y\frac{\partial}{\partial V_x}\right] f_L(\mathbf{V},t'). \end{aligned} \tag{3.40}$$

Here,

$$s(t) = \int_{t_0}^{t} dt'\, \nu(t') \tag{3.41}$$

is the average number of collisions per particle between the times t_0 and t. The action of the operator $\exp\left(atV_y\partial/\partial V_x\right)$ is similar to that of $\exp(-\Lambda t)$ in Eq. (3.7), namely

$$\exp\left(atV_y\frac{\partial}{\partial V_x}\right) X(\mathbf{V}) = X(\mathbf{V} + t\mathsf{a}\cdot\mathbf{V}). \tag{3.42}$$

Equations (3.39)–(3.41) give the solution to the BGK equation (3.34) for an arbitrary initial condition $f(\mathbf{V},t_0)$ and at any time $t > t_0$. However, we have already said that we are interested in obtaining the normal solution, which must be independent of the details of $f(\mathbf{V},t_0)$ and that applies beyond the so-called initial layer, i.e. in the limit $t - t_0 \to \infty$.[1] In order to explore the possible existence of a normal solution, it is convenient to introduce the reduced velocity

[1] In general, the existence of a normal solution also implies the consideration of points far enough from the boundaries. Since they are absent in the USF, we can focus on the initial layer only.

(1.88) and the reduced distribution (3.9). In terms of these quantities, Eq. (3.40) becomes

$$
\begin{aligned}
f^*(\boldsymbol{\xi},t) &= e^{-s(t)}\left[\frac{T(t)}{T(t_0)}\right]^{3/2} \exp\left[(t-t_0)a\xi_y\frac{\partial}{\partial\xi_x}\right] \\
&\quad \times f^*\left(\sqrt{T(t)/T(t_0)}\boldsymbol{\xi},t_0\right) \\
&\quad + \pi^{-3/2}\int_{t_0}^{t} dt'\,\nu(t')e^{-[s(t)-s(t')]}\left[\frac{T(t)}{T(t')}\right]^{3/2} \\
&\quad \times \exp\left[(t-t')a\xi_y\frac{\partial}{\partial\xi_x}\right]\exp\left[-\frac{T(t)}{T(t')}\xi^2\right]. \qquad (3.43)
\end{aligned}
$$

Due to the viscous heating, the details of the initial distribution are irrelevant in the limit $t-t_0\to\infty$. As a matter of fact (Santos and Brey, 1991a),

$$
\lim_{t-t_0\to\infty}\left[\frac{T(t)}{T(t_0)}\right]^{3/2} f^*\left(\sqrt{T(t)/T(t_0)}\boldsymbol{\xi},t_0\right)=\delta(\boldsymbol{\xi}), \qquad (3.44)
$$

where $\delta(\boldsymbol{\xi})$ is Dirac's delta distribution. On the other hand, the limit $t-t_0\to\infty$ must be taken with caution in the second term on the right side of Eq. (3.43) in order to avoid a trivial result. If one fixes an initial condition at a given time t_0 [which corresponds to a certain value of the initial reduced shear rate $a_0^*=a/\nu(t_0)$] and then takes the limit $t-t_0\to\infty$, it is clear that one has $a^*=a/\nu(t)\to 0$ and the solution reduces trivially to that of equilibrium. In order to get a normal solution at *finite* a^*, a different strategy is needed. Instead of fixing the "present" at t_0 and looking towards an infinitely far "future" ($t\to\infty$), we fix the "present" at t and push away the initial condition to the remote "past" ($t_0\to-\infty$). According to this point of view, we obtain the normal solution corresponding to a certain value of a^* by formally applying the initial condition for $a_0^*\to\infty$ or $T(t_0)\to 0$. Having all of this in mind, Eq. (3.43) can be rewritten as

$$
\begin{aligned}
f^*(\boldsymbol{\xi};a^*) &= e^{-s(a^*)}\delta(\boldsymbol{\xi})+\pi^{-3/2}\int_{\infty}^{a^*} d\beta\,\frac{\partial s(\beta)}{\partial\beta}e^{-[s(a^*)-s(\beta)]}\left(\frac{\beta}{a^*}\right)^{3/2q} \\
&\quad \times\exp\left\{[\tau(a^*)-\tau(\beta)]\,\xi_y\frac{\partial}{\partial\xi_x}\right\}\exp\left[-\left(\frac{\beta}{a^*}\right)^{1/q}\xi^2\right].
\end{aligned}
\qquad (3.45)
$$

Here, the time dependence is monitored by the reduced shear rate a^*, so that $\tau\equiv a(t-t_0)$ and s are seen as functions of a^*. The function $\tau(a^*)$ can be obtained as the solution of the equation

$$
\frac{\partial\tau}{\partial a^*}=\left(\frac{2}{3}qa^*\frac{P_{xy}}{p}\right)^{-1}, \qquad (3.46)
$$

where the a^*-dependence of P_{xy}/p is obtained from the following set of non-linear differential equations:

$$\frac{\partial}{\partial a^*}\frac{P_{xy}}{p} = \left(\frac{2}{3}qa^*\frac{P_{xy}}{p}\right)^{-1}\left[-\frac{1}{a^*}\frac{P_{xy}}{p} - \frac{P_{yy}}{p} + \frac{2}{3}\left(\frac{P_{xy}}{p}\right)^2\right], \tag{3.47}$$

$$\frac{\partial}{\partial a^*}\frac{P_{yy}}{p} = \left(\frac{2}{3}qa^*\frac{P_{xy}}{p}\right)^{-1}\left(-\frac{1}{a^*}\frac{P_{yy}}{p} + \frac{1}{a^*} + \frac{2}{3}\frac{P_{xy}}{p}\frac{P_{yy}}{p}\right). \tag{3.48}$$

Equations (3.46)–(3.48) are obtained from simple manipulations of Eqs. (3.35)–(3.37). In addition, the number of collisions $s(a^*)$ is the solution of

$$\frac{\partial s}{\partial a^*} = \frac{1}{a^*}\frac{\partial \tau}{\partial a^*}. \tag{3.49}$$

Equations (3.46)–(3.49) must be solved subjected to the "boundary" conditions $\tau, P_{xy}/p, P_{yy}/p, s \to 0$ when $a^* \to \infty$. Once their solutions are known, Eq. (3.45) provides the normal solution for any value of the reduced shear rate. This scheme is easy to implement numerically (Santos and Brey, 1991a).

3.2 Rheological properties

The knowledge of the distribution allows one to get, in particular, the viscosity function $F_\eta(a^*)$ defined by Eq. (2.54). Rather than an evaluation from Eq. (3.45), it is more useful to get F_η as the solution of a closed second-order differential equation stemming from (3.47) and (3.48). This equation reads (Zwanzig, 1979; Santos et al., 1986a)

$$\begin{aligned} &\tfrac{8}{9}q^2z^4F_\eta(F_\eta^2)'' - \tfrac{4}{3}qz^2\left[1+2z\left(\tfrac{1}{2}-q\right)F_\eta\right]\left(F_\eta^2\right)' \\ &+\left[1+\tfrac{2}{3}\left(2-q\right)zF_\eta + \tfrac{8}{9}\left(\tfrac{1}{2}-q\right)\left(1-q\right)z^2F_\eta^2\right]F_\eta = 1, \end{aligned} \tag{3.50}$$

where the primes denote derivatives with respect to $z \equiv a^{*2}$. Equation (3.50) still corresponds to the *general* solution, valid for arbitrary initial conditions. Given an initial velocity distribution function $f^*(\boldsymbol{\xi}, t_0)$, one gets the initial values of the square of the shear rate $z_0 = a_0^{*2}$, the shear viscosity $F_\eta(z_0)$, and the derivative $F_\eta'(z_0)$ (Santos and Brey, 1991a). With these initial conditions, Eq. (3.50) provides the viscosity function F_η for $a^* < a_0^*$. Thus, there are as many particular solutions as initial conditions. On the other hand, we recall that we are interested in obtaining the *normal* solution to (3.50), in which case the initial condition at $t_0 \to -\infty$ becomes the boundary conditions $\lim_{z\to\infty} F_\eta(z) = \lim_{z\to\infty} F_\eta'(z) = 0$. In order to get the normal solution to Eq. (3.50), the most natural route seems to be the Chapman–Enskog expansion, i.e. a series expansion in powers of z:

$$F_\eta(a^*) = 1 + \sum_{k=1}^{\infty} c_k z^k. \tag{3.51}$$

The coefficients c_k are recursively obtained by insertion of Eq. (3.51) into Eq. (3.50). The first few terms are $c_1 = -\frac{2}{3}(2-q)$, $c_2 = \frac{4}{9}(7-13q+4q^2)$, and $c_3 = -\frac{8}{9}(10-39q+39q^2-9q^3)$. In the limit of Maxwell molecules ($q=0$), these coefficients reduce to those already obtained in Section 2.3 [cf. Eq. (2.62)]. In fact, the differential equation (3.50) becomes a third-degree algebraic equation when $q \to 0$, whose real solution is given by Eqs. (2.57) and (2.60). In this Maxwell case, the expansion (3.51) is convergent. Nevertheless, when $q > 0$, a dominant-term analysis (Santos and Brey, 1991a) shows that the asymptotic behavior of the coefficients c_k for large k is $|c_k| \sim (4q/3)^k k!$.[2] Therefore, the series (3.51) does not converge for *any* nonzero value of $z = a^{*2}$, except in the special case of Maxwell molecules.

Since the Chapman–Enskog expansion for $q > 0$ is divergent, let us look for a representation of F_η as an expansion about the point at infinity. A simple analysis of the differential equation (3.50) for large z shows that, either

$$F_\eta \approx \overline{c}_0 z^{-2/3}, \quad \overline{c}_0 = \left[\tfrac{8}{9}\left(\tfrac{1}{3}q+1\right)\left(\tfrac{1}{3}q+\tfrac{1}{2}\right)\right]^{-1/3}, \tag{3.52}$$

or $F_\eta \approx Az^{(1-q)/2q}$, with arbitrary A. The behavior (3.52) is *universal* (in the sense that the power of z is independent of q) and defines the normal solution, namely the one determined by the boundary conditions $\lim_{z\to\infty} F_\eta(z) = \lim_{z\to\infty} F'_\eta(z) = 0$. On the other hand, the second possible behavior $F_\eta \sim z^{(1-q)/2q}$ is neither universal nor compatible with the boundary conditions of the normal solution; it corresponds to the general mathematical solution of Eq. (3.50). Note that Eq. (3.52) is consistent with Eq. (2.65) in the limit $q \to 0$. As in the Maxwell case, the expansion about the point at infinity of the normal solution can be written as

$$F_\eta = z^{-2/3} \sum_{k=0}^{\infty} \overline{c}_k z^{-k/3}, \tag{3.53}$$

where $\overline{c}_0$ is given by (3.52). The next two coefficients are

$$\overline{c}_1 = -\frac{1}{2}\frac{6+5q}{3+5q+2q^2}, \quad \overline{c}_2 = -\frac{9+4(6+7q)\overline{c}_1+4(3+7q-\frac{34}{9}q^2)\overline{c}_1^2}{4(3+7q+\frac{38}{9}q^2)\overline{c}_0}. \tag{3.54}$$

The numerical values of the coefficients $\overline{c}_k$ for large k indicate that the series (3.53) is convergent, in contrast to the series (3.51). In particular, for hard spheres, one has $|\overline{c}_k| \approx 10^{-4} 3^{-2k/3}$ (Santos et al., 1986a), which implies that (3.53) converges for $z^{-1/3} < 3^{2/3}$, i.e. for $a^* > 1/3$. In Table 3.1 we

[2] In the case of hard spheres ($q = \frac{1}{2}$), the numerical evaluation of c_k up to $k \approx 1800$ (Santos et al., 1986a) indicates that $|c_k| \simeq 5 \times 10^8 (\frac{2}{3})^k k!$.

compare the values of F_η, as given by the expansion (3.53), with those obtained from the truncated Chapman–Enskog series $F_{\eta;N} = 1 + \sum_{k=0}^{N} c_k z^k$ ($N = 1, 2, 4, 6, 8, 10$) for several values of the shear rate, in the case of hard spheres. Table 3.1 clearly shows that, although the Chapman–Enskog expansion (3.51)

Table 3.1. Viscosity function for hard spheres, F_η, as given by the series (3.53), as well as by the series (3.51) truncated at $k = N$, $F_{\eta;N}$.

a^{*2}	F_η	$F_{\eta;1}$	$F_{\eta;2}$	$F_{\eta;4}$	$F_{\eta;6}$	$F_{\eta;8}$	$F_{\eta;10}$
0.12	0.8907	0.8800	0.8896	0.8908	0.8907	0.8907	0.8907
0.20	0.8304	0.8000	0.8267	0.8316	0.8301	0.8303	0.8306
0.40	0.7166	0.6000	0.7067	0.7356	0.6770	0.7287	0.9156
0.60	0.6376	0.4000	0.6400	0.7024	0.1844	1.411	10.82
0.80	0.5792	0.2000	0.6267	0.6911	−1.801	9.959	165.0

is not convergent, it is *asymptotic*. This is supported by the fact that the several $F_{\eta;N}$ with $N = 2$–10 match well with the correct value of F_η at $a^{*2} = 0.12$. Moreover, at a given value of the shear rate, the accuracy of the partial sum $F_{\eta;N}$ does not necessarily improve as the number of terms N grows, but there exists an optimum number of terms which increases as the shear rate decreases.

The convergent character of the Chapman–Enskog and Hilbert methods is still an open issue (Cercignani et al., 1994). A general discussion by Grad (1963) indicates that the Chapman–Enskog expansion is at least asymptotic. This conclusion, however, does not preclude the convergence of the expansion. As a matter of fact, in the case of the *linearized* Boltzmann equation, McLennan (1965) proved for a general class of cutoff potentials the *convergence* of the Chapman–Enskog expansion, assuming an exponential dependence of the distribution function on position and time. On the other hand, the results discussed here for the USF provide a test example where the Chapman–Enskog method is only asymptotic. In principle, it is difficult to determine whether this divergence is an artifact of the BGK model or it is actually truly representative of the Boltzmann equation itself. The results obtained in Chapter 2 support the latter possibility. In the case of the Boltzmann equation for Maxwell molecules, we have seen that only a limited number of finite moments exist for any shear rate, while all the moments obtained from the solution of the BGK model for Maxwell molecules, Eq. (3.8), are finite. This suggests that the solutions of the BGK model are generally less singular than the solutions of the Boltzmann equation. Finally, note that McLennan's proof of convergence (McLennan, 1965) does not apply to the USF problem, since in this state all the corrections to the Navier–Stokes regime are *nonlinear*, i.e. $\partial^k u_x/\partial y^k = 0$ for $k \geq 2$.

Now we return to the study of the viscosity function F_η associated with the normal solution, in the particular case of hard spheres. This function can be represented by the expansion (3.53) for $a^* > 1/3$ and by $F_{\eta;N}$ with, for instance, $N = 6$ for $a^* \leq 1/3$, since both representations match well for $a^* \simeq 0.12$, as shown in Table 3.1. It can be checked (Santos and Brey, 1991a) that the different numerical solutions of Eq. (3.50) corresponding to particular initial conditions at $z = z_0$ tend to be "attracted" by the normal F_η as z decreases. Figure 3.10 shows the viscosity function versus $z = a^{*2}$ as obtained from the BGK equation for hard spheres and from the Boltzmann equation for Maxwell molecules [cf. Eq. (2.57)], which coincides with that given by the BGK model for the latter interaction. It is seen that both curves are hardly distinguishable, what means that the viscosity function is rather insensitive to the interaction model. The figure also shows simulation values for hard spheres obtained from

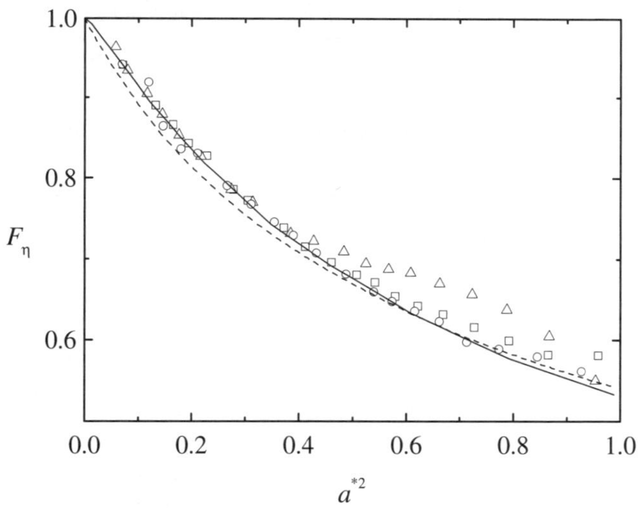

Figure 3.10. Viscosity function F_η versus the square of the reduced shear rate a^{*2}, as given by the BGK model for hard spheres (solid line) and the Boltzmann equation for Maxwell molecules (dashed line). The symbols refer to data obtained from the DSMC method for hard spheres with initial conditions corresponding to $a_0^* = 1.11$ (triangles), $a_0^* = 1.57$ (squares), and $a_0^* = 2.22$ (circles).

the Direct Simulation Monte Carlo (DSMC) method (Gómez Ordóñez et al., 1989). The simulations proceed as follows. The system is initially prepared in a local equilibrium state with a temperature $T(0) = T_0$ such that the initial value of the reduced shear rate is $a_0^* = a/\nu(0)$. As the system evolves and the temperature grows, the elements of the pressure tensor are evaluated as functions of time. This allows one to get the time evolution of the quantities $a^* = a/\nu(t)$ and $F_\eta = -[P_{xy}(t)/p(t)]/a^*$. From this one can finally get F_η

as a function of a^*. Figure 3.10 shows that, after a certain transient period, the data corresponding to each initial condition tend to collapse to a common curve. In addition, this common curve is very well described by the BGK prediction. This conclusion also holds for shear rates much larger than those considered in Fig. 3.10 (Gómez Ordóñez et al., 1990).

The fact that the viscosity function for hard spheres ($q = \frac{1}{2}$) is very close to that for Maxwell molecules ($q = 0$) suggests to consider the simple approximation

$$F_\eta(a^*) \simeq F_\eta^M(a^*) + q\Delta F_\eta(a^*), \tag{3.55}$$

where $F_\eta^M(a^*)$ is the Maxwell viscosity function, whose explicit expression is known [cf. Eqs. (2.57) and (2.60)]. Inserting Eq. (3.55) into Eq. (3.50) and neglecting terms nonlinear in q one gets

$$\Delta F_\eta = \frac{9(1 - F_\eta^M) - 10z\left(F_\eta^M\right)^2 + 8z^2\left(1 + zF_\eta^M\right)F_\eta^M\left(F_\eta^M\right)'}{3 + 8zF_\eta^M + 4z^2\left(F_\eta^M\right)^2}, \tag{3.56}$$

where the prime denotes a derivative with respect to $z = a^{*2}$. Figure 3.11

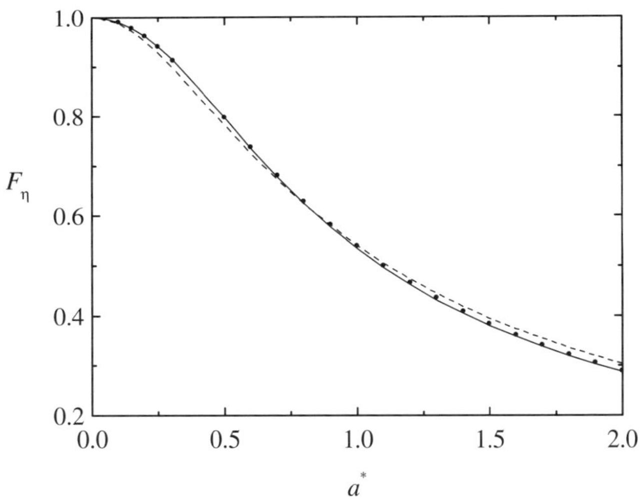

Figure 3.11. Shear-rate dependence of the viscosity function for hard spheres (solid line) and Maxwell molecules (dashed line). The dots represent the approximation (3.55) for hard spheres.

shows that the shear-rate dependence of F_η for hard spheres, as obtained from the expansion (3.53), is hardly distinguishable from that obtained by Eqs. (3.55) and (3.56). An expansion of the nonlinear viscosity in powers of the interaction parameter q was first proposed by Karlin et al. (1997, 1998) in a *longitudinal* viscous flow problem, while its comparison with the differential equation analogous to Eq. (3.50) was carried out by Santos (2000). In this longitudinal viscous flow problem, a good agreement requires to go to third order in ΔF_η. However, as seen in Fig. 3.11, the first-order approximation (3.55) is good enough in the USF problem.

Although the viscosity function F_η is the most relevant quantity associated with nonlinear transport in shearing problems, normal stresses are also important from a rheological point of view. These effects are characterized by the viscometric functions (2.55) and (2.56). First, note that the evolution equations for P_{yy} and P_{zz}, Eqs. (3.37) and (3.38), are formally the same. This implies that P_{zz} obeys an equation that can be obtained from Eq. (3.48) just by setting $P_{yy} \to P_{zz}$. Thus, the second viscometric function obeys the *homogeneous* equation

$$\left[-\frac{2}{3}qa^{*3}F_\eta(a^*)\frac{\partial}{\partial a^*}+1+\frac{2}{3}\left(1-2q\right)a^{*2}F_\eta(a^*)\right]\Psi_2=0. \qquad (3.57)$$

The hydrodynamic solution of this equation, i.e. the only one that is independent of the initial conditions, is simply $\Psi_2(a^*) = 0$. This result coincides with the one already obtained from the Boltzmann and BGK equations for Maxwell molecules. However, the Boltzmann equation for interactions other than Maxwell's leads to $\Psi_2 \neq 0$. In particular, the Chapman–Enskog method for hard spheres up to Burnett order (Chapman and Cowling, 1970; McLennan, 1989) yields $\Psi_2(0) = 0.172$. Simulation results by the DSMC method (Gómez Ordóñez et al., 1990) show that $\Psi_2(a^*)$ is a decreasing function of the shear rate. For instance, $\Psi_2 \simeq 0.03$ at $a^* \simeq 1$. This indicates that the BGK prediction $\Psi_2 = 0$ is not so crude after all.

Now, we analyze the shear-rate dependence of the first viscometric function. From Eq. (3.47) one easily gets

$$\Psi_1(a^*)=3\frac{F_\eta-1}{a^{*2}}+2\left(1-q\right)F_\eta^2-qa^*\frac{\partial}{\partial a^*}F_\eta^2, \qquad (3.58)$$

where we have made use of the relation $\Psi_1 = (3/a^{*2})(P_{yy}/p-1)$, since $P_{yy} = P_{zz}$. As expected, Eq. (3.58) reduces to Eq. (2.58) in the limit $q \to 0$. For small shear rates, the behavior of Ψ_1 can be obtained by inserting the expansion (3.51) into Eq. (3.58). The result is

$$\Psi_1(a^*)=-2\left[1-2(1-q)a^{*2}+8(\tfrac{1}{2}-q)(\tfrac{4}{3}-q)a^{*4}+\cdots\right]. \qquad (3.59)$$

In the limit of large shear rates, one gets from Eq. (3.52)

$$\Psi_1(a^*)=-a^{*-2}\left[3-2(1+\tfrac{1}{3}q)\overline{c}_0^2a^{*-2/3}+\cdots\right]. \qquad (3.60)$$

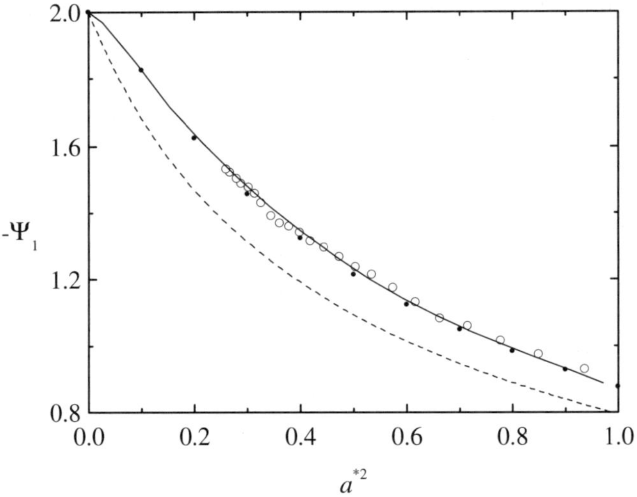

Figure 3.12. First viscometric function Ψ_1 versus the square of the reduced shear rate a^{*2}, as given by the BGK model for hard spheres (solid line) and the Boltzmann equation for Maxwell molecules (dashed line). The circles refer to data obtained from the DSMC method for hard spheres, while the dots correspond to the results obtained from the approximation (3.55).

The shear-rate dependence of Ψ_1 is shown in Fig. 3.12 for hard spheres and Maxwell molecules. Simulation results for hard spheres obtained by taking $a_0^* = 100$ are also presented (Gómez Ordóñez et al., 1990), as well as the approximate result for hard spheres obtained by inserting Eq. (3.55) into Eq. (3.58). We observe that the BGK prediction agrees very well with the DSMC data. In addition, in contrast to what happened with the viscosity function, a noticeable influence of the interaction potential is apparent, the normal stress difference $P_{xx} - P_{yy}$ being larger for hard spheres than for Maxwell molecules.

The good agreement between the rheological properties predicted by the BGK equation and the simulation results for hard spheres holds not only in the hydrodynamic regime but also in the transient regime. This is illustrated in Fig. 3.13, where simulation data for the viscosity function obtained starting from local equilibrium with two different values of a_0^* are compared with the corresponding numerical solutions of Eq. (3.50). A similar comparison for the first viscometric function also exhibits a good agreement (Gómez Ordóñez et al., 1990).

Apart from the transport properties, the comparison between the BGK solution and the DSMC simulations can also be made at the level of the velocity distribution function. Such a comparison (Gómez Ordóñez et al., 1990) shows a qualitatively good agreement in the small velocity region, although significant discrepancies exist for high velocities. This is not surprising since in the BGK

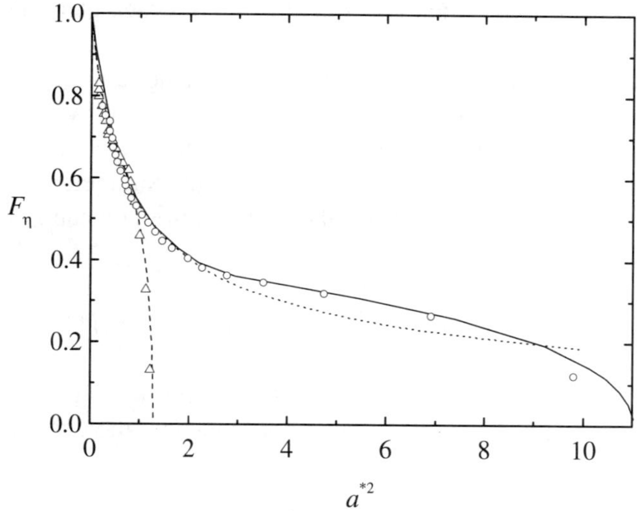

Figure 3.13. Viscosity function F_η as given by the BGK model (lines) and by the DSMC method (symbols) for hard spheres with initial conditions corresponding to $a_0^* = 3.33$ (solid line and circles) and $a_0^* = 1.11$ (dashed line and triangles). The dotted line represents the BGK hydrodynamic solution.

model the distribution is governed only by the first five moments, which probe the domain of velocities smaller than or of the order of the thermal velocity.

4. The thermostatted state

As said in this Chapter and in Chapter 2, the USF is a time-dependent state, due to the viscous heating [cf. Eq. (2.10)]. However, from the point of view of computer simulations, it is problematic to measure rheological properties in time-dependent states (Naitoh and Ono, 1979). Thus, it has become usual to introduce non-conservative external forces to compensate exactly for the viscous increase of temperature and achieve a steady state. This type of forces is familiarly referred to as "thermostats" (Evans and Morriss, 1984; Evans and Morriss, 1990).[3] Thermostatted systems have also attracted the attention of researchers in nonequilibrium statistical mechanics and dynamical systems theory (van Zhon, 1999; van Beijeren and Dorfman, 2000). The implicit assumption behind the introduction of these forces is that they play a neutral role in the transport properties, so that, for instance, the viscosity function $F_\eta(a^*)$ is the same with and without a thermostat. Nevertheless, this expectation is not

[3]Strictly speaking, the type of thermostat used in computer simulations keeps constant the kinetic energy of each microstate, not just the average kinetic energy (or temperature).

strictly true (Dufty et al., 1986a; Dufty et al., 1986b; Garzó et al., 1990), except in the Navier–Stokes regime.

Of course, the choice of the thermostat force **F** is not unique (Sarman et al., 1994; Montanero et al., 1995b). The simplest form is provided by Gauss's principle of least constraint (Evans and Morriss, 1990; Hoover, 1991), which gives the "drag" force $\mathbf{F} = -m\alpha\mathbf{V}$. The thermostat parameter α is a function of the shear rate adjusted as to keep the temperature constant. Under these conditions, the Boltzmann equation becomes

$$\frac{\partial}{\partial t}f - \frac{\partial}{\partial V_i}a_{ij}V_j f - \alpha\frac{\partial}{\partial \mathbf{V}}\cdot\mathbf{V}f = J[f,f]. \tag{3.61}$$

This equation differs from Eq. (2.6) by a term representing the effect of the thermostat force **F** [cf. Eq. (1.25)]. As a consequence, the evolution equation for the temperature is

$$\frac{dT(t)}{dt} = -\frac{2}{3nk_B}aP_{xy} + 2\alpha T, \tag{3.62}$$

instead of Eq. (2.10) By imposing the condition $dT/dt = 0$, one has

$$\alpha = -\frac{a}{3}\frac{P_{xy}}{p}. \tag{3.63}$$

In the long-time limit, it is expected that P_{xy} reaches a stationary value, so that α becomes independent of time. One could think that the only effect of the thermostat is equivalent to a rescaling of the velocities with respect to the thermal velocity. However, this is not true in general, since when one rescales the velocities starting from Eq. (2.6) one arrives at Eq. (2.17). Note that the latter is not equivalent to Eq. (3.61) due to the time-dependence of $\overline{a}$ and $\overline{\alpha}$ [cf. Eq. (2.19)]. An interesting exception is that of Maxwell molecules ($w = 4$), in which case $\overline{a} = a$ and $\overline{\alpha} = \alpha$, and Eq. (2.25) is totally equivalent to Eq. (3.61). Consequently, all the results derived in Chapter 2 referring to Maxwell molecules, such as rheological properties, high-degree moments, high-velocity tail, . . . , also hold in the thermostatted case for this interaction. This equivalence is closely related to the kinetic theory of a *dilute* Maxwell gas and in fact an influence of the thermostat should be expected at higher densities, even for Maxwell molecules.

In order to assess the influence of the thermostat on the properties of the system for interactions different from Maxwell's, it is convenient to consider the BGK model, since the Boltzmann equation is too complicated in the nonlinear regime for non-Maxwell interactions. As said before, all the dependence on the interaction potential appears through the effective collision frequency ν. In the thermostatted state, the temperature is constant and so is ν. Thus, according to the BGK model, the results are *universal*, i.e. they are independent of the

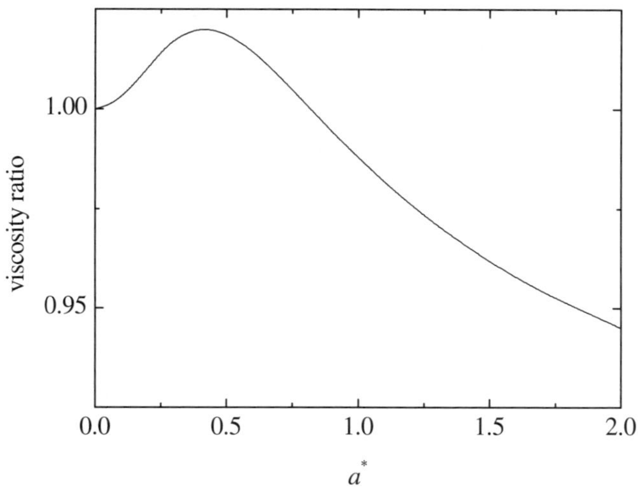

Figure 3.14. Shear-rate dependence of the ratio between the viscosity function for hard spheres without thermostat and the same quantity with thermostat.

potential, once the shear rate is reduced with respect to ν. As a consequence, all the results previously derived from the BGK model for Maxwell molecules apply to the thermostatted case for any potential. In particular, the viscosity function F_η for a thermostatted state is given by Eq. (2.57), while F_η for an unthermostatted state is obtained from the normal solution of the differential equation (3.50). As an illustration of the influence of the thermostat, Fig. 3.14 shows the ratio between F_η without thermostat and F_η with thermostat, in the case of hard spheres. We observe that at small shear rates the nonlinear viscosity is larger in the unthermostatted case, as expected from the fact that the super-Burnett coefficient is less negative in that situation. However, this behavior changes for sufficiently large shear rates, thus indicating that the thermostat enhances the shear-thinning effect. In any case, the quantitative discrepancies in F_η are smaller than about 5% in the range of shear rates $0 \leq a^* \leq 2$.

Since the BGK model provides quite a good picture of the rheology of USF far from equilibrium, it is fair to expect that this universality feature means that the rheological properties in real dilute systems are only weakly dependent on the interaction law, once the shear rate is nondimensionalized with an adequate collision frequency defined as $\nu = p/\eta_0$. This expectation is supported by a comparison with molecular dynamics simulations of a thermostatted Lennard–Jones gas (Loose and Hess, 1987; Loose and Hess, 1988; Loose, 1988). These authors computed the *kinetic* contributions to the pressure tensor for several values of the shear rate a at a reduced density $n\sigma^3 = 0.1$ and a reduced temperature $k_BT/\epsilon = 2.75$.[4] This leads to a Navier–Stokes shear viscosity equal to $\eta_0 = 0.276(m\epsilon)^{1/2}/\sigma^2$ and an effective collision frequency $\nu = p/\eta_0 = 0.996(\epsilon/m)^{1/2}\sigma^{-1}$ (Hirschfelder et al., 1964). Figure 3.15

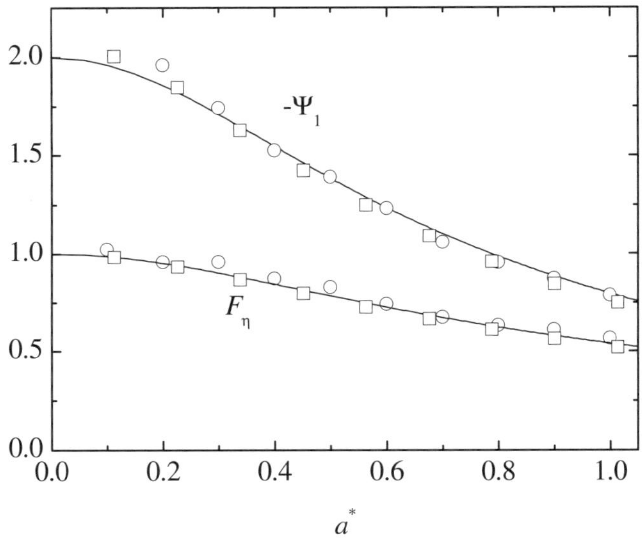

Figure 3.15. Viscosity function F_η and the first viscometric function Ψ_1 versus the shear rate a^*, as obtained from molecular dynamics simulations for a Lennard–Jones gas (circles), from the DSMC method for had spheres (squares), and from the Boltzmann equation for Maxwell molecules (lines).

shows the simulation data (Loose and Hess, 1988) for the viscosity function $F_\eta = -P_{xy}/a\eta_0$ and the first viscometric function $\Psi_1 = (P_{yy} - P_{xx})p/(a\eta_0)^2$ versus the reduced shear rate $a^* = a/(p/\eta_0)$. The figure also includes simulation data obtained by means of the DSMC method for hard spheres, as well as the theoretical curves for Maxwell molecules, Eqs. (2.57) and (2.58), which are identical in the Boltzmann and BGK descriptions. The excellent agreement observed confirms the expectation about the weak influence of the interaction on the transport properties, even in states far from equilibrium. In addition, the simulation data of Loose and Hess (1988) agree within statistical fluctuations with the BGK prediction $\Psi_2 = 0$. It is worthwhile remarking the relevance of a kinetic theory description beyond its expected range of applicability, since the simulated system represents a moderately dense gas.

[4]Here, σ and ϵ are the parameters of the Lennard–Jones potential. For Argon gas ($\sigma = 3.405$ Å, $\epsilon/k_B = 119.8$ K), the state $n\sigma^3 = 0.1$ and $k_BT/\epsilon = 2.75$ corresponds to a temperature $T \simeq 330$ K and a pressure $p \simeq 12$ MPa, i.e. a moderately dense gas.

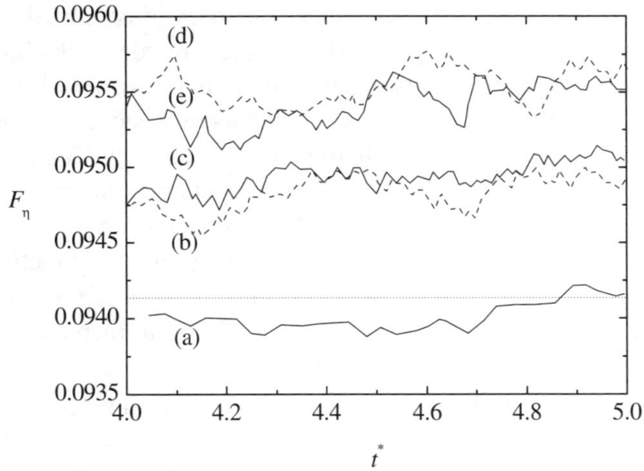

Figure 3.16. Viscosity function at $a^* = 6$ for repulsive potentials with (a) $w = 4$, (b) $w = 6$, (c) $w = 8$, (d) $w = 12$, and (e) $w = \infty$. The dotted line represents the value obtained from the exact solution for Maxwell molecules.

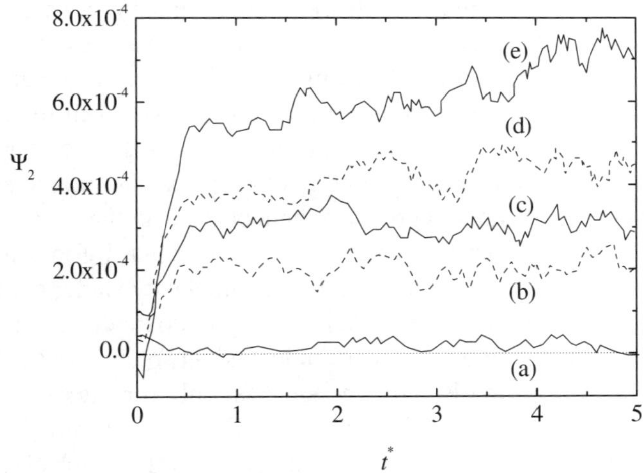

Figure 3.17. Time evolution of the second viscometric function at $a^* = 6$ for repulsive potentials with (a) $w = 4$, (b) $w = 6$, (c) $w = 8$, (d) $w = 12$, and (e) $w = \infty$. The dotted line represents the value obtained from the exact solution for Maxwell molecules.

It must be noted that a certain small influence of the potential on the rheological properties exists, as shown by careful DSMC simulations (Montanero et al., 1997a) of the thermostatted USF for repulsive potentials of the form $\phi(r) \propto r^{-w}$ with $w = 4$ (Maxwell molecules), 6, 8, 12, and ∞ (hard spheres). The viscosity function in the time window $4 \leq t^* = t\nu \leq 5$ (where the steady state has been reached) for a reduced shear rate $a^* = 6$ is plotted in Fig. 3.16. The differences among the curves are so small that, in order to distinguish them, the vertical scale has been widely magnified. Otherwise, we observe that the viscosity function tends to increase as the repulsion becomes harder. For instance, the value of $F_\eta(a^* = 6)$ for hard spheres is about 1.5% larger than that for Maxwell molecules. The first viscometric function presents a similar influence (Montanero et al., 1997a), the normal stress difference $P_{xx} - P_{yy}$ slightly decreasing as w increases. Regarding the second viscometric function, the simulation data show that, except for Maxwell molecules, $\Psi_2 \neq 0$. Nonetheless, its value is very small, as illustrated in Fig. 3.17. As w increases, the normal stress difference $P_{zz} - P_{yy}$ slightly increases.

Let us explore whether the practical insensitivity of the rheological properties on the interaction potential extends to higher-degree moments. Figure 3.18 shows the fourth-degree moment $\langle \xi^4 \rangle$ as a function of the shear rate as obtained from the exact solution of the Boltzmann equation for Maxwell molecules, from the DSMC method for hard spheres, and from molecular dynamics simulations for a Lennard–Jones gas (Loose and Hess, 1988). It is clear that, in contrast to what happens with the second-degree moments, the behavior of $\langle \xi^4 \rangle$ is significantly affected by the potential. While the simulation data for hard spheres and Lennard–Jones particles are quite close each other, the exact results for Maxwell molecules are clearly larger than the former ones. This is a signature of a high-velocity overpopulation much more pronounced for Maxwell molecules than in the case of harder potentials. We will further discuss this point at the end of this Section. Given that the BGK prediction for $\langle \xi^4 \rangle$ is common to any interaction, it is evident that it cannot agree simultaneously with the correct results for Maxwell molecules, for hard spheres, and for the Lennard–Jones potential. It turns out that the BGK prediction is much closer to the two latter than to the former case, as seen in Fig. 3.18. This might seem paradoxical, since the collision frequency in the BGK model is velocity-independent for any interaction potential, a feature shared by the Boltzmann equation only for Maxwell molecules, but not in general. On the other hand, the BGK model inhibits the formation of high-velocity tails because of the relaxation towards the local equilibrium distribution. This makes the BGK predictions for the moments exhibit a better agreement for hard interactions than for Maxwell molecules. The above agreement between the BGK results and the simulation data also holds for $\langle \xi^6 \rangle$ at a qualitative and semi-quantitative levels, as shown in Fig. 3.19.

As widely discussed in Chapter 2, the velocity moments of degree $k > 2$ diverge for shear rates larger than a critical value $a_c^{(k)}$ in the case of Maxwell molecules (cf. Table 2.1). Since we have just seen that those moments depend on the interaction potential, a natural question is whether the above singular behavior is also present for potentials different from Maxwell's. This question cannot be in principle answered by analytical means, since the hierarchy of

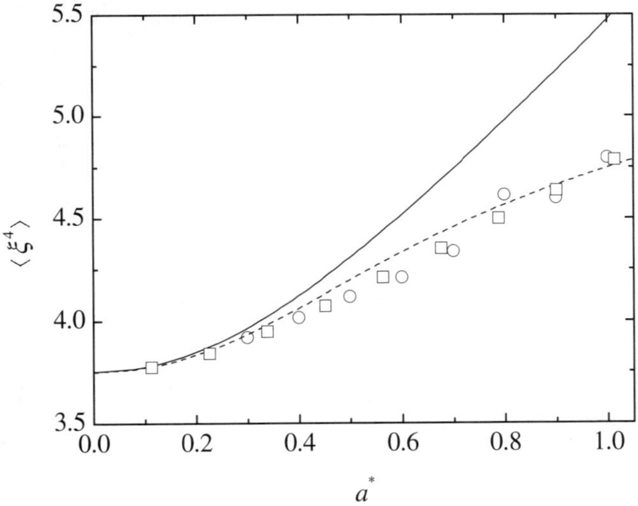

Figure 3.18. Fourth-degree moment $\langle \xi^4 \rangle$ versus the shear rate a^*, as obtained from molecular dynamics simulations for a Lennard–Jones gas (circles), from the DSMC method for hard spheres (squares), from the Boltzmann equation for Maxwell molecules (solid line), and from the BGK equation for any potential (dashed line).

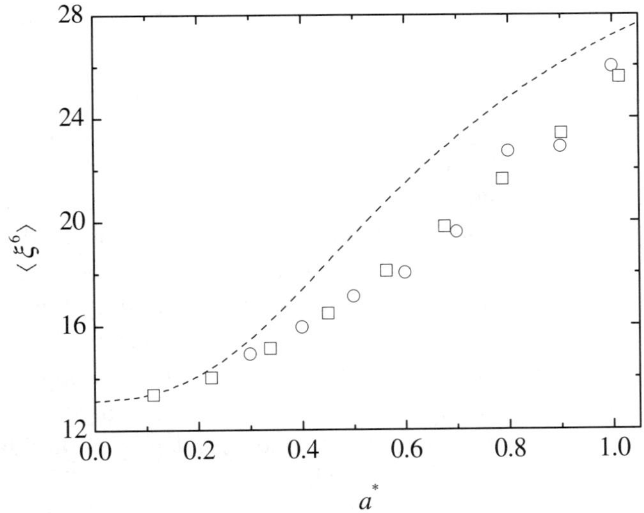

Figure 3.19. Sixth-degree moment $\langle \xi^6 \rangle$ versus the shear rate a^*, as obtained from molecular dynamics simulations for a Lennard–Jones gas (circles), from the DSMC method for hard spheres (squares), and from the BGK equation for any potential (dashed line).

moment equations derived from the Boltzmann equation is not closed, except for Maxwell molecules. As a consequence, it is helpful to resort to computer simulations by means of the DSMC method (Montanero et al., 1996a). Figure

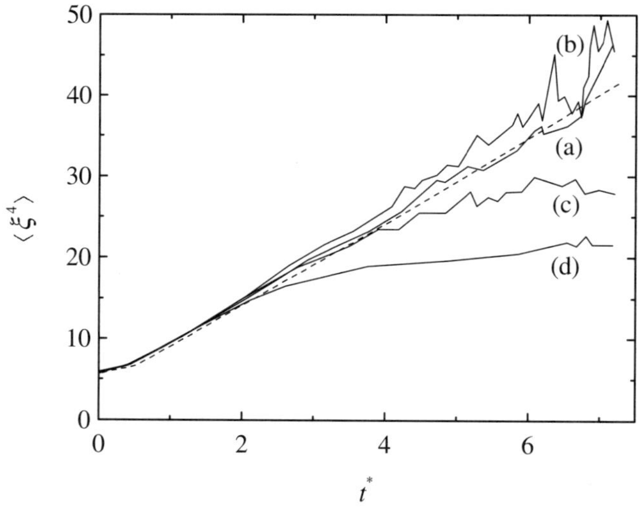

Figure 3.20. Time evolution of $\langle \xi^4 \rangle$, obtained by simulation, at a shear rate $a^* = 7.33$ and for (a) $w = 4$, (b) $w = 8$, (c) $w = 12$, and (d) $w = \infty$. The dashed line is the exact solution for Maxwell molecules ($w = 4$).

3.20 shows the time evolution of the fourth-degree moment $\langle \xi^4 \rangle$ at $a^* = 7.33$ and for $w = 4$, 8, 12, and ∞. The exact time evolution for Maxwell molecules, Eq. (2.A.3), is also plotted. First, it is worth noting the good agreement between exact and simulation results for Maxwell molecules. This can be seen as a rather stringent test of the simulation method (Montanero and Santos, 1995). The fact that the data for $w = 8$ (as well as for $w = 6$, not shown in Fig. 3.20 for the sake of clarity) practically overlap with those for $w = 4$ strongly suggests that the moment $\langle \xi^4 \rangle$ at $a^* = 7.33$ diverges for those potentials as well. Nevertheless, the data for $w = 12$ and $w = \infty$ seem to reach stationary values at the shear rate $a^* = 7.33$. An interesting question is whether this regular behavior for the hardest potentials is also true for moments of a degree higher than 4. As a matter of fact, in the particular case of Maxwell molecules, $\langle \xi^4 \rangle$ converges but $\langle \xi^6 \rangle$ diverges for shear rates within the range $2.346 < a^* < 6.846$, as shown in Table 2.1. The evolution of $\langle \xi^6 \rangle$ is shown in Fig. 3.21. This moment seems to diverge with time for $w = 12$, while it still reaches a stationary value for hard spheres. So far, the simulation results support the conclusion that the divergence of the moments found theoretically in the case of Maxwell molecules can be extended to other potentials with $4 \leq w \leq 12$. On the other hand, in the case of hard spheres ($w = \infty$), the moments $\langle \xi^4 \rangle$ and $\langle \xi^6 \rangle$ are not divergent, at least

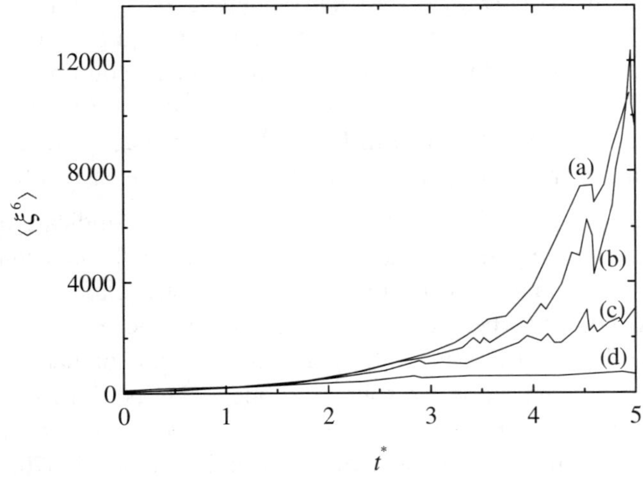

Figure 3.21. Time evolution of $\langle \xi^6 \rangle$, obtained by simulation, at a shear rate $a^* = 7.33$ and for (a) $w = 4$, (b) $w = 8$, (c) $w = 12$, and (d) $w = \infty$.

for $a^* \leq 7.33$. This might be due to the fact that the shear rate is not sufficiently large and/or the degree of the moments is not sufficiently high. Of course, a third possibility is that all the velocity moments are always convergent in the case of hard spheres. In order to shed light on this point, the moments $\langle \xi^k \rangle$, with $k = 4$,

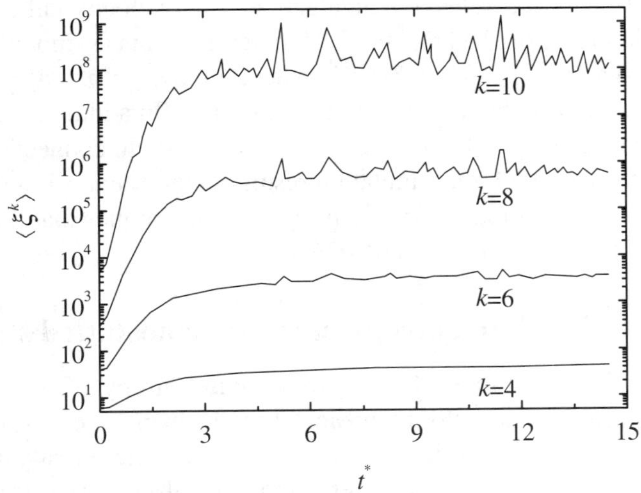

Figure 3.22. Time evolution of $\langle \xi^k \rangle$, $k = 4$, 6, 8, and 10 at a shear rate $a^* = 10$ for hard spheres ($w = \infty$).

6, 8, and 10, are plotted in Fig. 3.22 for hard spheres at $a^* = 10$. It is evident that the moments considered reach steady-state values after a few collision times. These values are, on the other hand, much larger than the equilibrium ones. For instance, $\langle\xi^{10}\rangle \sim 10^8$ at $a^* = 10$, while $\langle\xi^{10}\rangle = \frac{10\,395}{32} \simeq 325$ at equilibrium. On the basis of the results shown in Fig. 3.22, we can conjecture that *all* the velocity moments are convergent for *any* shear rate in the case of hard spheres.

The simulation results (Montanero et al., 1996a) thus indicate that the analytically known divergence of the moments for Maxwell molecules at sufficiently large shear rates is also present for other interactions. On the other hand, this phenomenon becomes less remarkable as the interaction becomes harder. In addition, it seems to disappear in the limit of hard spheres. A possible physical scenario is the following one. The inertial force $-maV_y\widehat{\mathbf{x}}$ in Eq. (3.61) acting during a time step δt tends to produce an increase in the magnitude of the x-component of the peculiar velocity of the particles: $V_x(t+\delta t) = V_x(t) - aV_y(t)\delta t$. This is not but the viscous heating effect. However, the thermostat force $\mathbf{F} = -m\alpha\mathbf{V}$ controls the mean kinetic energy by rescaling all the velocities with a common factor $e^{-\alpha\delta t}$. As a consequence, the *relatively small* high-velocity population may increase in time. This increase is counterbalanced by collisions taking place during the time interval δt, which tend to redistribute the energy. As the interaction becomes harder, the collision frequency for high-velocity particles increases and the energy redistribution is more efficient. Therefore, it is natural to expect that the high-velocity population decreases as the hardness of the potential increases.

In Chapter 2 we saw that the divergence of the moments in the case of Maxwell molecules was related to an algebraic high-velocity tail of the distribution function [cf. Eq. (2.111)]. For r^{-w}-potentials, this asymptotic form can be generalized as $f(\mathbf{V}) \sim V^{-5-\sigma(a^*,w)}$, where $\lim_{a^*\to 0}\sigma(a^*,w) = \infty$, i.e. all the velocity moments are convergent at equilibrium. In addition, according to Fig. 2.11, one can expect that $\lim_{a^*\to\infty}\sigma(a^*,w) = 0$. In connection with the influence of the potential, the simulation results suggest that σ is an increasing function of w, so that $\lim_{w\to\infty}\sigma(a^*,w) = \infty$. This implies that the algebraic decay disappears in the case of hard spheres.

5. Nonequilibrium entropy of the thermostatted state

In Section 2.7 we studied the nonequilibrium entropy of a dilute gas of Maxwell molecules under USF by means of the Boltzmann equation. Specifically, the exact first few terms in the series expansion of the velocity distribution function in powers of the shear rate were used to get the excess entropy to sixth order in the shear rate, Eq. (2.134). On the other hand, the truncated expansion (2.134) is not practical for intermediate or large shear rates, say for instance $a^* \approx 1$.

This limitation can in principle be overcome in the framework of the BGK model kinetic equation, since its explicit solution is known for arbitrary shear rates. It is given by Eq. (3.8) for Maxwell molecules without a thermostat, Eq. (3.45) for general repulsive potentials without a thermostat, and Eq. (3.8) for any interaction with a Gaussian thermostat. Henceforth, we will adopt the latter point of view. According to the definition of Boltzmann's entropy, Eq. (2.130), the reduced excess entropy $s_{\rm ex}^* = s_{\rm ex}/nk_B$ is given by

$$s_{\rm ex}^* = -\int d\boldsymbol{\xi}\, f^*(\boldsymbol{\xi}) \ln \frac{f^*(\boldsymbol{\xi})}{\pi^{-3/2} e^{-\xi^2}}. \tag{3.64}$$

Despite the fact that f^* is explicitly known (in terms of a quadrature), the velocity integral appearing in Eq. (3.64) does not seem to lend itself to an analytical evaluation. Consequently, one has to resort to a numerical computation (see Appendix 3.A). Alternatively, analytical information can be obtained in the form of a series expansion of $s_{\rm ex}^*$ in powers of a^*. The first nonzero nine terms are (Brey and Santos, 1992; Montanero and Santos, 1996)

$$\begin{aligned} s_{\rm ex}^* = & -\frac{1}{2}a^{*2} + \frac{1}{4}a^{*4} - \frac{5}{54}a^{*6} + \frac{153}{8}a^{*8} + \frac{192\,101}{270}a^{*10} \\ & + \frac{78\,664\,283}{2916}a^{*12} + \frac{2\,336\,070\,287}{10\,206}a^{*14} - \frac{7\,627\,231\,329\,077}{34\,992}a^{*16} \\ & - \frac{2\,347\,536\,574\,990\,979}{39\,366}a^{*18} + \mathcal{O}(a^{*20}). \end{aligned} \tag{3.65}$$

The fact that the coefficient of a^{*4} in the BGK model is smaller than in the Boltzmann equation [cf. Eq. (2.134)] suggests that the entropy in the BGK model is smaller than in the Boltzmann equation, at least in the domain of small shear rates. This seems to be counter-intuitive since the BGK model can be viewed as a "coarse-grained" version of the Boltzmann equation and contains less information than the latter one. According to this intuitive argument, one would expect the distribution function of the Boltzmann equation to be more "ordered" than the one of the BGK model, and hence the entropy of the former to be smaller than that of the latter. However, this expectation turns out to be incorrect, what shows that the nonequilibrium entropy is a quantity much more difficult to interpret than one might expect. As a matter of fact, the Navier–Stokes distribution $f_{\rm NS}^*(\boldsymbol{\xi}) = \pi^{-3/2}e^{-\xi^2}\,(1 - 2\xi_x\xi_y a^*)$ is simpler (i.e. less "ordered") than the solutions to the Boltzmann and BGK equations and yet its associated excess entropy $s_{\rm ex,NS}^* = -\frac{1}{2}a^{*2}$ is smaller than $s_{\rm ex}^*$ for both solutions.

It is apparent that the numerical coefficients appearing in Eq. (3.65) grow dramatically, what suggests that the series either converges with a very small radius of convergence or it is only asymptotic. A useful approximate representation is provided, at least for small or moderate shear rates, by the Padé approximant

$$s^*_{\rm ex} \simeq -\frac{a^{*2}}{2}\frac{1-\frac{7}{54}a^{*2}}{1+\frac{10}{27}a^{*2}}. \tag{3.66}$$

As seen in Section 2.7, information theory gives an upper bound of the excess entropy, Eq. (2.133). Since the pressure tensor given by the BGK model for any interaction potential (in the thermostatted USF) is the same as that given by the Boltzmann equation for Maxwell molecules, the expression of $s^*_{\rm ex,IT}$ is again given by Eq. (2.135).

Figure 3.23 shows the shear-rate dependence of $s^*_{\rm ex}$, as given by the numerical evaluation of Eq. (3.64), the Padé approximant (3.66), the Navier–Stokes approximation, and the information theory. We see that up to $a^* \simeq 0.2$ the Navier–Stokes approximation and the information-theory estimate practically coincide with the numerical evaluation. For larger shear rates, the entropy obtained numerically lies in between the Navier–Stokes value and the information-theory estimate. Despite its simplicity, the Padé approximant (3.66) presents a pretty good agreement with the numerical values, although tends to underestimate them as the shear rate increases. It is worthwhile noting that higher order Padé approximants constructed from the series (3.65) exhibit larger discrepancies from the numerical results than (3.66). This might be due to the possible asymptotic character of the expansion (3.65)

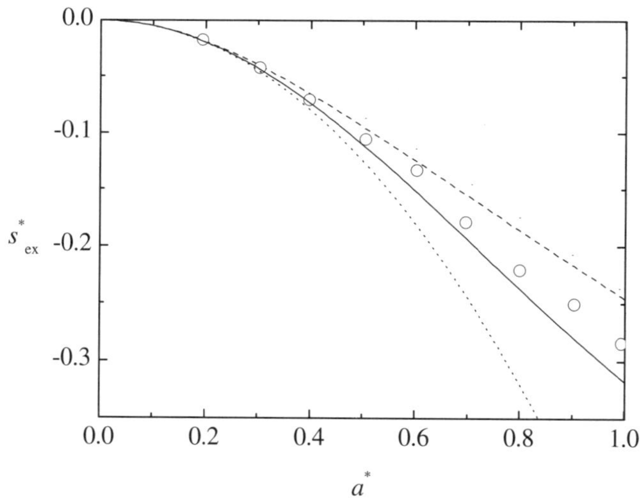

Figure 3.23. Shear-rate dependence of the reduced excess entropy, according to the numerical evaluation of Eq. (3.64) (circles), the Navier–Stokes approximation $s^*_{\rm ex,NS} = -\frac{1}{2}a^{*2}$ (dotted line), the Padé approximant (3.66) (solid line), and information theory (dashed line).

Once the nonequilibrium entropy is known, one can obtain associated "thermodynamic" quantities which are defined as the first derivatives of the entropy with respect to the basic variables, in a similar way as their analogous equilibrium quantities (Jou et al., 2000; Jou et al., 2001). As is well known, if a system is in equilibrium with an entropy per unit volume $s_0(n, e)$, where n is the number density and e is the internal energy per unit volume, then the temperature T and pressure p are defined by the first derivatives

$$T^{-1} = \left(\frac{\partial s_0}{\partial e}\right)_n, \tag{3.67}$$

$$p = -Tn^2 \left(\frac{\partial (s_0/n)}{\partial n}\right)_e - e. \tag{3.68}$$

For an ideal gas, the equilibrium entropy per unit volume is given by Eq. (1.49) and so Eqs. (3.67) and (3.68) give the equations of state $T = \frac{2}{3}(e/nk_B)$ and $p = \frac{2}{3}e$. Out of equilibrium, n and e become well defined local quantities, and the temperature and pressure are defined by the local versions of the above equations of state. On the other hand, one can define a nonequilibrium thermodynamic temperature θ and a nonequilibrium thermodynamic pressure π as the first derivatives of the *nonequilibrium* entropy, namely using Eqs. (3.67) and (3.68) with the replacements $s_0 \to s$, $T \to \theta$, and $p \to \pi$. Of course, θ and π differ from T and p, respectively, and depend on the hydrodynamic gradients (or fluxes). In the case of the USF, the thermodynamic temperature is given by (Brey and Santos, 1992)

$$\begin{aligned} \theta^{-1} &= T^{-1} + \left(\frac{\partial s_{\text{ex}}}{\partial e}\right)_{n,a} \\ &= T^{-1}\left[1 - \frac{2}{3}\left(\frac{\partial \ln \nu}{\partial \ln T}\right)_n a^* \frac{\partial s^*_{\text{ex}}(a^*)}{\partial a^*}\right], \end{aligned} \tag{3.69}$$

where in the last step we have taken into account that s_{ex} depends on e through the collision frequency ν entering in the reduced shear rate $a^* = a/\nu$. Analogously,

$$\begin{aligned} \frac{\pi}{\theta} &= \frac{p}{T} - n^2 \left(\frac{\partial (s_{\text{ex}}/n)}{\partial n}\right)_{e,a} - e\left(\theta^{-1} - T^{-1}\right) \\ &= nk_B\left[1 + a^* \frac{\partial s^*_{\text{ex}}(a^*)}{\partial a^*}\right]. \end{aligned} \tag{3.70}$$

In the BGK description of the thermostatted USF the reduced excess entropy $s^*_{\text{ex}}(a^*)$ is a *universal* function of the shear rate. However, the relationship between the nonequilibrium thermodynamic temperature and its kinetic counterpart is not universal due to the influence of the interaction potential through the dependence of the collision frequency on the temperature. For the sake of concreteness, in the case of interaction potentials of the form $\phi(r) = Kr^{-w}$, the collision frequency is $\nu \propto nT^q$ with $q = \frac{1}{2} - \frac{2}{w}$. Therefore, $\partial \ln n/\partial \ln T = q$. In the special case of Maxwell molecules ($q = 0$), Eq. (3.69) implies that $\theta = T$. This result not only holds in the BGK description but also applies to the Boltzmann equation for Maxwell molecules. In general, however, $\theta < T$ since the entropy is a monotonically decreasing function of the shear rate. At a given value of the reduced shear rate a^*, the harder the potential the smaller the ratio θ/T. It is interesting to remark that the ratio π/θ is a universal function of the shear rate, since the dependence of ν on T is absent in Eq. (3.70). In addition, one has $\pi/p < \theta/T \leq 1$, so π differs from p even in the case of Maxwell molecules.

The knowledge of the series expansion of the excess entropy, Eq. (3.65), allows us to get the series expansion of the thermodynamic quantities:

$$\frac{T}{\theta} = 1 + \frac{2}{3}qa^{*2}\left(1 - a^{*2} + \frac{5}{9}a^{*4} - 153a^{*6}\right) + \mathcal{O}(a^{*10}), \tag{3.71}$$

$$\frac{\pi}{p} = \frac{\theta}{T}\left[1 - a^{*2}\left(1 - a^{*2} + \frac{5}{9}a^{*4} - 153a^{*6}\right) + \mathcal{O}(a^{*10})\right]. \tag{3.72}$$

The possible asymptotic character of the series (3.65) is inherited by the above series. However, in view of the good performance of the Padé approximant (3.66) observed in Fig. 3.23, one can expect that a similar Padé approximant for the series appearing in Eqs. (3.71) and (3.72) may be useful, at least in the range $a^* \leq 1$. Thus, we approximate

$$\frac{T}{\theta} \approx 1 + \frac{2}{3}qa^{*2}\frac{1 - \frac{4}{9}a^{*2}}{1 + \frac{5}{9}a^{*2}}, \tag{3.73}$$

$$\frac{\pi}{p} \approx \frac{\theta}{T}\left(1 - a^{*2}\frac{1 - \frac{4}{9}a^{*2}}{1 + \frac{5}{9}a^{*2}}\right). \tag{3.74}$$

It can be checked that, for $a^* \leq 1$, the approximations (3.73) and (3.74) give practically the same results as those obtained from Eqs. (3.69) and (3.70) by using the Padé approximant (3.66) for the entropy.

In addition to the nonequilibrium thermodynamic temperature defined as $\theta^{-1} = \partial s/\partial e$, one can introduce an *effective* temperature $\widehat{T}$ defined as follows (Casas-Vázquez and Jou, 1994). Since the nonequilibrium entropy $s(n, e)$ is obviously smaller than the equilibrium one $s_0(n, e)$ at the same number density n and internal energy density e, one can ask, which is the value of an (effective) internal energy $\widehat{e}$ of a gas in *equilibrium* such that its entropy coincides with that of the nonequilibrium gas with an internal energy e? More explicitly,

$$s_0(n, \widehat{e}) = s(n, e) = s_0(n, e) + s_{\text{ex}}(n, e). \tag{3.75}$$

Of course, the value of $\widehat{e}$ depends on e, n, and the hydrodynamic gradients (or fluxes), and one must have $\widehat{e} < e$. The effective temperature $\widehat{T}$ is the equilibrium temperature associated with $s_0(n, \widehat{e})$, i.e. $\widehat{T}^{-1} = \partial s_0/\partial e|_{e=\widehat{e}}$. In the case of a dilute gas subjected to USF, the relationship between $\widehat{T}$ and T is

$$\frac{T}{\widehat{T}} = \exp\left[-\frac{2}{3}s_{\text{ex}}^*(a^*)\right], \tag{3.76}$$

where use has been made of Eq. (1.49). According to Casas–Vázquez and Jou (1994), the ratio $\widehat{T}/T$ represents the fraction of internal energy which actually contributes to the molecular disorder. In the BGK model the ratio $T/\widehat{T}$ is a universal function of the reduced shear rate. The series expansion (3.65) leads to the series

$$\frac{T}{\widehat{T}} = 1 + \frac{1}{3}a^{*2} - \frac{1}{9}a^{*4} + \frac{1}{81}a^{*6} - \frac{3092}{243}a^{*8} + \mathcal{O}(a^{*10}). \tag{3.77}$$

Comparison between Eqs. (3.71) and (3.77) shows that $\widehat{T} < \theta \leq T$ for any value of the interaction parameter q. For hard spheres ($q = \frac{1}{2}$), the difference between $\widehat{T}$ and θ is of order a^{*4}. The series (3.77) can be approximated by

$$\frac{T}{\widehat{T}} \simeq 1 + \frac{1}{3}a^{*2}\frac{1 - \frac{2}{9}a^{*2}}{1 + \frac{1}{9}a^{*2}}. \tag{3.78}$$

Again, this approximation yields results for $a^* \leq 1$ hardly distinguishable from those obtained by inserting the approximant (3.66) into Eq. (3.76).

Figure 3.24 shows the shear-rate dependence of the ratios θ/T, $\widehat{T}/T$, and π/p, both for Maxwell molecules and hard spheres. We observe that in the case of hard spheres the thermodynamic quantities θ and π differ from their kinetic counterparts T and p more than in the case of Maxwell molecules. The departure from equilibrium as measured by the ratios θ/T and π/p is more significant in the case of the latter quantity than in that of the former one. As a matter of fact, for Maxwell molecules $\theta = T$ while π/p is appreciably different from 1. Finally, it is interesting to note that the (universal) effective temperature $\widehat{T}$ practically coincides with the thermodynamic temperature θ in the case of hard spheres for $a^* \lesssim 0.4$.

The nonequilibrium thermodynamic temperature and pressure for a two-dimensional fluid of particles interacting with an r^{-12}-potential undergoing USF have been computed in molecular dynamics simulations (Evans, 1989). At a qualitative level, the behavior of θ/T and π/p agrees with the one analyzed here from the BGK equation. In particular, the thermodynamic quantities θ and π are smaller than T and p, respectively, the discrepancy being bigger in the case of the pressure. A more quantitative comparison cannot be carried out

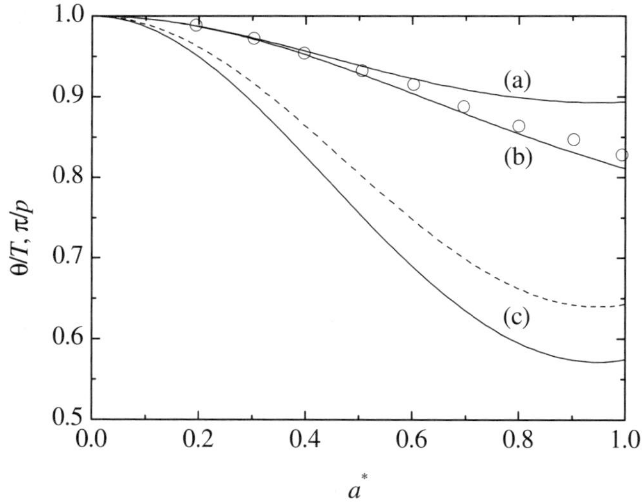

Figure 3.24. Shear-rate dependence of the ratios (a) θ/T, (b) $\widehat{T}/T$, and (c) π/p for Maxwell molecules (dashed line) and hard spheres (solid lines), as given by the Padé approximants (3.73), (3.74), and (3.78). Note that $\theta/T = 1$ for Maxwell molecules and that $\widehat{T}/T$ is the same for both potentials. The circles correspond to the values of $\widehat{T}/T$ as obtained from Eq. (3.76) by using the numerical values of the excess entropy.

because the simulations consider a system in two dimensions and, moreover, the densities are not small enough to allow for a description based on the Boltzmann equation.

It is worthwhile remarking that the thermodynamic quantities have been defined by Eqs. (3.69) and (3.70) by differentiating the entropy at constant shear rate a, but not at constant momentum flux (i.e. the irreversible part $\Pi_{ij} = P_{ij} - p\delta_{ij}$ of the pressure tensor). In fact, it is not possible to differentiate s with respect to e at constant Π_{ij}. While in the USF there is only one hydrodynamic gradient (the shear rate a), there are two independent fluxes, namely the shear stress P_{xy} and the normal stress difference $P_{xx} - P_{yy}$.[5] As a consequence, it is not possible to change $e \to e + de$ and $a \to a + da$ by fixing both P_{xy} and $P_{xx} - P_{yy}$. Let us see this point more specifically. According to Eq. (2.54),

$$P_{xy} = -\frac{2}{3} e a^* F_\eta(a^*). \tag{3.79}$$

[5]Strictly speaking, there is an additional third flux given by $P_{yy} - P_{zz}$. However, it vanishes in the BGK description, i.e. $\Psi_2(a^*) = 0$.

Thus, if we change the energy $e \to e + de$, the reduced shear rate must change as $a^* \to a^* + da^*$ with

$$\frac{da^*}{de} = -\left[e\frac{\partial \ln (a^* F_\eta)}{\partial a^*}\right]^{-1} \tag{3.80}$$

in order to keep P_{xy} fixed. On the other hand, if we keep $P_{xx} - P_{yy}$ constant, then da^*/de is given by Eq. (3.80), except that $a^* F_\eta(a^*)$ must be replaced by $a^{*2}\Psi_1(a^*)$. In general, both relationships are not compatible and cannot be satisfied simultaneously.

6. Small perturbations from the thermostatted state

Thus far, we have offered an extensive account of several physical properties of dilute gases under USF. In all the analyses (including DSMC simulations) the system has been assumed to be *homogeneous* in the local Lagrangian frame moving with the flow velocity $u_i(\mathbf{r}) = a_{ij} r_j$, $a_{ij} = a\delta_{ix}\delta_{jy}$ [cf. Eq. (2.5)]. In general, however, the USF state can be disturbed by small spatial perturbations. The response of the system to these perturbations gives rise to additional contributions to the fluxes, which can be characterized by generalized (shear-rate dependent) transport coefficients (Marchetti and Dufty, 1982). This section is devoted to the study of such small perturbations.

In order to analyze the above problem, we need to start from the kinetic equation with a general time and space dependence. It is convenient to keep using the relative velocity $\mathbf{V} = \mathbf{v} - \mathbf{u}_0$, where $\mathbf{u}_0(\mathbf{r}) = \mathsf{a} \cdot \mathbf{r}$ is the flow velocity in the *undisturbed* USF state. In the disturbed state, however, the true flow velocity $\mathbf{u}(\mathbf{r}, t)$ is in general different from $\mathbf{u}_0$, i.e. $\mathbf{u} = \mathbf{u}_0 + \delta\mathbf{u}$, so that the true peculiar velocity is $\mathbf{c} \equiv \mathbf{v} - \mathbf{u} = \mathbf{V} - \delta\mathbf{u}$. In the reference frame moving with velocity $\mathbf{u}_0$, the BGK kinetic equation (1.165) can be written as

$$\left[\frac{\partial}{\partial t} + \nu - 3\alpha + \Lambda + (\mathbf{V} + \mathbf{u}_0) \cdot \nabla + \alpha\delta\mathbf{u} \cdot \frac{\partial}{\partial \mathbf{V}}\right] f = \nu f_L, \tag{3.81}$$

where the operator Λ is defined in Eq. (3.5). In this equation, the thermostat force has been assumed to be proportional to the actual peculiar velocity, $\mathbf{F} = -m\alpha(\mathbf{V} - \delta\mathbf{u})$. It must be noted that now the collision frequency ν and the thermostat parameter α are in general functions of $\mathbf{r}$ and t through their functional dependence on $n(\mathbf{r}, t)$ and $T(\mathbf{r}, t)$. For repulsive potentials proportional to r^{-w}, one has $\nu(\mathbf{r}, t) \propto n(\mathbf{r}, t)[T(\mathbf{r}, t)]^{1/2-2/w}$. As for α, its generalization to the inhomogeneous case is essentially a matter of choice. For simplicity, it is convenient to take α as the local version of Eq. (2.46), i.e.

$$\alpha(\mathbf{r}, t) = \nu(\mathbf{r}, t)\frac{2}{3}\sinh^2\left\{\frac{1}{6}\cosh^{-1}\left[1 + 9\left(\frac{a}{\nu(\mathbf{r}, t)}\right)^2\right]\right\}. \tag{3.82}$$

The macroscopic balance equations are obtained by taking moments in Eq. (3.81):

$$\frac{\partial n}{\partial t} + \mathbf{u}_0 \cdot \nabla n + \nabla \cdot (n\delta\mathbf{u}) = 0, \tag{3.83}$$

$$\frac{\partial \delta\mathbf{u}}{\partial t} + \mathsf{a} \cdot \delta\mathbf{u} + (\mathbf{u}_0 + \delta\mathbf{u}) \cdot \nabla\delta\mathbf{u} + \frac{1}{mn}\nabla \cdot \mathsf{P} = 0, \tag{3.84}$$

$$\frac{3}{2}nk_B\frac{\partial T}{\partial t} + aP_{xy} + \frac{3}{2}nk_B(\mathbf{u}_0+\delta\mathbf{u})\cdot\nabla T + \nabla\cdot\mathbf{q} + \mathsf{P} : \nabla\delta\mathbf{u} + 3\alpha p = 0, \tag{3.85}$$

where P is the pressure tensor and $\mathbf{q}$ is the heat flux, which are defined by Eqs. (1.40) and (1.45), respectively, with $\mathbf{V} \to \mathbf{c}$.

Now we assume that the deviations from the USF state are small, which means that the spatial gradients of the hydrodynamic fields

$$A(\mathbf{r}, t) \equiv \{n(\mathbf{r}, t), T(\mathbf{r}, t), \delta\mathbf{u}(\mathbf{r}, t)\} \tag{3.86}$$

are small. Thus, we look for a normal solution to Eq. (3.81) by means of a generalization of the conventional Chapman–Enskog method outlined in Section 1.4, where now the reference state is not the local equilibrium one, but a *local* version of the USF distribution (3.8). More specifically, we write (Lee and Dufty, 1997)

$$f(A(\mathbf{r}, t), \mathbf{V}) = f^{(0)}(A(\mathbf{r}, t), \mathbf{V}) + \varepsilon f^{(1)}(A(\mathbf{r}, t), \mathbf{V}) + \cdots, \tag{3.87}$$

where ε is a formal uniformity parameter measuring the spatial gradients of the fields $A(\mathbf{r}, t)$. Upon writing Eq. (3.87), it has been indicated that all the space and time dependence of f occurs entirely through a functional dependence on the hydrodynamic fields, i.e. $f(\mathbf{r}, \mathbf{V}, t) = f(A(\mathbf{r}, t), \mathbf{V})$. The (reference) zeroth-order distribution is given by Eqs. (3.8) and (3.9), where it is understood that $n \to n(\mathbf{r}, t)$, $T \to T(\mathbf{r}, t)$, $\alpha^* \to \alpha(\mathbf{r}, t)/\nu(\mathbf{r}, t)$, $a^* \to a/\nu(\mathbf{r}, t)$, and $\boldsymbol{\xi} \to [m/2k_BT(\mathbf{r}, t)]^{1/2}[\mathbf{V} - \delta\mathbf{u}(\mathbf{r}, t)]$. Consequently, the zeroth-order heat flux $\mathbf{q}^{(0)}$ vanishes, while the zeroth-order pressure tensor, $\mathsf{P}^{(0)}$ is given by Eqs. (2.51)–(2.53), i.e.

$$P_{xx}^{(0)} = 3p - 2P_{yy}^{(0)}, \tag{3.88}$$

$$P_{yy}^{(0)} = P_{zz}^{(0)} = \frac{p}{1 + 2\alpha/\nu}, \tag{3.89}$$

$$P_{xy}^{(0)} = -p\frac{a/\nu}{[1 + 2\alpha/\nu]^2}, \tag{3.90}$$

$$P_{xz}^{(0)} = P_{yz}^{(0)} = 0. \tag{3.91}$$

In the same way as in the conventional Chapman–Enskog method, the time derivative is expanded in powers of the uniformity parameter as

$$\frac{\partial}{\partial t} = \partial_t^{(0)} + \varepsilon\partial_t^{(1)} + \cdots, \tag{3.92}$$

where the action of each operator $\partial_t^{(k)}$ is obtained from the hydrodynamic equations (3.83)–(3.85) by formally replacing the nabla operator as $\nabla \to \varepsilon\nabla$ and expanding the fluxes. In particular, to zeroth order, one gets

$$\partial_t^{(0)} n = 0, \tag{3.93}$$

$$\partial_t^{(0)} \delta\mathbf{u} + \mathsf{a} \cdot \delta\mathbf{u} = 0, \tag{3.94}$$

$$\partial_t^{(0)} T = 0, \tag{3.95}$$

where in the last equation we have taken into account that $aP_{xy}^{(0)} + 3\alpha p = 0$. The balance equations to first order are

$$\partial_t^{(1)} n + \mathbf{u}_0 \cdot \nabla n + \nabla \cdot (n\delta\mathbf{u}) = 0, \tag{3.96}$$

$$\partial_t^{(1)} \delta\mathbf{u} + (\mathbf{u}_0 + \delta\mathbf{u}) \cdot \nabla\delta\mathbf{u} + \rho^{-1}\nabla \cdot \mathsf{P}^{(0)} = 0, \tag{3.97}$$

$$\frac{3}{2} nk_B \partial_t^{(1)} T + aP_{xy}^{(1)} + \frac{3}{2} nk_B (\mathbf{u}_0 + \delta\mathbf{u}) \cdot \nabla T + \mathsf{P}^{(0)} : \nabla\delta\mathbf{u} = 0. \tag{3.98}$$

In Eq. (3.97) $\rho = mn$ is the mass density. Inserting Eqs. (3.87) and (3.92) into Eq. (3.81), where again the nabla operator is understood to be multiplied by ε, one gets the kinetic equations for the different contributions $f^{(k)}$. As expected, the equation for $f^{(0)}$ is identically satisfied. The equation for $f^{(1)}$ is

$$\left(\partial_t^{(0)} + \nu - 3\alpha + \Lambda + \alpha\delta\mathbf{u} \cdot \frac{\partial}{\partial \mathbf{V}} \right) f^{(1)} = -\left[\partial_t^{(1)} + (\mathbf{V} + \mathbf{u}_0) \cdot \nabla \right] f^{(0)}. \tag{3.99}$$

Since $f^{(1)}$ is linear in the hydrodynamic gradients, it must be of the form

$$f^{(1)} = \mathbf{X}_n \cdot \nabla n + \mathbf{X}_T \cdot \nabla T + \mathsf{X}_u : \nabla\delta\mathbf{u}, \tag{3.100}$$

where the vectors $\mathbf{X}_n$ and $\mathbf{X}_T$, and the tensor X_u are functions of the true peculiar velocity $\mathbf{c} = \mathbf{V} - \delta\mathbf{u}$, to be determined from Eq. (3.99). According to Eq. (3.100), the first-order corrections to the fluxes are

$$q_i^{(1)} = -\kappa_{ij} \frac{\partial T}{\partial r_j} - \mu_{ij} \frac{\partial n}{\partial r_j}, \tag{3.101}$$

$$P_{ij}^{(1)} = -\eta_{ijkl} \frac{\partial \delta u_k}{\partial r_l}, \tag{3.102}$$

where

$$\kappa_{ij} = -\frac{m}{2} \int d\mathbf{c}\, c^2 c_i X_{T,j}, \tag{3.103}$$

$$\mu_{ij} = -\frac{m}{2} \int d\mathbf{c}\, c^2 c_i X_{n,j}, \tag{3.104}$$

$$\eta_{ijkl} = -m \int d\mathbf{c}\, c_i c_j X_{u,kl} \tag{3.105}$$

are *generalized* transport coefficients. Note that $\eta_{ijkl} = \eta_{jikl} \neq \eta_{ijlk}$. Obviously, in the limit of zero shear rate, $\kappa_{ij} \to \kappa_0 \delta_{ij}$, $\mu_{ij} \to 0$ and $\eta_{ijkl} \to \eta_0(\delta_{ik}\delta_{jl} + \delta_{jk}\delta_{il} - \frac{2}{3}\delta_{ij}\delta_{kl})$ and we recover the usual Navier–Stokes transport coefficients [cf. Eqs. (1.103) and (1.108)]. On the other hand, κ_{xy}, κ_{yx}, μ_{xy}, and μ_{yx} are generalizations of Burnett coefficients that, for small shear rates, behave as (Garzó, 1988) $\kappa_{xy} \to -\frac{2}{5}(7-q)\kappa_0 a/\nu$, $\kappa_{yx} \to -\frac{2}{5}(2-q)\kappa_0 a/\nu$, $\mu_{xy} = \mu_{yx} \to \frac{2}{5}(T/n)\kappa_0 a/\nu$, where $q = \frac{1}{2} - \frac{2}{w}$, w being the potential parameter. Because of symmetry reasons, the off-diagonal elements xz, zx, yz, and zy of the tensors κ_{ij} and μ_{ij} are identically zero. This implies that if $\nabla T \parallel \widehat{\mathbf{z}}$, then $\mathbf{q}^{(1)} \parallel \widehat{\mathbf{z}}$, while if $\nabla T \perp \widehat{\mathbf{z}}$, then $\mathbf{q}^{(1)} \perp \widehat{\mathbf{z}}$. Analogously, many of the elements of η_{ijkl} are zero. For instance, if the only nonzero velocity gradient is $\partial \delta u_x/\partial z$, then $P_{ij}^{(1)} = P_{xz}^{(1)}(\delta_{ix}\delta_{jz} + \delta_{jx}\delta_{iz})$.

The right-hand side of Eq. (3.99) can be evaluated by using Eqs. (3.96)–(3.98) as

$$\left[\partial_t^{(1)} + (\mathbf{V} + \mathbf{u}_0)\cdot\nabla\right] f^{(0)} = \mathbf{Y}_n \cdot \nabla n + \mathbf{Y}_T \cdot \nabla T + \mathsf{Y}_u : \nabla \delta\mathbf{u}, \tag{3.106}$$

where

$$\mathbf{Y}_n = \frac{\partial f^{(0)}}{\partial n}\mathbf{c} - \rho^{-1}\frac{\partial f^{(0)}}{\partial \delta\mathbf{u}} \cdot \frac{\partial \mathsf{P}^{(0)}}{\partial n}, \tag{3.107}$$

$$\mathbf{Y}_T = \frac{\partial f^{(0)}}{\partial T}\mathbf{c} - \rho^{-1}\frac{\partial f^{(0)}}{\partial \delta\mathbf{u}} \cdot \frac{\partial \mathsf{P}^{(0)}}{\partial T}, \tag{3.108}$$

$$Y_{u,kl} = -n\frac{\partial f^{(0)}}{\partial n}\delta_{kl} - \frac{2}{3nk_B}\frac{\partial f^{(0)}}{\partial T}\left(P_{kl}^{(0)} - a\eta_{xykl}\right) + \frac{\partial f^{(0)}}{\partial \delta u_k}c_l. \tag{3.109}$$

Note that, due to the presence of $P_{xy}^{(1)}$ in Eq. (3.98), the unknown coefficients η_{xykl} appear in Eq. (3.109). The explicit knowledge of the reference state $f^{(0)}$ allows one to evaluate all the derivatives involved in Eqs. (3.107)–(3.109). The reader is referred to the paper by Lee and Dufty (1997) for specific results. Substitution of Eqs. (3.100) and (3.106) into Eq. (3.99) yields

$$\left(\nu - 3\alpha + \Lambda'\right)\mathbf{X}_n = -\mathbf{Y}_n, \tag{3.110}$$

$$\left(\nu - 3\alpha + \Lambda'\right)\mathbf{X}_T = -\mathbf{Y}_T, \tag{3.111}$$

$$\left(\nu - 3\alpha + \Lambda'\right)X_{u,kl} - aX_{u,xl}\delta_{ky} = -Y_{u,kl}, \tag{3.112}$$

where the operator Λ' is

$$\begin{aligned}\Lambda' &= \Lambda + a_{ij}\delta u_j \frac{\partial}{\partial c_i} + \alpha\delta\mathbf{u}\cdot\frac{\partial}{\partial \mathbf{c}} \\ &= -a_{ij}c_j\frac{\partial}{\partial c_i} - \alpha\mathbf{c}\cdot\frac{\partial}{\partial \mathbf{c}}.\end{aligned} \tag{3.113}$$

Upon writing Eqs. (3.110)–(3.112), we have made use of the property

$$\partial_t^{(0)} X = \frac{\partial X}{\partial \delta u_i} \partial_t^{(0)} \delta u_i = a_{ij} \delta u_j \frac{\partial X}{\partial c_i}, \tag{3.114}$$

where in the last step we made used Eq. (3.94) and the fact that X depends on $\delta \mathbf{u}$ through $\mathbf{c} = \mathbf{V} - \delta \mathbf{u}$. The solutions to Eqs. (3.110)–(3.112) are straightforward. For instance,

$$X_{u,kl} = -\int_0^\infty d\tau \, e^{-(\nu-3\alpha)\tau} e^{-\Lambda'\tau} \left(Y_{u,kl} - a X_{u,xl} \delta_{ky}\right). \tag{3.115}$$

The action of the operator $\exp(-\Lambda'\tau)$ is given by an equation analogous to Eq. (3.7) with $\mathbf{V} \to \mathbf{c}$. To close the evaluation of the functions $X_{u,kl}$, we need to get the coefficients η_{xykl}. This is done by inserting Eq. (3.115) into Eq. (3.105). As an illustration, let us consider the coefficient η_{xyxy}:

$$\begin{aligned} \eta_{xyxy} &= m \int_0^\infty d\tau \, e^{-(\nu+2\alpha)\tau} \int d\mathbf{c} \, (c_x - a\tau c_y) c_y \\ &\times \left[-\frac{2}{3nk_B} \frac{\partial f^{(0)}}{\partial T} \left(P_{xy}^{(0)} - a\eta_{xyxy} \right) - \frac{\partial f^{(0)}}{\partial c_x} c_y \right], \end{aligned} \tag{3.116}$$

where we have used the identity

$$\int d\mathbf{c} \, X_1(\mathbf{c}) e^{-\Lambda'\tau} X_2(\mathbf{c}) = e^{-3\alpha\tau} \int d\mathbf{c} \, X_2(\mathbf{c}) e^{\Lambda'\tau} X_1(\mathbf{c}). \tag{3.117}$$

By performing the integrals over $\mathbf{c}$ and over τ, one simply gets

$$\begin{aligned} \eta_{xyxy} &= -\frac{2}{3nk_B} \left(P_{xy}^{(0)} - a\eta_{xyxy} \right) \left[(\nu + 2\alpha)^{-1} \frac{\partial P_{xy}^{(0)}}{\partial T} \right. \\ & \left. -a(\nu + 2\alpha)^{-2} \frac{\partial P_{yy}^{(0)}}{\partial T} \right] + (\nu + 2\alpha)^{-1} P_{yy}^{(0)}. \end{aligned} \tag{3.118}$$

In general, $P_{ij}^{(0)}$ [cf. Eqs. (3.88)–(3.90)] depends on T through its dependence on p and ν. In the particular case of Maxwell molecules ν is independent of T, and so $\partial P_{ij}^{(0)}/\partial T = P_{ij}^{(0)}/T$. As a consequence, the solution of Eq. (3.118) is

$$\eta_{xyxy} = p \frac{\nu(\nu - 2\alpha)}{(\nu + 2\alpha)^2 (\nu + 6\alpha)}, \tag{3.119}$$

where we have used the relation $a^2 = 3(\alpha/\nu)(\nu + 2\alpha)^2$. The remaining coefficients η_{ijkl}, κ_{ij}, and μ_{ij} can be obtained in a similar way.

Let us reobtain Eq. (3.119) from a completely independent route. We consider the particular case $\delta\mathbf{u} = (\delta a)y\widehat{\mathbf{x}}$. Therefore, the shear stress is $P_{xy}^{(1)} = -\eta_{xyxy}\delta a$ and so

$$\begin{aligned} P_{xy} &= P_{xy}^{(0)} + P_{xy}^{(1)} \\ &= -\eta_0 \left[F_\eta(a)a + \frac{\eta_{xyxy}}{\eta_0}\delta a \right]. \end{aligned} \tag{3.120}$$

On the other hand, the above situation is equivalent to an unperturbed USF with a shear rate $a + \delta a$, so that

$$\begin{aligned} P_{xy} &= -\eta_0 F_\eta(a+\delta a)(a+\delta a) \\ &= -\eta_0 \left\{ F_\eta(a)a + \left[F_\eta(a) + a\frac{\partial F_\eta(a)}{\partial a} \right] \delta a \right\} + \mathcal{O}(\delta a^2). \end{aligned} \tag{3.121}$$

Comparison between Eqs. (3.120) and (3.121) implies that

$$\eta_{xyxy} = \eta_0 \left[F_\eta(a) + a\frac{\partial F_\eta(a)}{\partial a} \right]. \tag{3.122}$$

It is easy to check from Eq. (2.57) that Eqs. (3.119) and (3.122) are indeed equivalent. By using a similar argument we can get the coefficient η_{zyzy}. In this case, it is convenient to assume that the only perturbation is $\delta\mathbf{u} = (\delta a)y\widehat{\mathbf{z}}$. This is equivalent to an unperturbed USF state in which the flow velocity points in a direction x' lying on the plane xz and making an angle $\varphi = \tan^{-1}\delta a/a$ with the x-direction, and where the actual shear rate is $a' = \sqrt{a^2 + (\delta a)^2}$. Consequently,

$$P_{zy} = \sin\varphi P_{x'y} = -\sin\varphi\eta_0 F_\eta(a')a'. \tag{3.123}$$

The above result holds for arbitrary δa. On the other hand, if $\delta a \ll a$, then $a' \approx a$ and $\sin\phi \approx (\delta a)/a$, so that

$$P_{zy} = -\eta_0 F_\eta(a)\delta a + \mathcal{O}(\delta a^3). \tag{3.124}$$

This equation implies that

$$\eta_{zyzy} = \eta_0 F_\eta. \tag{3.125}$$

Analogously, one can obtain $\eta_{xzxz} = \eta_0 F_\eta$.

Figure 3.25 shows the shear-rate dependence of the coefficients η_{xyxy} and $\eta_{zyzy} = \eta_{xzxz}$. We observe that η_{xyxy}/η_0 presents a shear-thinning effect more pronounced than that of F_η, as expected from Eq. (3.122).

With all the shear-rate dependent generalized transport coefficients known, the new constitutive equations are completed and the corresponding set of closed

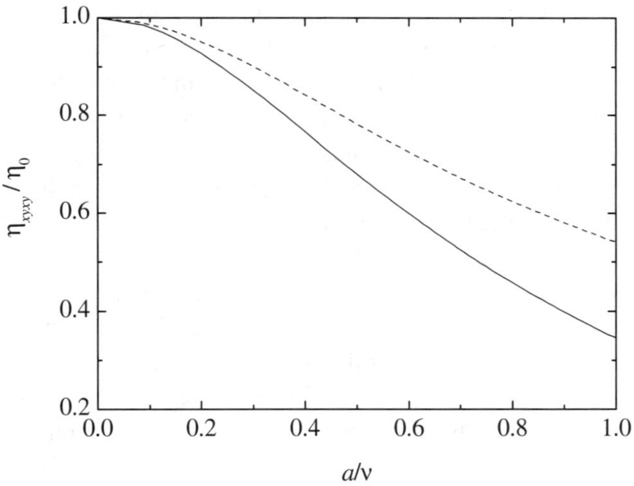

Figure 3.25. Transport coefficient η_{xyxy}/η_0 as a function of the shear rate (solid line). The viscosity function F_η, which coincides with $\eta_{zyzy}/\eta_0 = \eta_{xzxz}/\eta_0$ is also plotted (dashed line).

hydrodynamic equations can be derived. They are obtained from Eqs. (3.83)–(3.85) as

$$\frac{\partial n}{\partial t} + \mathbf{u}_0 \cdot \nabla n + \nabla \cdot (n \delta \mathbf{u}) = 0, \tag{3.126}$$

$$\frac{\partial \delta u_i}{\partial t} + a_{ij}\delta u_j + (\mathbf{u}_0 + \delta \mathbf{u}) \cdot \nabla \delta u_i + \frac{1}{mn}\frac{\partial}{\partial r_j}\left(P_{ij}^{(0)} - \eta_{ijkl}\frac{\partial \delta u_k}{\partial r_l}\right) = 0, \tag{3.127}$$

$$\frac{3}{2}nk_B\frac{\partial T}{\partial t} - a\eta_{xykl}\frac{\partial \delta u_k}{\partial r_l} + \frac{3}{2}nk_B(\mathbf{u}_0 + \delta \mathbf{u}) \cdot \nabla T$$
$$-\frac{\partial}{\partial r_k}\left(\kappa_{kl}\frac{\partial T}{\partial r_l} + \mu_{kl}\frac{\partial n}{\partial r_l}\right) + \left(P_{ij}^{(0)} - \eta_{ijkl}\frac{\partial \delta u_k}{\partial r_l}\right)\frac{\partial \delta u_i}{\partial r_j} = 0. \tag{3.128}$$

These equations are valid to second order in spatial gradients relative to the reference USF state, but there is no restriction on the value of the shear rate. The hydrodynamic equations (3.126)–(3.128) will be the starting point of the analysis on the stability of the USF in Section 3.8.

7. Heat transport under uniform shear flow

The results of the previous Section give the time evolution of the hydrodynamic fields when they are weakly perturbed from their values in the USF state. In particular, the heat flux induced in the system by density and temperature gradients is given by Eq. (3.101). In this Section, however, we will be interested in a perturbed *steady* state (Garzó, 1995a). Steady states are in general desirable for practical purposes, especially in computer simulations (Daivis and Evans, 1993). It is important to note that the fact that $\nabla T \neq 0$ and/or $\nabla n \neq 0$ implies that $\delta \mathbf{u} \neq 0$, in contrast to what occurs at equilibrium, where a thermal gradient can exist in the absence of convection. This is due to the term $\nabla \cdot \mathsf{P}$ in the momentum balance equation [cf. Eq. (3.84)], which is different from zero, even if the hydrostatic pressure is uniform, because of the dependence of P/p on the reduced shear rate $a^* = a/\nu(\mathbf{r})$. Thus, in order to achieve a steady state with $\delta \mathbf{u} = 0$, $\nabla p = 0$ and $\nabla T \neq 0$, we need to add an external field $m\mathbf{g}(\mathbf{r})$ that exactly compensates for the increase or decrease of momentum due to $\nabla \cdot \mathsf{P}$. This resembles the situation of a fluid in equilibrium under a gravity field, where the field cancels the term ∇p in the momentum balance equation. The addition of this external field does not modify the mass and energy balance equations (3.83) and (3.85). On the other hand, a term $\mathbf{g}$ appears then on the right-hand side of the momentum equation (3.84). Consequently, in the *steady* state with $\delta \mathbf{u} = 0$, Eqs. (3.83)–(3.85) become

$$\mathbf{u}_0 \cdot \nabla n = 0, \tag{3.129}$$

$$\frac{1}{mn} \nabla \cdot \mathsf{P} = \mathbf{g}, \tag{3.130}$$

$$a P_{xy} + \frac{3}{2} n k_B \mathbf{u}_0 \cdot \nabla T + \nabla \cdot \mathbf{q} + 3\alpha p = 0. \tag{3.131}$$

Equation (3.129) implies that $\nabla n \perp \mathbf{u}_0$, i.e. $\partial n/\partial x = 0$. Since we will assume that $\nabla p = 0$, then we also have $\partial T/\partial x = 0$, so the thermal gradient must be orthogonal to the direction of the flow velocity and $\mathbf{u}_0 \cdot \nabla T = 0$ in Eq. (3.131). The constraint $\partial T/\partial x = 0$ is a requirement imposed by the consistency of the perturbed state we are considering. Note that, due to the anisotropy of the USF, $\mathbf{g}$ and ∇T are in general not parallel.

7.1 Boltzmann description for Maxwell molecules

Under the above conditions, the stationary Boltzmann equation can be written as

$$\left(-3\alpha + \Lambda + \mathbf{V} \cdot \nabla + \mathbf{g} \cdot \frac{\partial}{\partial \mathbf{V}} \right) f = J[f, f]. \tag{3.132}$$

Equation (3.132) can be obtained from the Boltzmann equation (1.25) by following steps similar to those employed to get Eq. (3.81) from the BGK equation. In contrast to the results derived in the previous Section, here we will deal with the Boltzmann equation and not with a kinetic model. The price to be paid is the restriction to Maxwell molecules, in which case the USF state is well determined to the level of second- and fourth-degree moments (cf. Chapter 2). Now we assume that $\nabla T \sim \varepsilon$ is a small parameter. The goal is to solve Eq. (3.132) by means of a perturbation expansion of the form (3.87) around the steady USF. Similarly, the external field $\mathbf{g}$ must be also expanded, with $\mathbf{g}^{(0)} = 0$ and

$$\begin{aligned} \mathbf{g}^{(1)} &= \frac{1}{mn} \nabla \cdot \mathsf{P}^{(0)} \\ &= \frac{k_B T}{m} \frac{\partial \mathsf{P}^*}{\partial T} \cdot \nabla T \\ &= \frac{k_B a^*}{m} \frac{\partial \mathsf{P}^*}{\partial a^*} \cdot \nabla T, \end{aligned} \tag{3.133}$$

where $\mathsf{P}^* \equiv \mathsf{P}^{(0)}/p$ and in the last step we have taken into account that $\partial a^*/\partial T = a^*/T$ since $\nu \propto p/T$. The explicit expressions of the elements of $\mathbf{g}^{(1)}$ are

$$g_x^{(1)} = -\frac{a^*(1-2\alpha^*)}{(1+2\alpha^*)^2(1+6\alpha^*)} \frac{k_B}{m} \frac{\partial T}{\partial y}, \tag{3.134}$$

$$g_y^{(1)} = -\frac{4\alpha^*}{(1+2\alpha^*)(1+6\alpha^*)} \frac{k_B}{m} \frac{\partial T}{\partial y}, \tag{3.135}$$

$$g_z^{(1)} = -\frac{4\alpha^*}{(1+2\alpha^*)(1+6\alpha^*)} \frac{k_B}{m} \frac{\partial T}{\partial z}, \tag{3.136}$$

where $\alpha^* = \alpha/\nu$.

The kinetic equation for $f^{(1)}$ is

$$\left(-3\alpha + \Lambda + \mathcal{I}\right) f^{(1)} = -\left(\mathbf{V} \cdot \nabla + \mathbf{g}^{(1)} \cdot \frac{\partial}{\partial \mathbf{V}}\right) f^{(0)}. \tag{3.137}$$

Here, $\mathcal{I}$ is the linear operator defined as

$$\mathcal{I} f^{(1)} = -J[f^{(0)}, f^{(1)}] - J[f^{(1)}, f^{(0)}]. \tag{3.138}$$

Equation (3.137) is a linear integro-differential equation for the unknown $f^{(1)}$, provided that $f^{(0)}$ is known. In the BGK version of Eq. (3.137) the operator $\mathcal{I}$ is simply replaced by the effective collision frequency ν and one then has an equation similar to Eq. (3.99). Since the distribution function $f^{(0)}$ is not completely known in the case of the Boltzmann equation, the explicit solution of Eq. (3.137) cannot be obtained. On the other hand, Eq. (3.137) allows one to get the velocity moments of degree k of $f^{(1)}$ in terms of the moments of degree $k+1$ of $f^{(0)}$. The most relevant quantity in our problem is the heat flux $\mathbf{q}^{(1)}$, which requires the knowledge of the fourth-degree moments of $f^{(0)}$. As seen in Section 2.5, those moments are finite for shear rates $a^* < 6.846$. Their expressions are displayed in Appendix 2.B.

In order to get the heat flux, it is convenient to introduce the following set of 10 third-degree moments

$$\boldsymbol{\mathcal{M}}^{(1)} = \begin{pmatrix} M^{(1)}_{2|x} \\ M^{(1)}_{2|y} \\ M^{(1)}_{2|z} \\ M^{(1)}_{0|xxy} \\ M^{(1)}_{0|xxz} \\ M^{(1)}_{0|xyy} \\ M^{(1)}_{0|yyz} \\ M^{(1)}_{0|xzz} \\ M^{(1)}_{0|yzz} \\ M^{(1)}_{0|xyz} \end{pmatrix}, \tag{3.139}$$

where

$$M^{(1)}_{2|i} = \frac{1}{n} \int d\mathbf{V}\, Y_{2|i}(\mathbf{V}) f^{(1)}(\mathbf{V}), \tag{3.140}$$

$$M^{(1)}_{0|ijk} = \frac{1}{n} \int d\mathbf{V}\, Y_{0|ijk}(\mathbf{V}) f^{(1)}(\mathbf{V}), \tag{3.141}$$

and the polynomials $Y_{2|i}(\mathbf{V})$ and $Y_{0|ijk}(\mathbf{V})$are defined by Eqs. (2.71) and (2.72). The moments (3.139) are the first-order correction to the USF moments (2.73). The corresponding collisional moments are [cf. Eqs. (1.A.6) and (1.A.7)]

$$\frac{1}{n} \int d\mathbf{V}\, Y_{2|i}(\mathbf{V}) \mathcal{I} f^{(1)}(\mathbf{V}) = \frac{2}{3} \nu M^{(1)}_{2|i}, \tag{3.142}$$

$$\frac{1}{n} \int d\mathbf{V}\, Y_{0|ijk}(\mathbf{V}) \mathcal{I} f^{(1)}(\mathbf{V}) = \frac{3}{2} \nu M^{(1)}_{0|ijk}. \tag{3.143}$$

Taking moments in Eq. (3.137), we get a set of 10 inhomogeneous linear equations for the moments $\boldsymbol{\mathcal{M}}^{(1)}$. By using matrix notation, it can be written as

$$\mathcal{A}_{\sigma\sigma'} \mathcal{M}^{(1)}_{\sigma'} = \mathcal{N}_{\sigma}, \tag{3.144}$$

where $\boldsymbol{\mathcal{A}}$ is the matrix

$$\boldsymbol{\mathcal{A}} = \begin{pmatrix} c_1 & \frac{7}{5}a & 0 & 2a & 0 & 0 & 0 & 0 & 0 & 0 \\ \frac{2}{5}a & c_1 & 0 & 0 & 0 & 2a & 0 & 0 & 0 & 0 \\ 0 & 0 & c_1 & 0 & 0 & 0 & 0 & 0 & 0 & 2a \\ \frac{8}{25}a & 0 & 0 & c_2 & 0 & \frac{8}{5}a & 0 & 0 & 0 & 0 \\ 0 & 0 & 0 & 0 & c_2 & 0 & 0 & 0 & 0 & \frac{8}{5}a \\ 0 & \frac{8}{25}a & 0 & -\frac{7}{5}a & 0 & c_2 & 0 & 0 & -a & 0 \\ 0 & 0 & 0 & 0 & 0 & 0 & c_2 & 0 & 0 & -\frac{2}{5}a \\ 0 & -\frac{2}{25}a & 0 & -\frac{2}{5}a & 0 & 0 & 0 & c_2 & a & 0 \\ -\frac{2}{25}a & 0 & 0 & 0 & 0 & -\frac{2}{5}a & 0 & 0 & c_2 & 0 \\ 0 & 0 & \frac{1}{5}a & 0 & 0 & 0 & a & 0 & 0 & c_2 \end{pmatrix} \tag{3.145}$$

with $c_1 \equiv \frac{2}{3}\nu + 3\alpha$ and $c_2 \equiv \frac{3}{2}\nu + 3\alpha$, and $\boldsymbol{\mathcal{N}}$ is the column matrix of elements

$$\mathcal{N}_\sigma = -\frac{1}{n}\nabla \cdot \int d\mathbf{V}\, Y_\sigma(\mathbf{V})\mathbf{V} f^{(0)}(\mathbf{V}) + \frac{1}{n}\mathbf{g}^{(1)} \cdot \int d\mathbf{V} \left[\frac{\partial}{\partial \mathbf{V}} Y_\sigma(\mathbf{V})\right] f^{(0)}(\mathbf{V}). \tag{3.146}$$

Here, $Y_\sigma(\mathbf{V})$ is the polynomial associated with the moment $M_\sigma^{(1)}$. The velocity integrals in the above equation can be expressed in terms of moments of $f^{(0)}$. Simple algebra leads to

$$\int d\mathbf{V}\, Y_{2|i}(\mathbf{V}) V_j f^{(0)}(\mathbf{V}) = n\left(M_{2|ij}^{(0)} + \frac{1}{3}M_{4|0}^{(0)}\delta_{ij}\right), \tag{3.147}$$

$$\begin{aligned} \int d\mathbf{V}\, Y_{0|ijk}(\mathbf{V}) V_l f^{(0)}(\mathbf{V}) \quad = \quad & n\bigg[\frac{1}{7}\left(M_{2|ij}^{(0)}\delta_{kl} + M_{2|ik}^{(0)}\delta_{jl} + M_{2|jk}^{(0)}\delta_{il}\right) \\ & -\frac{2}{35}\left(M_{2|il}^{(0)}\delta_{jk} + M_{2|jl}^{(0)}\delta_{ik} + M_{2|kl}^{(0)}\delta_{ij}\right) \\ & + M_{0|ijkl}^{(0)}\bigg], \end{aligned} \tag{3.148}$$

$$m\int d\mathbf{V} \left[\frac{\partial}{\partial V_j} Y_{2|i}(\mathbf{V})\right] f^{(0)}(\mathbf{V}) = 3p\delta_{ij} + 2P_{ij}^{(0)}, \tag{3.149}$$

$$\begin{aligned} m\int d\mathbf{V} \left[\frac{\partial}{\partial V_l} Y_{0|ijk}(\mathbf{V})\right] f^{(0)}(\mathbf{V}) \quad = \quad & P_{ij}^{(0)}\delta_{kl} + P_{ik}^{(0)}\delta_{jl} + P_{jk}^{(0)}\delta_{il} \\ & -\frac{2}{5}\left(P_{il}^{(0)}\delta_{jk} + P_{jl}^{(0)}\delta_{ik} + P_{kl}^{(0)}\delta_{ij}\right) \\ & -\frac{3}{5}p\left(\delta_{il}\delta_{jk} + \delta_{jl}\delta_{ik} + \delta_{kl}\delta_{ij}\right). \end{aligned} \tag{3.150}$$

The fourth-degree moments $M_{4|0}^{(0)}$, $M_{2|ij}^{(0)}$, and $M_{0|ijkl}^{(0)}$ are given in Appendix 2.B [cf. Eqs. (2.B.1) and (2.B.8)]. They depend on temperature both explicitly and implicitly through a^*. Taking into account that

$$\frac{1}{n}\nabla = \frac{k_BT}{p}(\nabla T)\frac{\partial}{\partial T} \tag{3.151}$$

and the expression (3.133) for $\mathbf{g}^{(1)}$, the elements $\mathcal{N}_\sigma$ can be written as

$$\mathcal{N}_\sigma = \mathbf{B}_\sigma \cdot \nabla T, \tag{3.152}$$

where $\mathbf{B}_\sigma$ is a known function of the shear rate (Garzó, 1995a). The solution to Eq. (3.144) is then

$$\mathcal{M}_\sigma^{(1)} = \left[(\boldsymbol{\mathcal{A}})_{\sigma\sigma'}^{-1}\mathbf{B}_{\sigma'}\right]\cdot\nabla T. \tag{3.153}$$

The heat flux is $q_i^{(1)} = (mn/2)M_{2|i}^{(1)}$ and can be cast into the form of a generalized Fourier's law:

$$q_i^{(1)} = -\widetilde{\kappa}_{ij}\frac{\partial T}{\partial r_j}, \tag{3.154}$$

where $\widetilde{\kappa}_{ij}$ represents a generalized thermal conductivity tensor, which is a highly nonlinear function of the shear rate. Note that this tensor differs from the tensor κ_{ij} introduced in Eq. (3.101) since the former incorporates contributions coming from both ∇T and $\nabla n = -(n/T)\nabla T$, as well as contributions associated with the external field $\mathbf{g}^{(1)}$. Nevertheless, at zero shear rate, $\mathbf{g}^{(1)}$ and μ_{ij} vanish, and so $\widetilde{\kappa}_{ij} = \kappa_{ij} = \kappa_0\delta_{ij}$. The behaviors for small shear rates of the relevant elements of $\widetilde{\kappa}_{ij}$ are $\widetilde{\kappa}_{yy}/\kappa_0 \approx 1 + 3.04a^{*2}$, $\widetilde{\kappa}_{zz}/\kappa_0 \approx 1 - 1.18a^{*2}$ and $\widetilde{\kappa}_{xy}/\kappa_0 \approx -3.90a^*$. It must be remarked that the tensor $\widetilde{\kappa}_{ij}$ becomes meaningless for shear rates such that the unperturbed fourth-degree moments diverge, namely for $a^* \geq a_c^{(4)} \simeq 6.846$. This is associated with the fact that the high-velocity tail of $f^{(0)}$, Eq. (2.111), is transferred to $f^{(1)}$ as $f^{(1)}(\mathbf{V}) \sim V^{-4-\sigma(a^*)}$. Consequently, the third-degree moments of $f^{(1)}$ diverge whenever the fourth-degree moments of $f^{(0)}$ do [cf. Eq. (3.137)].

7.2 BGK description for general interactions

The above study can be extended to general interaction potentials by using the BGK model (Garzó, 1993). The equation for $f^{(1)}$ is still Eq. (3.137), with $\mathbf{g}^{(1)}$ given by Eq. (3.133), except that $\mathcal{I} \to \nu$. Although an explicit expression for $f^{(1)}$ can be obtained, a moment method is enough to get the heat flux. The closed equation for the third-degree moments $\mathcal{M}^{(1)}$ is again given by Eq. (3.144), except that $c_1 = c_2 = \nu + 3\alpha$ in Eq. (3.145) and the fourth-degree moments $M_{4|0}^{(0)}, \ldots$ appearing in $\mathcal{N}_\sigma$ are different from those of the Boltzmann equation (cf. Sec. 3.2). While the shear-rate dependence of the unperturbed moments is universal, the temperature dependence of the reduced shear rate a^* is not the same for different interactions. As a consequence, the thermal conductivity tensor defined by Eq. (3.154) presents a different shear-rate dependence for each possible potential. To fix ideas, in the case of repulsive potentials $\phi(r) \propto r^{-w}$, one has $a^* \propto p^{-1}T^{1-q}$, where $q = \frac{1}{2} - \frac{2}{w}$, so that $\partial a^*/\partial T = (1-q)(a^*/T)$. Further details can be found in the paper by Garzó (1993).

For small shear rates, the elements of the generalized thermal conductivity tensor behave as $\widetilde{\kappa}_{yy}/\kappa_0 \approx 1 + \frac{1}{15}(5+14q)a^{*2}$, $\widetilde{\kappa}_{zz}/\kappa_0 \approx 1 - \frac{1}{15}(19-2q)a^{*2}$ and $\widetilde{\kappa}_{xy}/\kappa_0 \approx -\frac{2}{5}(8-q)a^*$. It is interesting to note that, to first order in a^*, $\widetilde{\kappa}_{xy} = \kappa_{xy} - (n/T)\mu_{xy}$, where κ_{ij} and μ_{ij} are the generalized transport coefficients defined by Eq. (3.101). The results show that the qualitative dependence of $\widetilde{\kappa}_{ij}$ on a^* is the same for the different values of the interaction parameter q. Differences appear, however, at a quantitative level, especially for $\widetilde{\kappa}_{yy}$ and $\widetilde{\kappa}_{xy}$ (Garzó, 1993). For instance, at a given value of a^*, $\widetilde{\kappa}_{yy}$ increases as the potential becomes harder. Of course, the hardest physical interaction corresponds to hard spheres ($w \to \infty$, $q = 1/2$). On the other hand, a mathematical model (not associated with any physical interaction law) that can be interpreted as representing an interaction "harder" than that of hard spheres was introduced by Ernst and Hendriks (Ernst and Hendriks, 1979; Ernst, 1981) and, independently, by Bobylev (Bobylev, 1984; Bobylev, 1988). This model is called VHP ("Very Hard Particles") and is defined so that the collision rate $gB(g,\theta)$ is proportional to g^2, while $gB(g,\theta) \propto g$ for hard spheres. The main advantage of the VHP model is that it allows to solve exactly the Cauchy or initial-value problem of the two-dimensional Boltzmann equation for homogeneous and isotropic states (Ernst, 1981; Bobylev, 1988). In the BGK equation, where the collision frequency is independent of the velocity, the VHP model translates into $\nu \propto nT$, i.e. $q = 1$. In this case, $\nu \propto p$ is a constant, even if a thermal gradient exists, so that Eq. (3.130) implies that the external field g vanishes. Consequently, it is easy to show that the heat flux obeys the generalized Fourier's law (3.154) to *any* order in the thermal gradient, provided that $\nabla T = \text{const}$.

In Section 3.2 we saw that the fourth-degree moments of $f^{(0)}$ are different in the Boltzmann and BGK descriptions. Since the shear-rate dependence of the thermal conductivity tensor $\widetilde{\kappa}_{ij}$ involves the unperturbed fourth-degree moments and their derivatives, it can be expected that $\widetilde{\kappa}_{ij}$, as predicted by both descriptions, exhibits significant discrepancies. Of course, a comparison can only be carried out for Maxwell molecules, since only in that case the exact solution of the Boltzmann equation is known. Figure 3.26 compares the Boltzmann and BGK predictions for the element $\widetilde{\kappa}_{yy}$ in the region $0 \leq a^* \leq 1$. It is quite apparent that the BGK value widely differs from the Boltzmann one. While the Boltzmann equation predicts a monotonically increasing coefficient

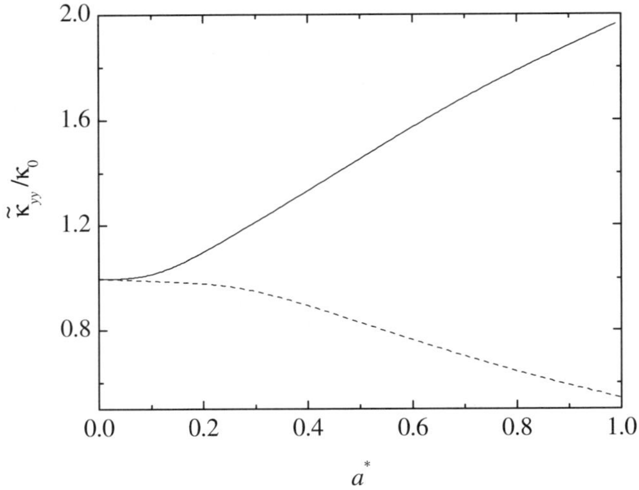

Figure 3.26. Shear-rate dependence of the reduced element $\widetilde{\kappa}_{yy}/\kappa_0$ for Maxwell molecules, as predicted by the Boltzmann equation (solid line) and the BGK equation (dashed line).

$\widetilde{\kappa}_{yy}$, the BGK equation yields a decreasing coefficient, except in a narrow region of shear rates smaller than $a^* \simeq 0.2$. This discrepancy is a reflection of the disagreement between the predictions of the Boltzmann and BGK equation for the fourth-degree moments, especially those where the y-component of the velocity plays an important role [cf. Fig. 3.7]. Actually, the thermal conductivity tensor $\widetilde{\kappa}_{ij}$ probes the population of particles beyond the region of thermal velocities. The fact that $\widetilde{\kappa}_{yy}$ increases with a^* implies that the shearing enhances the transport of energy along the direction of the gradient of the flow velocity, produced by a thermal gradient along the same direction. This effect agrees with computer simulations of a strongly sheared Lennard–Jones fluid (Daivis and Evans, 1993). The other diagonal element, $\widetilde{\kappa}_{zz}$, is plotted in Fig. 3.27. In this case, the BGK values agree qualitatively well with the Boltzmann ones. In fact, the behavior for small shear rates is quite similar in both cases: $\widetilde{\kappa}_{zz}/\kappa_0 \approx 1-1.18a^{*2}$ (Boltzmann equation) and $\widetilde{\kappa}_{zz}/\kappa_0 \approx 1 - \frac{19}{15}a^{*2}$ (BGK equation). Moreover, both theories predict that $\widetilde{\kappa}_{zz}$ decreases as the shear rate increases, so that the transport of energy along the direction orthogonal to the shearing plane is inhibited by the shear flow. It is interesting to note that the BGK model gives a larger deviation from Fourier's law than the true Boltzmann equation. Of course, these comments are restricted to shear rates sufficiently smaller than the critical value $a_c^* \simeq 6.846$. The off-diagonal element $\widetilde{\kappa}_{xy}$ measures cross-effects in the thermal conduction, namely a transport of energy parallel to the flow direction due to a thermal gradient along the velocity gradient. As shown in Fig. 3.28, this coefficient is negative and its magnitude increases with the shear rate. This

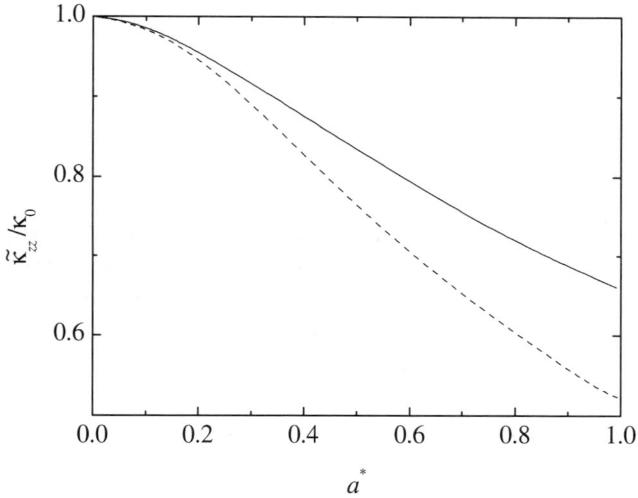

Figure 3.27. Shear-rate dependence of the reduced element $\widetilde{\kappa}_{zz}/\kappa_0$ for Maxwell molecules, as predicted by the Boltzmann equation (solid line) and the BGK equation (dashed line).

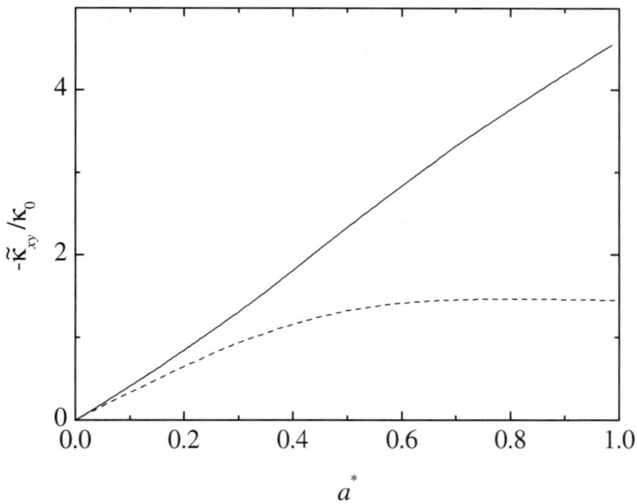

Figure 3.28. Shear-rate dependence of the reduced element $-\widetilde{\kappa}_{xy}/\kappa_0$ for Maxwell molecules, as predicted by the Boltzmann equation (solid line) and the BGK equation (dashed line).

behavior is also predicted by the BGK equation for a^* smaller than about 0.8, after which it gives a decreasing value of $|\widetilde{\kappa}_{xy}|$ (Garzó, 1993). The initial slope is similar in both theories: $-\widetilde{\kappa}_{xy}/\kappa_0 \approx -3.90a^*$ (Boltzmann equation) and $-\widetilde{\kappa}_{xy}/\kappa_0 \approx -\frac{16}{5}a^*$ (BGK equation). As in the case of $\widetilde{\kappa}_{yy}$, the qualitative shear-rate dependence of $\widetilde{\kappa}_{zz}$ and $\widetilde{\kappa}_{xy}$ obtained from the Boltzmann equation agrees with the one observed in molecular dynamics simulations (Daivis and Evans, 1993).

7.3 Heat flux induced by an external force

The natural way of creating heat transport is by means of a thermal gradient. On the other hand, a heat flux can also be induced by the action of an external non-conservative force. This method was originally proposed as an efficient way of measuring the Navier–Stokes thermal conductivity in computer simulations of homogeneous systems (Evans, 1982; Gillan and Dixon, 1983; Sarman and Evans, 1992a; Sarman and Evans, 1992b). The equilibrium state is disturbed by the action of a force

$$\mathbf{F}_\epsilon = -\frac{1}{2T}\left(mV^2 - 3k_BT\right)\boldsymbol{\epsilon}, \tag{3.155}$$

where the field strength $\boldsymbol{\epsilon}$ plays the role of a thermal gradient ∇T. This force generates a heat flux $\mathbf{q} \parallel \boldsymbol{\epsilon}$ and the Navier–Stokes thermal conductivity coefficient is obtained by extrapolating the ratio $|\mathbf{q}|/|\boldsymbol{\epsilon}|$ to zero-field limit, i.e. $\kappa_0 = -\lim_{\epsilon\to 0}|\mathbf{q}|/|\boldsymbol{\epsilon}|$.

When the above force is applied on a gas subjected to the *nonequilibrium* state of USF, the induced heat flux is related to the field strength by means of a field susceptibility tensor λ_{ij} as

$$q_i = -\lambda_{ij}\epsilon_j, \tag{3.156}$$

where the weak-field limit is implicitly assumed. By expanding $f = f^{(0)} + f^{(1)} + \cdots$, where $f^{(k)}$ is of order k in $\boldsymbol{\epsilon}$, one gets the following Boltzmann equation for $f^{(1)}$:

$$\left(-3\alpha + \Lambda + \mathcal{I}\right)f^{(1)} = \frac{k_B}{2m}\boldsymbol{\epsilon}\cdot\frac{\partial}{\partial\mathbf{V}}\left(\frac{mV^2}{k_BT} - 3\right)f^{(0)}, \tag{3.157}$$

which is the counterpart of Eq. (3.137). By following the same mathematical steps as done previously, one gets an equation similar to Eq. (3.144), except that now

$$\begin{aligned}\mathcal{N}_\sigma &= -\frac{k_B}{mn}\boldsymbol{\epsilon}\cdot\int d\mathbf{V}\left[\frac{\partial}{\partial\mathbf{V}}Y_\sigma(\mathbf{V})\right]\left(\frac{mV^2}{k_BT} - 3\right)f^{(0)} \\ &= \mathbf{B}'_\sigma\cdot\boldsymbol{\epsilon},\end{aligned} \tag{3.158}$$

where $\mathbf{B}'_\sigma$ involves second- and fourth-degree moments of $f^{(0)}$ and is different from the quantity $\mathbf{B}_\sigma$ appearing in Eq. (3.152). The *susceptibility* tensor λ_{ij} is then obtained analogously to the case of the thermal conductivity tensor $\widetilde{\kappa}_{ij}$. The explicit expressions derived from the Boltzmann equation for Maxwell molecules can be found elsewhere (Garzó, 1995b). For small shear rates, the nonzero elements of the tensor λ_{ij} are $\lambda_{xx}/\kappa_0 \approx 1+59.4a^{*2}$, $\lambda_{yy}/\kappa_0 \approx 1+3.05a^{*2}$, $\lambda_{zz}/\kappa_0 \approx 1+0.72a^{*2}$, $\lambda_{xy}/\kappa_0 \approx -2.90a^*$, and $\lambda_{yx}/\kappa_0 \approx -1.40a^*$. In the case of the BGK model, the results are universal, i.e. independent of the interaction potential. For small shear rates, the behaviors predicted by the BGK model are (Garzó, 1994) $\lambda_{xx}/\kappa_0 \approx 1+\frac{79}{15}a^{*2}$, $\lambda_{yy}/\kappa_0 \approx 1+\frac{43}{15}a^{*2}$, $\lambda_{zz}/\kappa_0 \approx 1+\frac{7}{15}a^{*2}$, $\lambda_{xy}/\kappa_0 \approx -\frac{11}{5}a^*$, and $\lambda_{yx}/\kappa_0 \approx -\frac{6}{5}a^*$. Except at zero shear rate, the tensors λ_{ij} and $\widetilde{\kappa}_{ij}$ are different. This means that, because of the complexity of the USF problem, the response of the system to a thermal gradient is different from the response to the external force $\mathbf{F}_\epsilon$. Comparison between the shear-rate dependence of λ_{ij} and $\widetilde{\kappa}_{ij}$ shows that, while the yy and xy elements present a qualitative good agreement, the zz elements are quite different (Garzó, 1995b).

8. Stability of the uniform shear flow

In the two previous Sections we have analyzed the response of a gas undergoing USF to perturbations in the form of small hydrodynamic gradients ∇n, ∇T, and $\nabla \delta\mathbf{u}$. Implicitly, the stability of the USF against these spatial perturbations has been taken for granted. However, since the pioneering work of Erpenbeck (1984), it has been known by molecular dynamics simulations of dense systems (Loose and Hess, 1989) that the USF undergoes a transition to an ordered state ("string" phase) at large shear rates, which has been attributed to a *short wavelength* hydrodynamic instability (Kirkpatrick and Nieuwoudt, 1986; Lutsko and Dufty, 1987; Lutsko et al., 1989). On the other hand, this shear-induced ordering transition is related to excluded volume effects and seems to be preceded by a vanishing (in certain directions) of the anisotropic pair correlation function at contact (Lutsko, 1996). Consequently, this transition is not expected to appear in the low-density limit, which is the regime we are mainly interested in. This, however, does not preclude the existence of instabilities in the USF state, even for dilute gases, different from the above ordering transition. As a matter of fact, the BGK kinetic model shows that the USF state presents a *long wavelength* instability for any value of the shear rate (Lee et al., 1996; Lee and Dufty, 1997; Montanero et al., 1998). Here we analyze the onset of this instability.

8.1 Theoretical analysis

Equations (3.126)–(3.128) describe the time evolution of spatial perturbations to the USF with a finite shear rate a, under the assumption that the gradients ∇n, ∇T, and $\nabla \delta\mathbf{u}$ are small. Let us suppose that the initial perturbations $\delta n(\mathbf{r},0) = n(\mathbf{r},0) - n_0$, $\delta T(\mathbf{r},0) = T(\mathbf{r},0) - T_0$, and $\delta\mathbf{u}(\mathbf{r},0) = \mathbf{u}(\mathbf{r},0) - \mathbf{u}_0$, are also small with respect to a given set of reference values n_0, T_0, and

$\mathbf{u}_0 = \mathsf{a} \cdot \mathbf{r}$. As a consequence, Eqs. (3.126)–(3.128) can be linearized with respect to

$$\{\delta A_\alpha(\mathbf{r}, t)\} \equiv \{n_0^{-1}\delta n(\mathbf{r}, t), T_0^{-1}\delta T(\mathbf{r}, t), (m/k_B T_0)^{1/2}\delta \mathbf{u}(\mathbf{r}, t)\}. \quad (3.159)$$

The resulting set of five linear equations defines the hydrodynamic modes, namely, linear response excitations to small perturbations. If at least one of these modes grows in time, the reference state is linearly unstable. For the sake of simplicity, we consider here perturbations along the y-direction only, i.e. along a direction parallel to the velocity gradient of the reference state. In this case, the resulting linear hydrodynamic balance equations are

$$\frac{\partial \delta n}{\partial t} + n_0 \frac{\partial \delta u_y}{\partial y} = 0, \quad (3.160)$$

$$\frac{3}{2} n_0 k_B \frac{\partial \delta T}{\partial t} + (P_{iy} - a\eta_{xyiy}) \frac{\partial \delta u_i}{\partial y} - \kappa_{yy} \frac{\partial^2 \delta T}{\partial y^2} - \mu_{yy} \frac{\partial^2 \delta n}{\partial y^2} = 0, \quad (3.161)$$

$$\frac{\partial \delta u_i}{\partial t} + a\delta_{ix}\delta u_y + \frac{1}{mn_0}\left(\frac{\partial P_{iy}}{\partial n_0}\frac{\partial \delta n}{\partial y} + \frac{\partial P_{iy}}{\partial T_0}\frac{\partial \delta T}{\partial y} - \eta_{iyjy}\frac{\partial^2 \delta u_j}{\partial y^2}\right) = 0. \quad (3.162)$$

Here, the pressure tensor P_{ij} and the transport coefficients $\eta_{ijk\ell}$, κ_{ij}, and μ_{ij}, defined in Eqs. (3.101) and (3.102), are understood to be evaluated at the reference state. As in the usual case of the analysis of hydrodynamic modes around equilibrium (Résibois and de Leener, 1977), it is convenient to use a Fourier representation

$$\delta \widetilde{A}_\alpha(k, t) = \int_{-L/2}^{L/2} dy\, e^{iky} \delta A_\alpha(y, t), \quad (3.163)$$

where the Lees–Edwards periodic boundary conditions (2.7) are applied at $y = \pm L/2$, so that the wavenumber k is restricted to $k = 2\pi/L, 4\pi/L, 6\pi/L, \ldots$. Therefore, Eqs. (3.160)–(3.162) can be written in the matrix form

$$\frac{\partial \delta \widetilde{A}_\alpha}{\partial t} + F_{\alpha\beta}\delta \widetilde{A}_\beta = 0, \quad (3.164)$$

where

$$F_{\alpha\beta} = a\delta_{\alpha 3}\delta_{\beta 4} - ikB_{\alpha\beta} + k^2 D_{\alpha\beta} \quad (3.165)$$

with the matrices

$$\mathsf{B} = \sqrt{\frac{k_B T_0}{m}} \begin{pmatrix} 0 & 0 & 0 & 1 & 0 \\ 0 & 0 & \frac{2}{3}\frac{P_{xy} - a\eta_{xyxy}}{p_0} & \frac{2}{3}\frac{P_{yy} - a\eta_{xyyy}}{p_0} & 0 \\ \frac{n_0}{p_0}\frac{\partial P_{xy}}{\partial n_0} & \frac{T_0}{p_0}\frac{\partial P_{xy}}{\partial T_0} & 0 & 0 & 0 \\ \frac{n_0}{p_0}\frac{\partial P_{yy}}{\partial n_0} & \frac{T_0}{p_0}\frac{\partial P_{yy}}{\partial T_0} & 0 & 0 & 0 \\ 0 & 0 & 0 & 0 & 0 \end{pmatrix}, \quad (3.166)$$

$$\mathsf{D} = \begin{pmatrix} 0 & 0 & 0 & 0 & 0 \\ \frac{2n_0\mu_{yy}}{3p_0} & \frac{2T_0\kappa_{yy}}{3p_0} & 0 & 0 & 0 \\ 0 & 0 & \frac{\eta_{xyxy}}{mn_0} & \frac{\eta_{xyyy}}{mn_0} & 0 \\ 0 & 0 & \frac{\eta_{yyxy}}{mn_0} & \frac{\eta_{yyyy}}{mn_0} & 0 \\ 0 & 0 & 0 & 0 & \frac{\eta_{zyzy}}{mn_0} \end{pmatrix}. \tag{3.167}$$

The eigenvalues $\omega_\alpha(k,a)$ of the matrix $\mathsf{F}(k,a)$ determine the time evolution of $\delta\widetilde{A}_\alpha(k,t)$. The reference USF state will be linearly stable if the real parts of those eigenvalues are positive. The structure of the matrix $\mathsf{F}(k,a)$ shows that the perturbation $\delta\widetilde{A}_5 \propto \delta\widetilde{u}_z$ is decoupled from the remaining ones. This is due to the choice of gradients along the y-direction only. The eigenvalue associated with this mode is simply $\omega_5(k,a) = [\eta_0 F_\eta(a)/mn_0]k^2$, where we have made use of the relation (3.125). At equilibrium ($a = 0$), the hydrodynamic modes of the Navier–Stokes equations are recovered (Résibois and de Leener, 1977). They correspond to a heat mode, two sound modes, and a twofold-degenerate shear mode. To second order in the wavenumber they are given by

$$\omega_\alpha(k,0) = \begin{pmatrix} D_T k^2 \\ \mathrm{i}c_s k + \Gamma k^2 \\ -\mathrm{i}c_s k + \Gamma k^2 \\ (\eta_0/mn_0)k^2 \\ (\eta_0/mn_0)k^2 \end{pmatrix}. \tag{3.168}$$

Here, $D_T = \kappa_0/(5nk_B/2)$ is the heat diffusion coefficient, $c_s = \sqrt{5k_BT_0/3m}$ is the sound velocity, and $\Gamma = D_T/3 + 2\eta_0/3mn_0$ is the sound damping constant. These coefficients are positive, and so the excitations around equilibrium are damped. On the other hand, for finite shear rates the modes are more complicated and the behavior at long wavelength is qualitatively different:

$$\omega_\alpha(k,a) = \begin{pmatrix} C_1(a)k^2 \\ -e^{\mathrm{i}\pi/3}C_2(a)k^{2/3} + e^{-\mathrm{i}\pi/3}C_3(a)k^{4/3} + C_4(a)k^2 \\ -e^{-\mathrm{i}\pi/3}C_2(a)k^{2/3} + e^{\mathrm{i}\pi/3}C_3(a)k^{4/3} + C_4(a)k^2 \\ C_2(a)k^{2/3} + C_3(a)k^{4/3} + C_4(a)k^2 \\ \left[\eta_0 F_\eta(a)/mn_0\right]k^2 \end{pmatrix}, \tag{3.169}$$

where the coefficients $C_i(a)$ are nonlinear functions of the shear rate whose explicit expressions are given by Lee and Dufty (1997). As in the Navier–Stokes case, there are two oscillating modes and three purely damped modes. However, Eqs. (3.168) and (3.169) present two important qualitative differences. First, the long-wavelength behavior at finite shear rate is not analytic in k, as given by the terms of order $k^{2/3}$. Thus, the purely damped mode ω_4 does not represent diffusive behavior. The origin of this non-analytic behavior can be traced back to the fact that at $k = 0$ the matrix F becomes $F_{\alpha\beta} = a\delta_{\alpha 3}\delta_{\beta 4}$, which is not diagonalizable (Lee and Dufty, 1997).

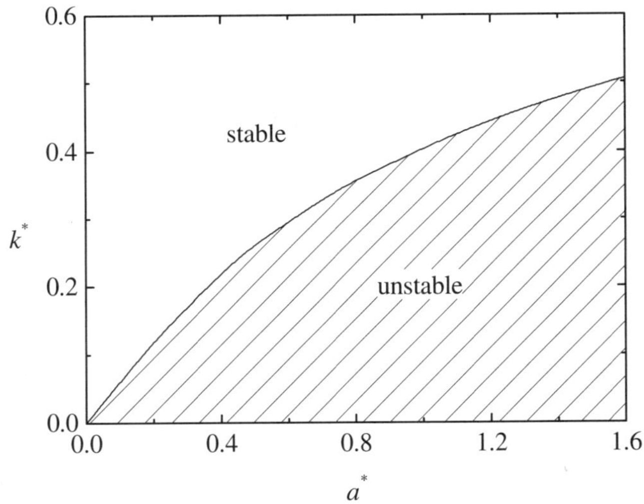

Figure 3.29. Stability line $k_s^*(a^*)$ in the case of Maxwell molecules. The region above the curve corresponds to the stable domain, while the region below the curve corresponds to the unstable domain.

The second important difference between (3.168) and (3.169) is that the two oscillatory modes $\omega_{2,3}$ of (3.169) are unstable in the long-wavelength limit due to the fact that the coefficient $C_2(a)$ is positive definite (Lee and Dufty, 1997). This implies that there are excitations growing in time. In other words, the reference USF state becomes *unstable* under long wavelength perturbations. It might be possible, however, that the hydrodynamic modes decay in the case of wavenumbers k larger than those considered in the small k expansion (3.169). The analysis of this requires the evaluation of the eigenvalues $\omega_\alpha(k, a)$ with its full nonlinear dependence on k. It turns out that, at a given value of the shear rate a, $\mathrm{Re}\,\omega_{2,3}(k, a) > 0$ if k is larger than a certain threshold value $k_s(a)$. Consequently, the USF state is linearly stable against excitations with a wavenumber $k > k_s(a)$. The stability line $k_s(a)$ is plotted in Fig. 3.29 in the case of Maxwell molecules (Lee et al., 1996). Dimensionless variables are used, $k^* = k\ell_0$ and $a^* = a/\nu_0$, where $\ell_0 = \sqrt{2k_BT_0/m}/\nu_0$ is the mean free path and ν_0 is the collision frequency in the reference state. The dynamics of the stable domain in the case $a \to 0$ and $k \to 0$ with $k/a = \mathrm{const}$ has been carefully studied by Lee and Dufty (1997).

8.2 Monte Carlo simulations

So far, the results presented in this Section are limited to (a) the hydrodynamic regime (since terms of order higher than k^2 are neglected in the hydrodynamic equations), (b) small perturbations (since nonlinear terms in $\{\delta A_\alpha\}$ are neglected), and (c) the use of the BGK kinetic model. In order to overcome these limitations one needs to resort to numerical methods. In this context, Montanero et al. (1995a) solved (by means of the DSMC method) the *inhomogeneous* Boltzmann equation for Maxwell molecules with Lees–Edwards boundary conditions (Lees and Edwards, 1972), starting from initial conditions far from those consistent with the USF. The numerical solution goes beyond the limitations of a hydrodynamic regime and small perturbations. The initial conditions were of the local equilibrium form, namely

$$f(y,\mathbf{V};0)=n(y,0)\left[\frac{m}{2\pi k_BT(y,0)}\right]^{3/2}\exp\left\{-\frac{m\left[\mathbf{V}-\delta\mathbf{u}(y,0)\right]^2}{2k_BT(y,0)}\right\}, \tag{3.170}$$

where $\mathbf{V}=\mathbf{v}-\mathsf{a}\cdot\mathbf{r}$ is the peculiar velocity in the USF state. The deviations of $\delta\mathbf{u}(y,t)$ from zero and of the density and temperature profiles from their average values

$$\overline{n}=\frac{1}{L}\int_{-L/2}^{L/2}dy\, n(y,t),\quad \overline{T}(t)=\frac{1}{\overline{n}L}\int_{-L/2}^{L/2}dy\, n(y,t)T(y,t), \tag{3.171}$$

are a measure of the departure from the USF state. In the simulations (Montanero et al., 1995a), the size of the system was taken as $L=10\sqrt{2k_B\overline{T}(0)/m}/\overline{\nu}$ and two shear rates were considered: $a=5\overline{\nu}$ and $a=10\overline{\nu}$, where $\overline{\nu}\propto\overline{n}$ is the average collision frequency. In both cases the simulations show that the global temperature $\overline{T}(t)$ reaches a constant value $\overline{T}_\infty$ after a few collisions per particle (Montanero et al., 1995a). This fact, however, is not sufficient to imply that the system achieves a homogeneous state in the long-time limit. Figure 3.30 shows the velocity and temperature profiles for $a=5\overline{\nu}$ and at times $t=0$, $11\overline{\nu}^{-1}$, and $110\overline{\nu}^{-1}$, starting from the initial condition (3.170) with

$$\begin{aligned} n(y,0) &= \overline{n}\left(1+\frac{3}{5}\frac{y}{L}\right),\quad T(y,0)=\frac{50}{49}\overline{T}(0)\left(1-\frac{2}{5}\frac{y}{L}\right),\\ \delta\mathbf{u}(y,0) &= -\frac{2}{5}\sqrt{2k_B\overline{T}(0)/m}\frac{y}{L}\widehat{\mathbf{x}}. \end{aligned} \tag{3.172}$$

Figure 3.30 clearly indicates that the system evolves towards a state with vanishing velocity perturbation $\delta\mathbf{u}(y,t)\to\mathbf{0}$ and constant temperature $T(y,t)\to\overline{T}_\infty$. This final state is a USF state with a temperature $T\simeq1.26\overline{T}(0)$ slightly different from the initial average value $\overline{T}(0)$ (Montanero et al., 1995a). It

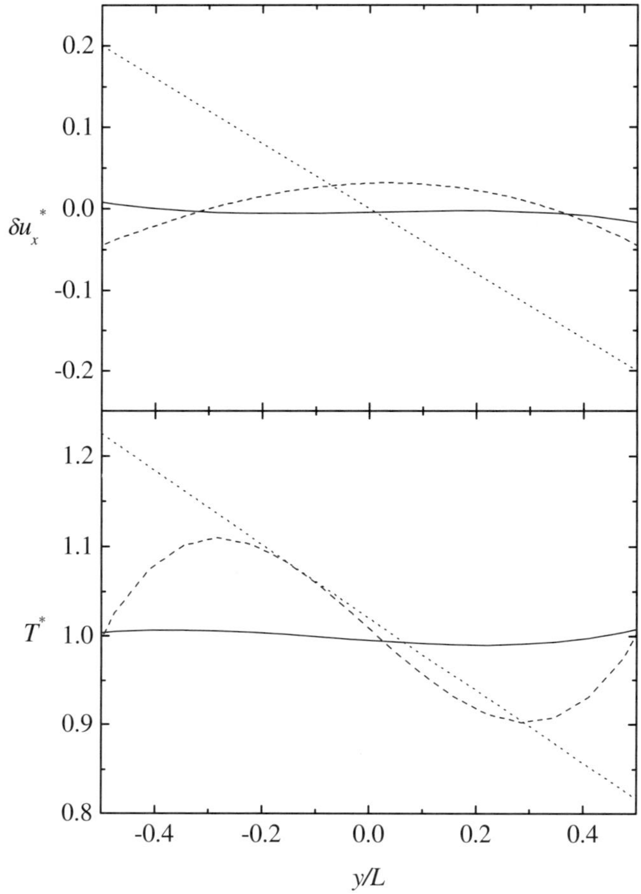

Figure 3.30. Profiles of $\delta u_x^*(y,t) = \delta u_x(y,t)/\sqrt{2k_B\overline{T}(0)/m}$ and $T^*(y,t) = T(y,t)/\overline{T}(t)$ for $a = 5\overline{\nu}$ and at times $t = 0$ (dotted lines), $11\overline{\nu}^{-1}$ (dashed lines), and $110\overline{\nu}^{-1}$ (solid lines), starting from the initial condition (3.172).

is worth noticing that the relaxation time associated with the *local* temperature $T(y,t)$ is much longer than that associated with the *global* temperature $\overline{T}(t)$. Thus, for instance, at $t = 11\overline{\nu}^{-1}$ the local temperature is still evolving, while the global temperature has already reached its stationary value. The results displayed in Fig. 3.30 show that the final USF state, which corresponds to $a^* = 5$, is stable under perturbations characterized by a wavenumber $k^* = (2\pi/L)\ell = (2\pi/10)\sqrt{\overline{T}/\overline{T}(0)} \simeq 0.71$, where $\ell = \sqrt{2k_B\overline{T}/m}/\overline{\nu}$ is the mean free path corresponding to the final state. This conclusion is consistent with the extrapolation of the BGK stability line plotted in Fig. 3.29, although

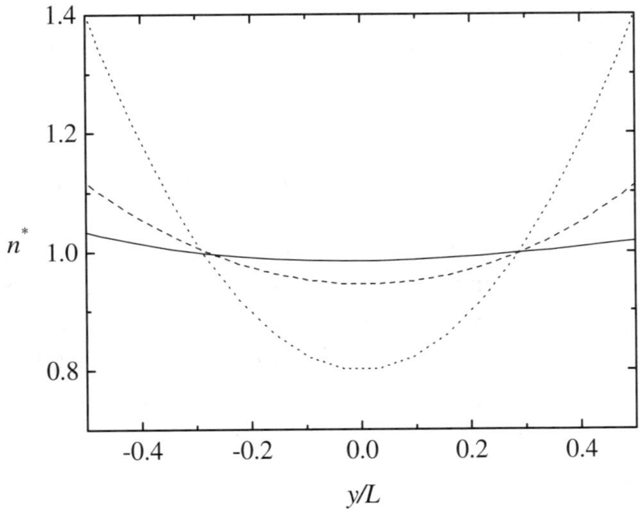

Figure 3.31. Profile of $n^*(y,t) = n(y,t)/\overline{n}$ for $a = 10\overline{\nu}$ and at times $t = 0$ (dotted line), $73\overline{\nu}^{-1}$ (dashed line), and $220\overline{\nu}^{-1}$ (solid line), starting from the initial condition (3.173).

it must be borne in mind that the latter has been obtained in the framework of a hydrodynamic theory for small k^*.

A similar behavior is found in the case $a = 10\overline{\nu}$, as illustrated in Fig. 3.31 for the density profile at times $t = 0$, $73\overline{\nu}^{-1}$, and $220\overline{\nu}^{-1}$, starting from the initial condition

$$\begin{aligned} n(y,0) &= \overline{n}\left(\frac{4}{5}+\frac{12}{5}\frac{y^2}{L^2}\right), \quad T(y,0) = \overline{T}(0)\left(1-\frac{2}{5}\frac{y}{L}\right), \\ \delta\mathbf{u}(y,0) &= \mathbf{0}. \end{aligned} \tag{3.173}$$

Now, the final temperature is $T \simeq 1.75\overline{T}(0)$ and the relaxation time is longer than in the previous case. Nevertheless, the hydrodynamic quantities eventually reach the USF state, and so we can conclude that the USF with $a^* = 10$ is stable when $k^* \simeq 0.83$.

The above cases do not develop instabilities because the minimum wavenumber $k = 2\pi/L$ is not small enough. However, according to the linear stability analysis discussed before, the USF state is expected to be unstable if the size L of the system is large enough as to allow for long wavelength perturbations. In addition, the stability analysis does not provide information about the state of the system for asymptotically long times, where deviations from the USF state can be quite large. In order to overcome the latter limitation, numerical solutions of the kinetic equations are needed again. This was carried out by Montanero et al. (1998), who solved the BGK kinetic equation by means of a DSMC-like method. In the simulations a size system corresponding to $k^* = 0.1$ at a shear rate $a^* = 0.5$ was considered. This case is clearly inside the unstable region of Fig. 3.29. Taking as a starting point the velocity distribution function (3.8) corresponding to USF, a perturbed initial condition was prepared by displacing and rescaling velocities so that

$$\delta u_{x,y}(y,0) = -\delta\widetilde{u}_{x,y}(0)\sin ky, \quad \delta u_z(y,0) = \delta n(y,0) = \delta T(y,0) = 0, \tag{3.174}$$

with $\delta\widetilde{u}_x(0) = -0.03v_0$ and $\delta\widetilde{u}_y(0) = 0.1v_0$, where $v_0 = \sqrt{2k_BT_0/m}$ is the thermal velocity of the unperturbed USF state. First we consider the short-time dependence at a fixed position $y = -L/4$, which corresponds to the maximum initial deviation. Figure 3.32 shows $\delta\widetilde{u}_x(t) = \delta u_x(-L/4,t)$ and

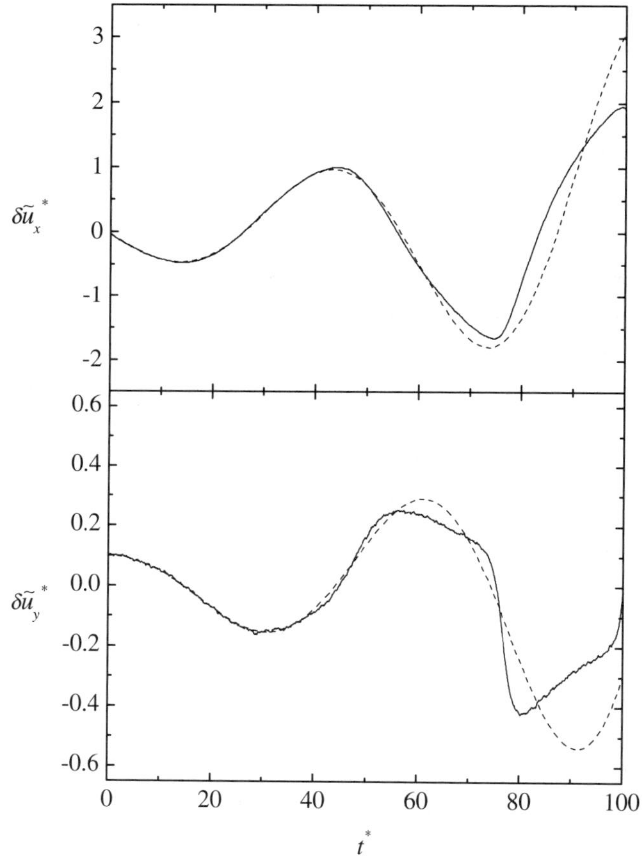

Figure 3.32. Plot of $\delta\widetilde{u}_x^*(t^*) = \delta\widetilde{u}_x(t^*)/v_0$ and $\delta\widetilde{u}_y^*(t) = \delta\widetilde{u}_y(t^*)/v_0$. The solid lines are simulation results and the dashed lines correspond to the linear stability analysis.

$\delta\widetilde{u}_y(t) = \delta u_y(-L/4,t)$. The dashed lines are the results from the hydrodynamic analysis of the BGK model near USF (Lee and Dufty, 1997) summarized in the previous Subsection. The good agreement up to $t^* \equiv t\nu_0 \approx 50$ shows that the instability is not just a consequence of the assumptions behind the hydrodynamic description. The subsequent differences between simulation and theory are due to the fact that the latter is limited to small deviations from USF.

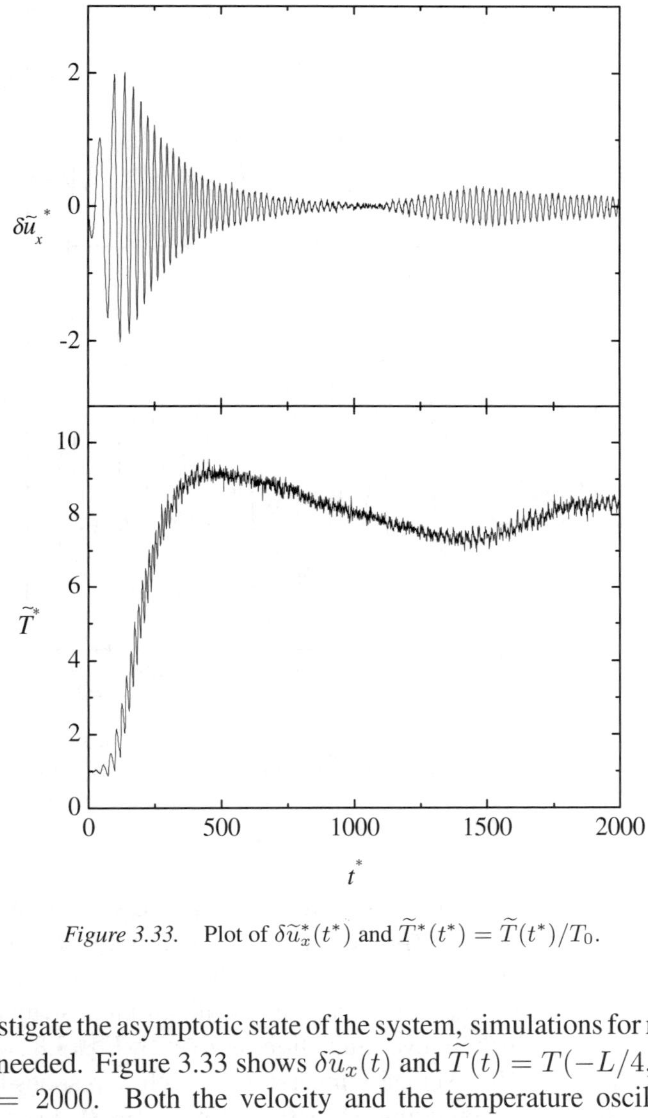

Figure 3.33. Plot of $\delta\widetilde{u}_x^*(t^*)$ and $\widetilde{T}^*(t^*) = \widetilde{T}(t^*)/T_0$.

To investigate the asymptotic state of the system, simulations for much longer times are needed. Figure 3.33 shows $\delta\widetilde{u}_x(t)$ and $\widetilde{T}(t) = T(-L/4, t)$ for times up to $t^* = 2000$. Both the velocity and the temperature oscillate in time and are modulated by a slowly varying amplitude relative to their asymptotic average values. The maximum value of $\widetilde{T}$ is $\widetilde{T}_{\text{max}} \approx 9T_0$. Expressed in units of the mean free path associated with this temperature, the wavenumber is $k \approx 0.3\nu_0/\sqrt{2k_B\widetilde{T}_{\text{max}}/m}$, which is close to the value $k_s(a)$ at which the USF with $a^* = 0.5$ would become marginally stable (cf. Fig. 3.29). Thus the initial

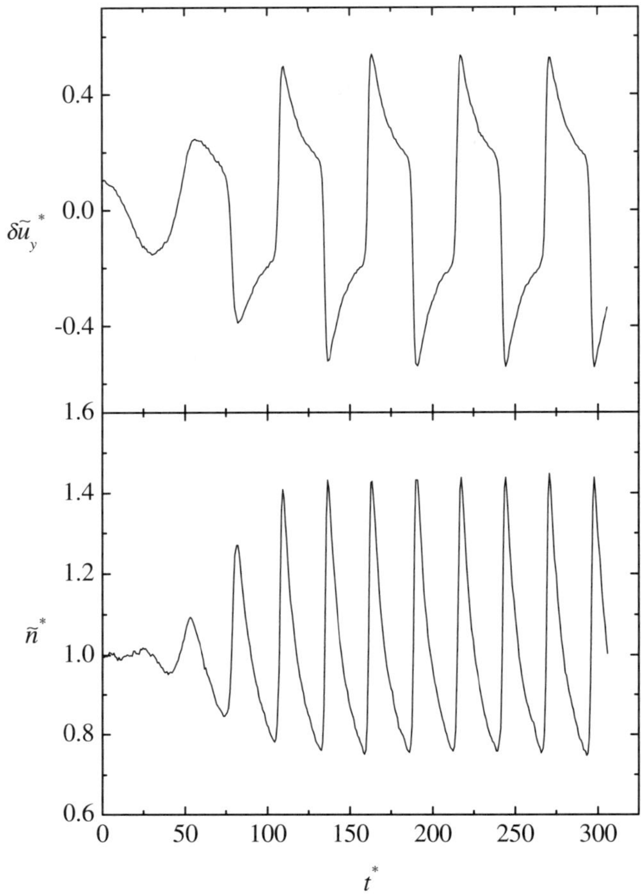

Figure 3.34. Plot of $\delta\widetilde{u}_y^*(t^*)$ and $\widetilde{n}^*(t^*) = \widetilde{n}(t^*)/n_0$.

dynamics tends towards stabilization but does not ever cross over into the stable domain. Consequently, the asymptotic dynamics is not simply that of stationary uniform shear flow at a different temperature, as happened in the cases of Figs. 3.30 and 3.31, but rather a quasi-stationary state with different spatial structure.

The thermostat used in the above simulations is given by Eq. (3.82) and allows for a global change in the average temperature, which is responsible for the amplitude modulations at long times (Montanero et al., 1998). In order to have a more controlled asymptotic state, it is convenient to consider a variation of this thermostat that maintains the average temperature constant. The quantities $\delta\widetilde{u}_y(t)$ and $\widetilde{n}(t) = n(-L/4, t)$ obtained by using this second more efficient thermostat are plotted in Fig. 3.34. After a transient period of length $t^* \approx 100$, a stable oscillatory behavior of the velocity appears with a period $\tau_p \simeq 54$. The

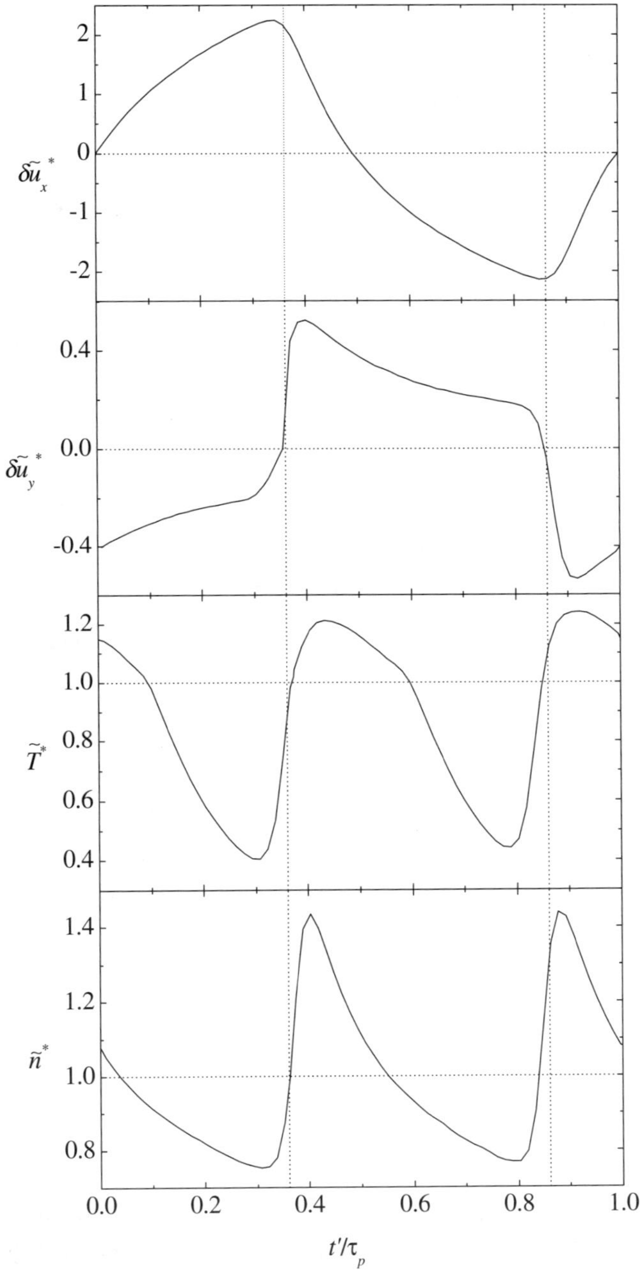

Figure 3.35. Plot of $\delta\widetilde{u}_x^*(t')$, $\delta\widetilde{u}_y^*(t')$, $\widetilde{T}^*(t')$, and $\widetilde{n}^*(t')$ over one cycle τ_p. Here $t' = t^* - j - t_w^*$, where $t_w^* = 463.4$, and the results are averaged over $j = 1, 2, \ldots, 20$ successive cycles. The vertical dotted lines indicates the times $t' = 0.36\tau_p$ and $t' = 0.86\tau_p$, approximately at which $\delta\widetilde{u}_x^*(t')$ have extrema and $\delta\widetilde{u}_y^*(t')$, $\widetilde{T}^*(t')$, and $\widetilde{n}^*(t')$ have nodes.

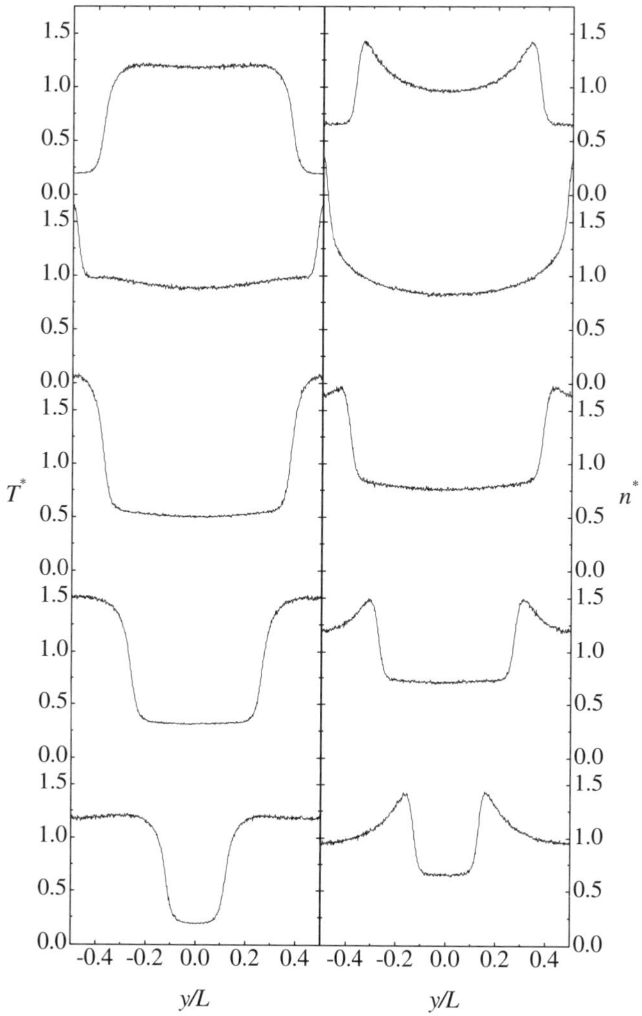

Figure 3.36. Plot of $T^*(y,t')$ (left) and $n^*(y,t')$ (right) as functions of y for (from top to bottom) $t' = 0, 0.14\tau_p, 0.25\tau_p, 0.36\tau_p$, and $0.5\tau_p$.

shape of $\delta\widetilde{A}_\alpha$ over one cycle is shown in Fig. 3.35. Here $t' = t^* - t_w^*$, where $t_w^* = 463.4$, and the curves are averages over 20 successive cycles. The value of the waiting time t_w^* is chosen as to assure that the transient time is over and also with the criterion that $\delta\widetilde{u}_x = 0$ at $t' = 0$. Inspection of the results in Fig. 3.35 shows several regularities. First, the following symmetry relation appears:

$$\delta\widetilde{A}_\alpha(t' + \tau_p/2) = \pm\delta\widetilde{A}_\alpha(t'), \tag{3.175}$$

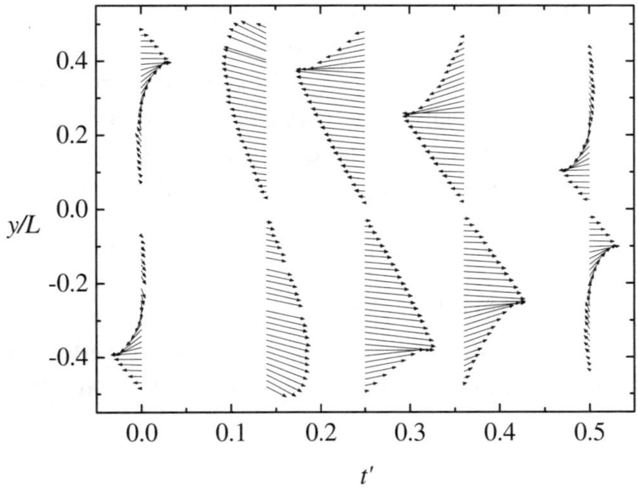

Figure 3.37. Vector plot representing $\delta u_x(y,t')$ and $\delta u_y(y,t')$ at $t' = 0, 0.14\tau, 0.25\tau, 0.36\tau$, and 0.5τ.

where the minus sign applies to the velocity and the plus sign applies to the density and the temperature. Next, at times $t' \approx 0.36\tau_p$ and $t' \approx 0.86\tau_p$, $\delta\widetilde{u}_x$ have extrema, while $\delta\widetilde{u}_y$, $\delta\widetilde{T}$, and $\delta\widetilde{n}$ have nodes. Note also that the vector $\delta\widetilde{\mathbf{u}}$ rotates anti-clockwise and that most of the time $\delta\widetilde{T} > 0$ is correlated to $\delta\widetilde{n} > 0$ and $\delta\widetilde{T} < 0$ is correlated to $\delta\widetilde{n} < 0$.

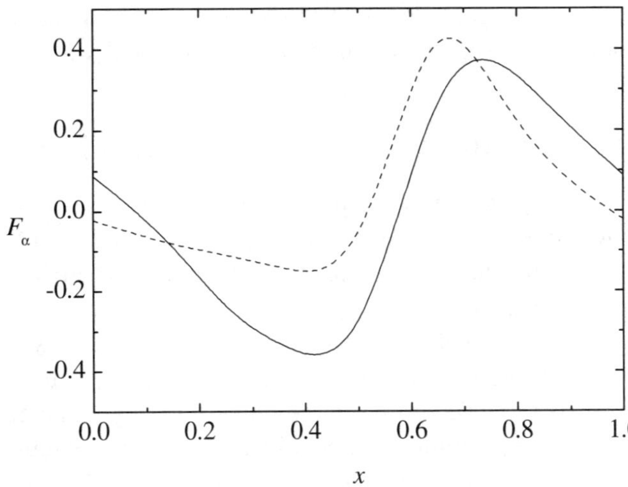

Figure 3.38. Functions $F_T(x)$ (solid line) and $F_n(x)$ (dashed line) determining the wave behavior of the temperature and the density, respectively.

The above results indicate the wave character of the asymptotic state. In Figs. 3.32–3.35 we have focused on the time evolution of the hydrodynamic quantities at the representative point $y = -L/4$. To confirm the wave nature of the state, it is convenient to analyze the spatial dependence of the hydrodynamic fields at several times. It turns out that the simulation results are consistent with two independent invariance relations (Montanero et al., 1998):

$$\delta A_\alpha(y, t') = \delta A_\alpha(y + L/2, t' + \tau_p/2), \tag{3.176}$$

$$\delta A_\alpha(y, t') = \pm\delta A_\alpha(-y, t'), \tag{3.177}$$

where, as in Eq. (3.175), the sign + refers to the density and temperature and the sign – refers to the velocity. Combining Eqs. (3.176) and (3.177), one gets

$$\delta A_\alpha(y, t') = \pm\delta A_\alpha(-y - L/2, t' + \tau_p/2). \tag{3.178}$$

In particular, setting $y = -L/4$, Eq. (3.175) is recovered.

Figure 3.36 shows the spatial variation of $n(y, t')$ and $T(y, t')$ at $t' = 0$, $0.14\tau_p$, $0.25\tau_p$, $0.36\tau_p$, and $0.5\tau_p$. The profiles for the interval $0.5\tau_p \leq t' \leq \tau_p$ can be determined from those for $0 \leq t' \leq 0.5\tau_p$ by applying the property (3.176). As observed in Fig. 3.35 for the special point $y = -L/4$, it is seen that a high (low) temperature is generally correlated to a high (low) density. The spatial distribution of the temperature is highly non-uniform even though the thermostat maintains a constant average temperature. Figure 3.37 shows a vector representation of the components of $\delta\mathbf{u}$ at different values of t'. As anticipated from Fig. 3.35, $\delta\mathbf{u}$ rotates anti-clockwise throughout the system. The layers $y = 0, \pm L/2$ are always nodes of $\delta\mathbf{u}$ and extrema of δT and δn. The pattern indicated by Figs. 3.36 and 3.37 can be described as a periodic standing wave represented by the superposition of two symmetrical waves travelling in opposite directions:

$$\delta A_\alpha(y, t') = F_\alpha(t'/\tau_p - y/L) \pm F_\alpha(t'/\tau_p + y/L), \tag{3.179}$$

where the functions $F_\alpha(x)$ verify the periodicity relation $F_\alpha(x) = F_\alpha(x + 1)$. Figure 3.38 shows the functions $F_T(x)$ and $F_n(x)$ associated with δT^* and δn^*, respectively. It confirms the high degree of correlation between the temperature and density oscillations.

In summary, for initial values of the reduced wavenumber k^* and the reduced shear rate a^* in the predicted unstable domain, small perturbations of the hydrodynamic fields grow according to the linear hydrodynamic equations for $t^* < 50$. Subsequently, nonlinear effects invalidate this theoretical analysis. The simulations show a transient period up to about $t^* \approx 100$, after which a periodic state with a wave pattern is observed for $100 < t^* \leq 2000$. In this asymptotic state the vector quantities oscillate at a period approximately twice that of the scalar fields.

APPENDIX 3.A: Numerical evaluation of the entropy

The reduced excess entropy for the USF is given by Eq. (3.64), that can be rewritten as

$$s_{\text{ex}}^* = -\pi^{-3/2} \int d\boldsymbol{\xi}\, e^{-\xi^2} R(\boldsymbol{\xi}) \ln R(\boldsymbol{\xi}), \tag{3.A.1}$$

where $R(\boldsymbol{\xi}) \equiv f^*(\boldsymbol{\xi})/(\pi^{-3/2} e^{-\xi^2})$ can be obtained from Eq. (3.8). In principle, despite the fact that it is easy to evaluate $R(\boldsymbol{\xi})$ numerically, the triple velocity integral in Eq. (3.A.1) is quite difficult to compute by conventional methods. An alternative route is based on the interpretation of the integral in (3.A.1) as an *average* of the quantity $R(\boldsymbol{\xi}) \ln R(\boldsymbol{\xi})$ over the Maxwell–Boltzmann distribution $\pi^{-3/2} e^{-\xi^2}$. Consequently, s_{ex}^* can be estimated by a Monte Carlo integration as

$$s_{\text{ex}}^* \simeq -\frac{1}{N} \sum_{s=1}^{N} R(\boldsymbol{\xi}_s) \ln R(\boldsymbol{\xi}_s), \tag{3.A.2}$$

where $\boldsymbol{\xi}_s$ is a random velocity sampled from the Maxwell–Boltzmann distribution, namely

$$\xi_{s,i} = \sqrt{-\ln x_{s,i}} \cos(\pi y_{s,i}), \tag{3.A.3}$$

where $x_{s,i}$ and $y_{s,i}$ ($s = 1, \ldots, N$, $i = 1, 2, 3$) are random numbers uniformly distributed between 0 and 1. The data shown in Fig. 3.23 have been numerically evaluated with $N \sim 10^6$, their error bars being typically less than 1%.

Chapter 4

UNIFORM SHEAR FLOW IN A MIXTURE

1. Introduction

In Chapters 2 and 3 we have offered a thorough account of the transport properties and velocity distribution function of a *single* dilute gas under the uniform shear flow (USF) state far from equilibrium. Now, the objective is to extend the analysis to the case of a multi-component gas.

Needless to say, the analysis of transport phenomena in fluid systems is certainly more complicated when one deals with a mixture than when one considers the case of a one-component system. Not only the number of transport coefficients is higher but also they are functions of parameters such as the mole fractions, the mass ratios, and the size ratios. At a kinetic level, the description of a multi-component system made of N species is provided by the N distribution functions $f_s(\mathbf{r}, \mathbf{v}; t)$, $s = 1, \ldots, N$, obeying a set of coupled N Boltzmann equations, as discussed in Section 1.4. The aim of this Chapter is to analyze the transport properties of a dilute binary mixture under USF. First, we will consider a situation where the mole fractions are constant so that no mutual diffusion exists in the mixture and the transport of momentum is the only relevant phenomenon. This physical state is consistent with the hydrodynamic balance equations without prejudicing how far the system is from equilibrium. Therefore, the description holds for arbitrary values of the masses, concentrations, and force constants, the transport coefficients being nonlinear functions of the shear rate as well as of the parameters of the mixture. In the context of the Boltzmann equation, explicit results will be obtained in the particular case of Maxwell molecules. As happens in the single-gas case, the collision operator for mixtures of Maxwell molecules is mathematically more tractable than for other potentials. For instance, it can be proved that the coupled set of Boltzmann equations for a binary mixture of Maxwell gases reduces to a *single* Boltzmann equation in the limit of small mass ratio, at least for isotropic distributions (Bobylev and Spiga, 1992). In the USF state, the infinite hierarchy for the moments of the velocity distribution functions can be solved exactly in the case of a Maxwell mixture (Marín et al., 1995). In order to evaluate the effect of the interaction potential on the rheological properties, the same problem is formulated using a generalization (Garzó and López de Haro, 1995) of the Gross–Krook (GK) kinetic model (Gross and Krook, 1956).

An interesting point is the investigation of the tracer limit, namely, when the mole fraction of one of the components tends to zero. In this limit, the results obtained from the Boltzmann equation for Maxwell molecules (Marín et al., 1996c) and from the kinetic model for general repulsive potentials (Marín et al., 1997) show a surprising effect: the partial contribution of the tracer species to the total properties of the mixture becomes *finite* if the shear rate is larger than a certain critical value a_c, which is a function of the parameters of the mixture and the interaction model considered. This phenomenon can be interpreted as a *nonequilibrium phase transition* in velocity space. While for non-Maxwell molecules the corresponding order parameter is discontinuous at the critical point (first-order transition), the transition becomes continuous at a_c in the special case of Maxwell molecules.

The chapter ends with an analysis of transport of mass and energy in a binary mixture subjected to the USF. This allows us to get generalizations of Fick's law and Dufour's effect, leading to a diffusion tensor and a Dufour tensor, as nonlinear functions of the shear rate.

2. Maxwell molecules

Let us consider a gas mixture of N species under the USF state depicted in Fig. 2.1. At a macroscopic level, the number densities n_s are constant, the temperature T is uniform, and the flow velocity $\mathbf{u}$ has the linear profile $\mathbf{u} = ay\widehat{\mathbf{x}}$. Microscopically, the USF state is characterized by a uniform distribution in the Lagrangian frame, namely $f_s(\mathbf{r}, \mathbf{v}; t) = f_s(\mathbf{V}, t)$, where $\mathbf{V} = \mathbf{v} - \mathbf{u}$. As in the one-component case [cf. Eq. (2.6)] the Boltzmann equation in this frame becomes

$$\frac{\partial}{\partial t} f_s - \frac{\partial}{\partial V_i} a_{ij} V_j f_s = \sum_{r=1}^{N} J_{sr}[f_s, f_r], \tag{4.1}$$

where the binary collision operator $J_{sr}[f_s, f_r]$ is defined by Eq. (1.129). Equation (4.1) must be complemented by the Lees–Edwards periodic boundary conditions (2.7).

In order to get explicit results, we particularize to Maxwell molecules, i.e. the interaction potential is $\phi_{sr}(r) = K_{sr} r^{-4}$. In this case, Eq. (4.1) has a scaling property similar to the one seen in Section 2.2 for the one-component gas, namely

$$J_{sr}[\overline{\mathbf{V}}|\overline{f}_s, \overline{f}_r] = e^{3\alpha t} J_{sr}[\mathbf{V}|f_s, f_r], \tag{4.2}$$

where

$$\overline{\mathbf{V}} = e^{-\alpha t}\mathbf{V}, \quad \overline{f}_s(\overline{\mathbf{V}}) = e^{3\alpha t} f_s(\mathbf{V}), \tag{4.3}$$

and α is an arbitrary constant. As a consequence, Eq. (4.1) is mathematically equivalent to

$$\frac{\partial}{\partial t}\overline{f}_s - \frac{\partial}{\partial \overline{V}_i}\left(a_{ij}\overline{V}_j + \alpha \overline{V}_i\right)\overline{f}_s = \sum_{r=1}^{N} J_{sr}[\overline{f}_s, \overline{f}_r]. \tag{4.4}$$

This equation can be interpreted as the Boltzmann equation for USF in the presence of a non-conservative external force of the form $\mathbf{F}_s = -m_s\alpha\overline{\mathbf{V}}$. Without loss of generality, we will choose the constant α as a function of the shear rate a by the condition that the (scaled) temperature $\overline{T} = e^{-2\alpha t}T$ reaches a constant value in the long-time limit. Henceforth, for the sake of simplicity, we drop the bars and understand that all quantities are scaled according to Eq. (4.3).

The first non-trivial flux in a multi-component gas is the mass flux $\mathbf{j}_s = \rho_s(\mathbf{u}_s - \mathbf{u})$ defined by Eq. (1.138), i.e.

$$\mathbf{j}_s = m_s \int d\mathbf{v}\,(\mathbf{v} - \mathbf{u}) f_s(\mathbf{v}). \tag{4.5}$$

The evolution equation for this flux is obtained from Eq. (4.4) by multiplying by $m_s\mathbf{V}$ and integrating over velocity. It is given by

$$\frac{\partial}{\partial t}\mathbf{j}_s + \mathsf{a}\cdot\mathbf{j}_s + \alpha\mathbf{j}_s = -\sum_{r=1}^{N}\frac{\lambda_{sr}}{m_s m_r}\left(\rho_r\mathbf{j}_s - \rho_s\mathbf{j}_r\right), \tag{4.6}$$

where use has been made of Eq. (1.B.1) and where λ_{sr} is defined by Eq. (1.B.3). In general, due to the constraint $\sum_s \mathbf{j}_s = \mathbf{0}$, one has to deal with a set of $N-1$ linear independent equations. Equation (4.6) shows that the mass fluxes tend to vanish after a few collision times. As an illustration, let us consider the binary case ($N = 2$), which yields

$$\mathbf{j}_1(t) = (\mathsf{I} - t\mathsf{a})\cdot\mathbf{j}_1(0)e^{-(\alpha+\lambda_{12}\rho/m_1m_2)t}. \tag{4.7}$$

Here, I is the 3×3 identity tensor. From now on we will take the steady-state values $\mathbf{j}_s = \mathbf{0}$ for the mass fluxes and will focus on the transient regime and the steady-state values of the pressure tensor, which is the most relevant quantity in a shearing problem.

2.1 Transient regime

Although the transport properties are defined in the long-time limit (hydrodynamic regime), it is instructive to study the transient regime starting from an arbitrary (but uniform in the Lagrangian frame) initial condition. Multiplying both sides of Eq. (4.4) by $m_s\mathbf{V}\mathbf{V}$ and integrating over the velocity space, one gets

$$
\begin{aligned}
\frac{\partial}{\partial t}P_{s,ij} + a_{ik}P_{s,kj} + a_{jk}P_{s,ki} + 2\alpha P_{s,ij} &= \sum_{r=1}^{N}\frac{\lambda'_{sr}}{(m_s+m_r)m_s} \\
\times\left[(\rho_s p_r + \rho_r p_s)\,\delta_{ij} - \rho_s P_{r,ij} - \rho_r P_{s,ij}\right] \\
-\sum_{r=1}^{N}\frac{\lambda_{sr}}{(m_s+m_r)m_s}2\left(\frac{m_s}{m_r}\rho_r P_{s,ij} - \rho_s P_{r,ij}\right),
\end{aligned}
\tag{4.8}
$$

where $P_{s,ij}$ is the *partial* pressure tensor

$$
P_{s,ij} = m_s\int d\mathbf{V}\, V_i V_j f_s(\mathbf{V}) \tag{4.9}
$$

and $p_s = \frac{1}{3}\mathrm{tr}\,\mathsf{P}_s$ is the partial hydrostatic pressure. From it one can define a partial temperature T_s associated with species s as $T_s = p_s/n_s k_B$. The deviations of the partial temperatures T_s from the global temperature defined as $T = \sum_s x_s T_s$ [cf. Eq. (1.190)], where x_s are the mole fractions, are a measure of the breakdown of energy equipartition out of equilibrium. Upon writing Eq. (4.8), use has been made of Eq. (1.B.4), the coefficient λ'_{sr} being given by Eq. (1.B.5). Taking the trace in Eq. (4.8) and summing for $s = 1, 2, \ldots, N$, one gets the evolution equation

$$
\frac{\partial}{\partial t}p + \frac{2a}{3}P_{xy} + 2\alpha p = 0, \tag{4.10}
$$

where $p = \sum_s p_s$ is the total hydrostatic pressure and $\mathsf{P} = \sum_s \mathsf{P}_s$ is the total pressure tensor. The above equation coincides with that of the single-gas case, Eq. (2.34), and expresses the balance equation for the total energy [cf. Eq. (1.137)] particularized to the USF.

Equation (4.8) represents a linear set of $6N$ equations, whose solution gives the time evolution of the partial pressure tensors for any choice of α and for arbitrary values of the physical parameters of the problem, namely the shear rate a, the masses m_s, the densities n_s, and the force constants K_{sr}. In dimensionless units, the parameter space has $N(N+5)/2-2$, corresponding to the shear rate, $N-1$ mass ratios, $N-1$ mole fractions, and $N(N+1)/2-1$ force constant ratios. In a similar way as what happens in the one-component case [cf. Eqs. (2.31)–(2.33)], the elements $P_{s,xz}$ and $P_{s,yz}$, and the difference $P_{s,yy} - P_{s,zz}$

are decoupled from the remaining elements, so that the original set of $6N$ equations becomes three sets of N equations each (with the same eigenvalues) plus a coupled set of $3N$ equations. The latter, involving $P_{s,xx}$, $P_{s,yy}$, and $P_{s,xy}$, is the relevant one for long times. As said before, we choose the scaling parameter α by the thermostat condition, namely the temperature must reach a constant value in the long-time limit. This condition yields a $3N$-degree equation for α.

For the sake of concreteness, henceforth we will consider the case of a binary mixture ($N = 2$). Even in this case, the complexity of the problem with respect to the single-gas case increases significantly (Marín et al., 1995). As pointed out above, $P_{1,yz}$, $P_{2,yz}$, $P_{1,xz}$, $P_{2,xz}$, $P_{1,yy} - P_{1,zz}$, and $P_{2,yy} - P_{2,zz}$ are decoupled from the remaining elements. They tend to zero as a linear combination of terms of the form $e^{-\ell_\sigma t}$, with characteristic times ℓ_σ^{-1} given by the roots of the quadratic equation

$$(2\alpha - \ell)^2 + (B_{11} + B_{22})(2\alpha - \ell) + B_{11}B_{22} - B_{12}B_{21} = 0. \qquad (4.11)$$

Here, we have introduced the coefficients

$$B_{11} = 3A_{11} + 2\frac{\lambda_{12}}{m_1 + m_2}n_2, \qquad (4.12)$$

$$B_{12} = 3A_{12} - 2\frac{\lambda_{12}}{m_1 + m_2}n_1, \qquad (4.13)$$

where

$$A_{11} = \frac{\lambda'_{11}}{m_1}n_1 + \frac{\lambda'_{12}}{m_1 + m_2}\frac{m_2}{m_1}n_2, \qquad (4.14)$$

$$A_{12} = \frac{\lambda'_{12}}{m_1 + m_2}n_1. \qquad (4.15)$$

Similar expressions define B_{21}, B_{22}, A_{21}, and A_{22} just by making the changes $1 \leftrightarrow 2$.

The relevant elements are

$$\boldsymbol{\mathcal{P}} = \begin{pmatrix} P_{1,xx} \\ P_{2,xx} \\ P_{1,yy} \\ P_{2,yy} \\ P_{1,xy} \\ P_{2,xy} \end{pmatrix}. \qquad (4.16)$$

Their evolution equation is

$$\left(\frac{\partial}{\partial t} + \boldsymbol{\mathcal{L}}\right) \cdot \boldsymbol{\mathcal{P}} = 0, \qquad (4.17)$$

where $\boldsymbol{\mathcal{L}} = 2\alpha\boldsymbol{\mathcal{I}} + \boldsymbol{\mathcal{L}}' - \frac{1}{3}\boldsymbol{\mathcal{L}}''$ is the corresponding evolution operator, $\boldsymbol{\mathcal{I}}$ being the 6×6 identity matrix, and

$$\boldsymbol{\mathcal{L}}' = \begin{pmatrix} B_{11} & 0 & 2a & B_{12} & 0 & 0 \\ 0 & B_{11} & 0 & 0 & B_{12} & 0 \\ 0 & a & B_{11} & 0 & 0 & B_{12} \\ B_{21} & 0 & 0 & B_{22} & 0 & 2a \\ 0 & B_{21} & 0 & 0 & B_{22} & 0 \\ 0 & 0 & B_{21} & 0 & a & B_{22} \end{pmatrix}, \tag{4.18}$$

$$\boldsymbol{\mathcal{L}}'' = \begin{pmatrix} A_{11} & 2A_{11} & 0 & A_{12} & 2A_{12} & 0 \\ A_{11} & 2A_{11} & 0 & A_{12} & 2A_{12} & 0 \\ 0 & 0 & 0 & 0 & 0 & 0 \\ A_{21} & 2A_{21} & 0 & A_{22} & 2A_{22} & 0 \\ A_{21} & 2A_{21} & 0 & A_{22} & 2A_{22} & 0 \\ 0 & 0 & 0 & 0 & 0 & 0 \end{pmatrix}. \tag{4.19}$$

The solution of Eq. (4.17) is expressed in terms of the roots ℓ_σ of the corresponding characteristic polynomial, i.e.

$$\det(\boldsymbol{\mathcal{L}} - \ell\boldsymbol{\mathcal{I}}) = 0. \tag{4.20}$$

Equation (4.20) is a sixth-degree equation that, in general, must be solved numerically to get $2\alpha - \ell_\sigma$ as functions of a, m_1, m_2, n_1, n_2, K_{11}, K_{12}, and K_{22}.

At this level, it is convenient to choose a time unit. In general there are several characteristic times in a binary mixture, so that the choice of the adequate time unit depends on the case of interest (tracer limit, disparate-mass mixture, ...). Since our description applies for arbitrary masses, concentrations, and force constants, we take, for simplicity, an effective collision frequency ν defined by

$$\nu = 2n\frac{\lambda'_{12}}{m_1 + m_2}. \tag{4.21}$$

This choice is symmetrical with respect of the change of labels $1 \leftrightarrow 2$. In addition, in the limit of particles mechanically equivalent, Eq. (4.21) reduces to the effective collision frequency defined in the case of a single gas [cf. Eqs. (1.A.2) and (2.28)]. Next, we use the dimensionless quantities $t^* = \nu t$, $a^* = a/\nu$, and $\alpha^* = \alpha/\nu$. Additionally, we introduce the reduced parameters

$$\mu = m_1/m_2, \quad x_1 = n_1/(n_1 + n_2), \tag{4.22}$$

$$\gamma_{11} = \lambda'_{11}/\lambda'_{12} = \sqrt{(K_{11}/K_{12})(1 + \mu)/2}, \tag{4.23}$$

$$\gamma_{22} = \lambda'_{22}/\lambda'_{12} = \sqrt{(K_{22}/K_{12})(1 + \mu)/2\mu}. \tag{4.24}$$

This implies that, apart from the shear rate, the solution of the problem in dimensionless form involves four independent parameters: the mass ratio μ, the mole fraction x_1, and the force constant ratios K_{11}/K_{12} and K_{22}/K_{12}.

Before solving the general problem, it is instructive to consider some limit cases for which Eq. (4.20) factorizes into two cubic equations. First, let us analyze the case of a binary mixture made of particles mechanically equivalent ($\mu = 1, \gamma_{11} = \gamma_{22} = 1$) in which the only distinction between both species is a tag or "color." In that case, Eq. (4.20) reduces to

$$\frac{3}{2}(2\alpha^* - \ell)[1 + (2\alpha^* - \ell)]^2 = a^{*2}, \tag{4.25}$$

$$\frac{3}{4}(2\alpha^* - \ell + \varpi)[1 + 2(2\alpha^* - \ell) + 2\varpi]^2 = a^{*2}. \tag{4.26}$$

Here, the eigenvalues ℓ_σ are understood in units of ν. In Eq. (4.26) $\varpi = \lambda_{sr}/\lambda'_{sr} = 0.645$ is a pure number. Equation (4.25) coincides with the one obtained in the case of a single gas, Eq. (2.44). Its smallest root gives the dominant behavior of the *total* pressure tensor $\mathsf{P} = \mathsf{P}_1 + \mathsf{P}_2$ and its expression is

$$\ell_{\text{min}} = 2\alpha^* - G(a^*), \tag{4.27}$$

where $G(x) \equiv \frac{4}{3}\sinh^2[\frac{1}{6}\cosh^{-1}(1 + 9x^2)]$. On the other hand, Eq. (4.26) is related to the time evolution of the *difference* between P_1/n_1 and P_2/n_2. The smallest root is now

$$\ell'_{\text{min}} = 2\alpha^* + \varpi - \frac{1}{2}G(2a^*). \tag{4.28}$$

It can be easily proved that $\ell'_{\text{min}} > \ell_{\text{min}}$ for any value of the shear rate. This means that the relative temperature difference $(T_1 - T_2)/T$ tends to zero over a time scale shorter than the time scale of variation of the global temperature $T(t)$.

Another interesting limit corresponds to the so-called tracer limit, i.e. $x_1 \to 0$. In this situation, the excess component (species 2) is not disturbed by the presence of the tracer particles (species 1), and the collisions among tracer particles can also be neglected. As a consequence, f_2 obeys a closed Boltzmann equation, while f_1 verifies a linear Boltzmann–Lorentz equation (Garzó and López de Haro, 1991; Garzó and López de Haro, 1993). In the tracer limit, Eq. (4.20) factorizes into

$$\frac{3}{(1+\mu)\gamma_{22}}(2\alpha^* - \ell)\left[\frac{(1+\mu)\gamma_{22}}{2} + (2\alpha^* - \ell)\right]^2 = a^{*2}, \tag{4.29}$$

$$3\mu(2\alpha^* - \ell + \varpi)\left[\frac{1}{2\mu} + (2\alpha^* - \ell) + \varpi\right]^2 = a^{*2}. \tag{4.30}$$

Equation (4.29) is associated with the time evolution of the excess component. In fact, Eq. (4.29) is equivalent to that of a single gas, Eq. (4.25), with the adequate change of time unit.[1] Its smallest root is

$$\ell_{\rm min} = 2\alpha^* - \frac{(1+\mu)\gamma_{22}}{2} G\left(\frac{2}{(1+\mu)\gamma_{22}} a^*\right), \tag{4.31}$$

so that the dominant behavior of the temperature is

$$T(t) = T_2(t) \sim e^{-\ell_{\rm min} t^*}. \tag{4.32}$$

On the other hand, Eq. (4.30) gives the transient behavior of the tracer component. Its smallest root is

$$\ell'_{\rm min} = 2\alpha^* + \varpi - \frac{1}{2\mu} G(2\mu a^*). \tag{4.33}$$

Consequently, the asymptotic behavior of the temperature ratio is

$$\frac{T_1(t)}{T(t)} \sim e^{-(\ell'_{\rm min} - \ell_{\rm min})t^*} + \text{const}. \tag{4.34}$$

Whether the ratio T_1/T reaches a constant value for long times or monotonically grows in time depends on the sign of the difference $\ell'_{\rm min} - \ell_{\rm min}$. A detailed analysis of this point will be postponed to Section 4.4.

An additional limit case in which Eq. (4.20) factorizes into two cubic equations corresponds to $K_{22} \gg K_{12}$ (i.e. $\gamma_{22} \gg 1$), in which case particles of species 2 are again unaffected by collisions with particles of species 1, even though the mole fractions are comparable. If we formally take the limit $\gamma_{22} \to \infty$, we get

$$\frac{3}{(1+\mu)x_2\gamma_{22}}(2\alpha^* - \ell)\left[\frac{(1+\mu)x_2\gamma_{22}}{2} + (2\alpha^* - \ell)\right]^2 = a^{*2}, \tag{4.35}$$

$$\frac{3\mu(2\alpha^* - \ell + x_2\varpi)}{x_2 + (1+\mu)x_1\gamma_{11}}\left[\frac{x_2 + (1+\mu)x_1\gamma_{11}}{2\mu} + (2\alpha^* - \ell) + x_2\varpi\right]^2 = a^{*2}. \tag{4.36}$$

As expected, Eq. (4.35) is again equivalent to that of a single gas with the adequate change of time unit. Its smallest root is

$$\ell_{\rm min} = 2\alpha^* - \frac{(1+\mu)x_2\gamma_{22}}{2} G\left(\frac{2}{(1+\mu)x_2\gamma_{22}} a^*\right), \tag{4.37}$$

[1] More specifically, Eq. (4.29) comes from Eq. (4.25) with the changes $a^* \to a^*(\nu/\nu_{22})$, $\alpha^* \to \alpha^*(\nu/\nu_{22})$, $\ell_\sigma \to \ell_\sigma(\nu/\nu_{22})$, where $\nu_{22} = n_2\lambda'_{22}/m_2$, so that $\nu/\nu_{22} = 2/(1+\mu)\gamma_{22}$ in the tracer limit.

which characterizes the time evolution of the partial temperature T_2. Equation (4.36) incorporates the influence on the evolution of T_1 of collisions both of type 1–1 and of type 1–2. Its smallest root is

$$\ell'_{\rm min} = 2\alpha^* + x_2\varpi - \frac{x_2 + (1+\mu)x_1\gamma_{11}}{2\mu} G\left(\frac{2\mu a^*}{x_2 + (1+\mu)x_1\gamma_{11}}\right). \quad (4.38)$$

If one assumes that, in addition to $K_{22} \gg K_{12}$, one has $K_{11} \gg K_{12}$ (i.e. $\gamma_{11} \gg 1$), Eq. (4.36) reduces to that of a single gas. Of course, this describes a trivial situation in which both species evolve independently of each other. More interesting is the case $K_{11} \ll K_{12}$ (i.e. $\gamma_{11} \ll 1$), where the situation is formally equivalent to that of the tracer limit and Eqs. (4.36) and (4.38) become Eqs. (4.30) and (4.33), respectively, except that $a^* \to a^*/x_2$, $\alpha^* \to \alpha^*/x_2$, and $\ell_\sigma \to \ell_\sigma/x_2$.

With the exception of the limit cases just discussed, Eq. (4.20) must be solved numerically. Figure 4.1 shows the real part of the roots $(2\alpha^* - \ell)$ of Eq. (4.20) versus a^* for the case $\mu = \frac{1}{5}$, $x_1 = \frac{1}{6}$, and $K_{11} = K_{12} = K_{22}$. Obviously, exactly the same curves are obtained in the equivalent case $\mu = 5$, $x_1 = \frac{5}{6}$. At a given value of the shear rate, the difference between the two largest values of $2\alpha^* - \ell_\sigma$ gives the inverse of the relaxation time of the transient regime. It can be proved that this difference does not vanish if $x_s \neq 0$.

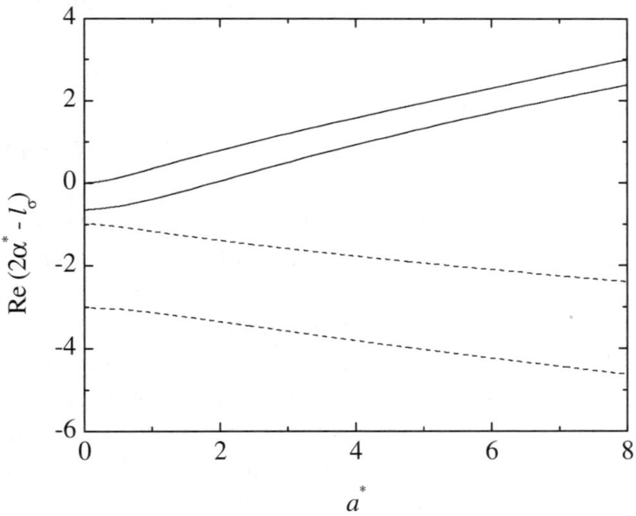

Figure 4.1. Shear-rate dependence of the real part of the roots of Eq. (4.20) for the case $\mu = \frac{1}{5}$, $x_1 = \frac{1}{6}$, and $K_{11} = K_{12} = K_{22}$. The solid lines refer to the real roots, while the dashed lines refer to the complex roots.

As said before, we take $\alpha^*(a^*)$ under the condition that $T(t)$ reaches a stationary value in the long-time limit. This is equivalent to choosing α^* so that

$\ell = 0$ is the smallest root of Eq. (4.20). In other words, $\alpha^*(a^*)$ is the largest root of the sixth-degree equation

$$\det \boldsymbol{\mathcal{L}} = 0. \tag{4.39}$$

Thus, the upper curve in Fig. 4.1 represents $2\alpha^*(a^*)$ for the case $\mu = \frac{1}{5}$, $x_1 = \frac{1}{6}$, and $K_{11} = K_{12} = K_{22}$. In the limit of small shear rates, $\alpha^* \approx \alpha_0 a^{*2}$, where

$$\alpha_0 = \frac{2}{3(1+\mu)} \frac{C}{C'} \tag{4.40}$$

with

$$C = (x_2 - \mu x_1)^2 + x_1 x_2 (\mu + 1)(\gamma_{11} + \mu\gamma_{22}) + 2\mu\varpi, , \tag{4.41}$$

$$C' = x_1^2 \gamma_{11}(2\varpi + \mu) + x_1 x_2(\mu+1)(\gamma_{11}\gamma_{22} + 2\varpi) + x_2^2 \gamma_{22}(2\varpi\mu + 1). \tag{4.42}$$

On the other hand, in the limit of large shear rates, $\alpha \approx \alpha_\infty a^{*2/3}$, where

$$\alpha_\infty = \frac{1}{2 \times 3^{1/3}} \left[A_{11}^* + A_{22}^* + \sqrt{(A_{11}^* - A_{22}^*)^2 + x_1 x_2} \right]^{1/3}, \tag{4.43}$$

and $A_{ss}^* = A_{ss}/\nu$, A_{sr} being defined by Eqs. (4.14) and (4.15). In the special case of particles mechanically equivalent (i.e. $\gamma_{11} = \gamma_{22} = \mu = 1$), we recover the values $\alpha_0 = \frac{1}{3}$ and $\alpha_\infty = \frac{1}{2}\left(\frac{2}{3}\right)^{1/3}$, in agreement with Eqs. (2.61) and (2.64).

2.2 Rheological properties

Now we compute the main transport properties of the problem for arbitrary values of the shear rate a^*, as well as of the ratios μ, x_1, γ_{11}, and γ_{22}. These properties are related to the elements of the pressure tensor P in the long-time limit, i.e. in the stationary regime according to our choice of α^* as a thermostat parameter. In the USF problem, the relevant transport coefficients are the shear viscosity function $F_\eta(a^*)$ and the first and second viscometric functions $\Psi_{1,2}(a^*)$ defined by Eqs. (2.54)–(2.56), respectively. These quantities refer to the properties of the mixture as a whole (total momentum flux). It is also interesting to study the partial contribution corresponding to each species. In particular, we will pay attention to the temperature ratio T_1/T_2, which provides information about how the kinetic energy is distributed between both species. Given that the mixture is arbitrarily far from equilibrium, the equipartition theorem does not apply and so it is expected that $T_1 \neq T_2$.

The stationary forms of P_1 and P_2 are obtained from Eq. (4.17) by solving the homogeneous equation $\boldsymbol{\mathcal{L}} \cdot \boldsymbol{\mathcal{P}} = 0$. This equation has a non-trivial solution due to Eq. (4.39). Such a solution can be written as

$$\widetilde{\boldsymbol{\mathcal{P}}} = \widetilde{\boldsymbol{\mathcal{L}}}^{-1} \cdot \boldsymbol{\mathcal{Q}}, \tag{4.44}$$

where

$$\widetilde{\mathcal{P}} = \begin{pmatrix} p_1 \\ P_{1,yy} \\ P_{2,yy} \\ P_{1,xy} \\ P_{2,xy} \end{pmatrix}, \tag{4.45}$$

$\widetilde{\mathcal{L}} = 2\alpha^* \mathcal{I} + \widetilde{\mathcal{L}}'$ with

$$\widetilde{\mathcal{L}}' = \begin{pmatrix} B_{11}^* - A_{11}^* + A_{12}^* - B_{12}^* & 0 & 0 & \frac{2}{3}a^* & 0 \\ A_{12}^* - A_{11}^* & B_{11}^* & B_{12}^* & 0 & 0 \\ A_{22}^* - A_{21}^* & B_{21}^* & B_{22}^* & 0 & 0 \\ 0 & a^* & 0 & B_{11}^* & B_{12}^* \\ 0 & 0 & a^* & B_{21}^* & B_{22}^* \end{pmatrix}, \tag{4.46}$$

and

$$\mathcal{Q} = \begin{pmatrix} A_{12}^* - B_{12}^* \\ A_{12}^* \\ A_{22}^* \\ 0 \\ 0 \end{pmatrix} p. \tag{4.47}$$

Equation (4.44) provides explicit expressions for p_s/p, $P_{s,yy}/p$, and $P_{s,xy}/p$ in terms of the thermostat parameter α^*, the shear rate a^*, and the parameters characterizing the mixture, namely μ, x_1, γ_{11}, and γ_{22} (Marín et al., 1995). To close the problem one needs to make use of the largest root of the sixth-degree equation (4.39), which gives the value of α^* corresponding to the shear rate and the mixture parameters of interest.

As said before, the temperature ratio $T_1/T_2 = x_2 p_1 / x_1(p - p_1)$ is a measure of the lack of equidistribution of the kinetic energy. Obviously, $T_1 = T_2$ for any value of the shear rate in the case of mechanically equivalent particles. In general, the behavior for small shear rates is

$$\frac{T_1}{T_2} \approx 1 + \frac{4}{3\varpi} \frac{\chi_0}{C'} a^{*2}, \tag{4.48}$$

with

$$\chi_0 = x_1(\mu - \gamma_{11}) + x_2(\mu\gamma_{22} - 1) \tag{4.49}$$

and C' given by Eq. (4.42) On the other hand, for large shear rates the temperature ratio tends asymptotically to a constant value

$$\begin{aligned} \frac{T_1}{T_2} - 1 &\approx \frac{x_2}{2(12\alpha_\infty^3 - A_{11}^*)} - 1 \\ &= \frac{\sqrt{[\chi_0(\mu+1) + \mu]^2 - 4x_1\chi_0\mu(\mu+1)} - \chi_0(\mu+1) - \mu}{2x_1\mu}. \end{aligned} \tag{4.50}$$

According to Eq. (4.48), the ratio T_1/T_2 starts increasing with the shear rate if the parameters of the mixture are such that $\chi_0 > 0$. In this case, Eq. (4.50) shows that the asymptotic value of T_1/T_2 is smaller than 1. The opposite happens if $\chi_0 < 0$. Therefore, the temperature ratio T_1/T_2 does not present a monotonic behavior and reaches a maximum (minimum) if χ_0 is positive (negative). Consequently, both temperatures coincide at a certain value of the shear rate, which depends on the parameters of the mixture. For the sake of illustration, the temperature ratio T_1/T_2 is plotted in Fig. 4.2 versus a^* for several values of μ and x_1. In this figure, we have assumed that the force constants have a mass dependence of the form $K_{sr} \propto (m_s m_r)^{1/2}$. Such a dependence has been proposed to model the cross section observed in disparate-mass binary mixtures (Johnson, 1978). We observe that, for small shear rates, the kinetic energy per particle of the solute component (species 1) is larger than that of the solvent component (species 2) if the former is heavier than the latter. However, the opposite happens for sufficiently large shear rates.

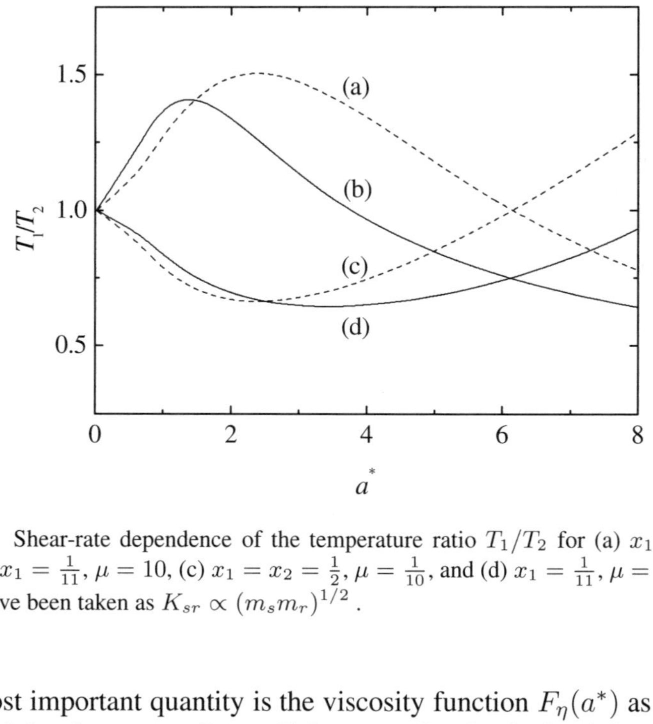

Figure 4.2. Shear-rate dependence of the temperature ratio T_1/T_2 for (a) $x_1 = x_2 = \frac{1}{2}$, $\mu = 10$, (b) $x_1 = \frac{1}{11}$, $\mu = 10$, (c) $x_1 = x_2 = \frac{1}{2}$, $\mu = \frac{1}{10}$, and (d) $x_1 = \frac{1}{11}$, $\mu = \frac{1}{10}$. The force constants have been taken as $K_{sr} \propto (m_s m_r)^{1/2}$.

The most important quantity is the viscosity function $F_\eta(a^*)$ as a nonlinear function of the shear rate. Its explicit expression is obtained from Eq. (2.54) with $P_{xy} = P_{1,xy} + P_{2,xy}$, where $P_{s,xy}/p$ are determined from Eq. (4.44) (Marín et al., 1995). Equivalently, Eq. (4.10) implies that

$$F_\eta(a^*) = \frac{3\alpha^*(a^*)}{a^{*2}}, \tag{4.51}$$

where $\alpha^*(a^*)$ is the largest root of Eq. (4.39). In the limit of zero shear rate one recovers the well-known Navier–Stokes expression (Chapman and Cowling, 1970) $F_\eta(0) = \eta_0(\nu/p)$, where the shear viscosity η_0 is given by Eq. (1.151). More explicitly,

$$F_\eta(0) = 3\alpha_0, \tag{4.52}$$

with α_0 given by Eq. (4.40). Because of the choice of units, $F_\eta(0) \neq 1$, except in the limit of particles mechanically equivalent. Figure 4.3 shows $F_\eta(a^*)/F_\eta(0)$ for $x_1 = \frac{1}{11}$, $\mu = \frac{1}{20}$, $\frac{1}{10}$, 1, 5, 10, and for $K_{sr} \propto (m_s m_r)^{1/2}$. The known shear-thinning effect for a single gas [cf. Fig. 2.3] extends as well to binary mixtures, i.e. $F_\eta(a^*)$ decreases as a^* increases. An interesting feature that can be observed from the figure is that, at a given value of the reduced shear rate, the relative viscosity is larger than that of the single gas ($\mu = 1$), regardless of the values of the mole fractions and the mass ratio. Furthermore, at given values of the mole fraction $x_1 < x_2$ and of the shear rate, $F_\eta(a^*)/F_\eta(0)$ decreases as the mass ratio increases if $\mu < 1$, while the opposite happens if $\mu > 1$.

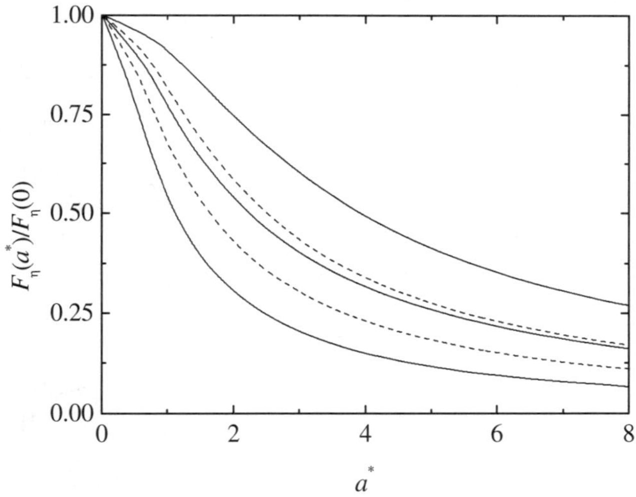

Figure 4.3. Shear-rate dependence of the relative viscosity function $F_\eta(a^*)/F_\eta(0)$ for $x_1 = \frac{1}{11}$ and (from top to bottom) $\mu = \frac{1}{20}$, $\frac{1}{10}$, 10, 5, and 1. Here, $K_{sr} \propto (m_s m_r)^{1/2}$.

The viscometric functions, defined by Eqs. (2.55) and (2.56), provide information about normal stress differences in the system. As seen in the previous Section, $P_{zz} - P_{yy}$ tends to zero in the long-time limit. Consequently, the second viscometric function is zero regardless of the values of a^*, x_1, μ, γ_{11}, and γ_{22}. This property is peculiar of the Maxwell interaction, as was illustrated in Fig. 3.17. The expression for $\Psi_1(a^*)$ can be obtained from Eq. (4.44) by taking into account that $P_{s,xx} = 3p_s - 2P_{s,yy}$ (Marín et al., 1995). In the limit

$a^* \to 0$, one gets

$$\Psi_1(0) = -\frac{8}{(1+\mu)^2}\frac{C''}{C'^2}, \tag{4.53}$$

where

$$\begin{aligned} C'' &= x_1^3\mu^2(\mu+2\varpi)^2 + x_1^2 x_2[\mu^2(1-2\mu) + 4\mu^2\varpi(3\varpi+\mu-2) \\ &\quad +(1+\mu)^2\gamma_{11}^2 + 2\mu(1+\mu)(2\gamma_{11}\varpi - \gamma_{11} + \mu^2\gamma_{22} + 2\mu\gamma_{22}\varpi)] \\ &\quad +x_1x_2^2[\mu(\mu-2) + 4\mu\varpi(3\mu\varpi+1-2\mu) + \mu^2(1+\mu)^2\gamma_{22}^2 \\ &\quad +2(1+\mu)(\gamma_{11} + 2\mu\gamma_{11}\varpi - \mu^2\gamma_{22} + 2\mu^2\varpi\gamma_{22})] \\ &\quad +x_1^3(1+2\mu\varpi)^2, \end{aligned} \tag{4.54}$$

and C' is given by Eq. (4.42). Notice that $\Psi_1(0)$ is a Burnett transport coefficient. It reduces to $\Psi_1(0) = -2$ in the case of identical particles. In Fig. 4.4 we plot $\Psi_1(a^*)/\Psi_1(0)$ as a function of a^* for the same cases as in Fig. 4.3. As happens in a single gas, $\Psi_1(a^*)$ is a decreasing function. In addition, the dependence of Ψ_1 on the parameters of the mixture is qualitatively similar to that of the viscosity function.

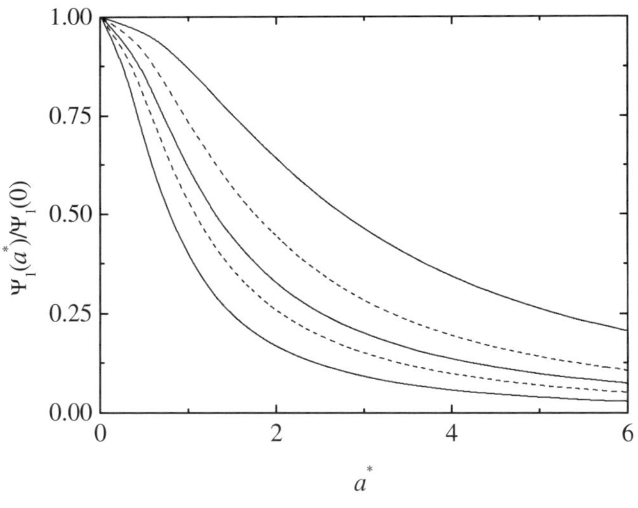

Figure 4.4. Shear-rate dependence of the reduced first viscometric function $\Psi_1(a^*)/\Psi_1(0)$ for $x_1 = \frac{1}{11}$ and (from top to bottom) $\mu = \frac{1}{20}, \frac{1}{10}$, 10, 5, and 1. Here, $K_{sr} \propto (m_s m_r)^{1/2}$.

3. General repulsive interactions

As in the single-gas case, the solution described in the previous Section has two limitations. First, it is restricted to the special case of Maxwell molecules, and so the results obtained before cannot be extrapolated *in principle* to more general interaction potentials. Second, and due to the mathematical difficulties embodied in the Boltzmann collision operators, the explicit expressions for the velocity distribution functions are not known, even for Maxwell molecules, although indirect information about them can be obtained through the velocity moments. An alternative to overcome analytically both limitations is to use a kinetic model. As discussed in Section 1.7, the idea behind a kinetic model is to replace the Boltzmann collision integrals (which contain a detailed description of the two-body interactions) by simple collision terms retaining the main physical properties of the true collision operators.

In this Section we re-examine the USF problem for a dilute binary mixture starting from a generalization of the well-known Gross–Krook (GK) kinetic model (Gross and Krook, 1956) [cf. Eqs. (1.182), (1.184), (1.193), and (1.195)] where the effective collision frequencies depend on the "temperature" of each species (Garzó and López de Haro, 1995) [cf. Eq. (1.198)]. This allows for the consideration of general repulsive interaction potentials of the form $\phi_{sr}(r) = K_{sr}r^{-w}$, where $w = 4, \ldots, \infty$. In Section 1.7 we discussed that the GK model violates in general the exact condition (1.199), *unless* $T_s = T$ in the limit of mechanically equivalent particles. Since the latter condition does apply in the steady-state USF we want to describe, the above drawback is not present and so we can use the GK model for this problem.

As indicated by Eq. (4.7), after a transient period, the flow velocities of all the species coincide with the barycentric velocity $\mathbf{u} = \mathsf{a} \cdot \mathbf{r}$, i.e. $\mathbf{u}_s = \mathbf{u}$. Consequently, the collision term of the GK model becomes

$$J_{sr}^{\mathrm{GK}}[f_s, f_r] = -\nu_{sr}(f_s - f_{sr}), \tag{4.55}$$

where ν_{sr} is defined by Eq. (1.198) with $\overline{T}_s = T_s$ and the reference distribution f_{sr} is

$$f_{sr} = n_s \left(\frac{m_s}{2\pi k_B T_{sr}}\right)^{3/2} \exp\left(-\frac{m_s}{2k_B T_{sr}} V^2\right) \tag{4.56}$$

with

$$T_{sr} = T_s + \frac{2m_s m_r}{(m_s + m_r)^2}(T_r - T_s). \tag{4.57}$$

In the case of Maxwell molecules, the scaling properties (4.3) and (4.4) are preserved by the GK model. For non-Maxwell interactions, however, such a scaling does not hold and, therefore, the properties of the system with and without the thermostat are different, as was discussed in Section 3.3 for the one-component case. Here we will assume the presence of the thermostat term to achieve a steady state in the long-time limit. Therefore, the GK model kinetic equation for this USF state is

$$\frac{\partial}{\partial t} f_s - \frac{\partial}{\partial V_i}\left(a_{ij} V_j + \alpha V_i\right) f_s = -\sum_{r=1}^{N} \nu_{sr}(f_s - f_{sr}). \tag{4.58}$$

This equation is the GK version of Eq. (4.4). Taking second-degree moments in Eq. (4.58), one gets

$$\frac{\partial}{\partial t}P_{s,ij} + a_{ik}P_{s,kj} + a_{jk}P_{s,ki} + 2\alpha P_{s,ij} = -\sum_{r=1}^{N} \nu_{sr}\left(P_{s,ij} - n_s k_B T_{sr}\delta_{ij}\right). \tag{4.59}$$

In the following, we restrict ourselves to binary mixtures ($N = 2$). In that case, Eq. (4.59) yields a closed set of 12 equations. However, as happened in the Boltzmann description for Maxwell molecules, the quantities $P_{s,xz}$, $P_{s,yz}$, and $P_{s,yy} - P_{s,zz}$ are decoupled from the other six moments given by Eq. (4.16) and tend to zero for long times. The time evolution of the column matrix $\mathcal{P}$ has the form of Eq. (4.17), except that in the expressions (4.18) and (4.19) for the matrices $\mathcal{L}'$ and $\mathcal{L}''$ the elements A_{sr} and B_{sr} are replaced by

$$A_{11} = \nu_{11} + \nu_{12}\frac{m_1^2 + m_2^2}{(m_1 + m_2)^2}, \quad A_{12} = 2\nu_{21}\frac{m_1 m_2}{(m_1 + m_2)^2}, \tag{4.60}$$

$$B_{11} = \nu_{11} + \nu_{12}, \quad B_{12} = 0. \tag{4.61}$$

For general potentials, the corresponding Eq. (4.17) is a *nonlinear* differential equation due to the temperature dependence of the collision frequencies. In the case of Maxwell molecules ($w = 4$), ν_{sr} is independent of time and one has to deal with a linear differential equation. The shear-rate dependence of the dominant eigenvalues are quite close to those obtained from the Boltzmann equation (Marín and Garzó, 1997).

3.1 Rheological properties

Now we focus on the analysis of the steady state. The elements of $\mathcal{P}$ are then obtained from $\mathcal{L} \cdot \mathcal{P} = 0$. This gives the elements of the reduced partial pressure tensors P_s/p in terms of the thermostat parameter α, the shear rate a, and the parameters of the mixture. To close the solution, one needs to know the shear-rate dependence of α. This is obtained from the consistency condition $(P_{1,xx} + 2P_{1,yy} + P_{2,xx} + 2P_{2,yy})/p = 3$, which is equivalent to the condition (4.39). More specifically, one gets the following closed set of two equations for α and the temperature ratio T_1/T_2 (Marín and Garzó, 1996):

$$(\widetilde{A}_{11} - 3)(\widetilde{A}_{22} - 3) - \widetilde{A}_{12}\widetilde{A}_{21} = 0, \tag{4.62}$$

$$\frac{T_1}{T_2} = \frac{3 - \widetilde{A}_{22}}{\widetilde{A}_{12}}, \tag{4.63}$$

where

$$\widetilde{A}_{sr} = \frac{A_{sr}}{\nu_r + 2\alpha}\left[3 + \frac{2a^2}{(\nu_r + 2\alpha)^2}\right]. \tag{4.64}$$

Here, $\nu_s = \sum_r \nu_{sr}$ is the total collision frequency for particles of species s. While in general one has to solve the set of nonlinear algebraic equations (4.62) and (4.63) numerically, the behavior for small shear rates can be obtained analytically. The result is

$$\alpha \approx \frac{1}{3}\left(\frac{x_1}{\nu_1^{(0)}} + \frac{x_2}{\nu_2^{(0)}}\right) a^2, \tag{4.65}$$

$$\frac{T_1}{T_2} \approx 1 + \frac{(m_1+m_2)^2}{3m_1m_2}\frac{x_2}{\nu_{12}^{(0)}}\left(\frac{1}{\nu_1^{(0)}} - \frac{1}{\nu_2^{(0)}}\right) a^2. \tag{4.66}$$

In these equations, $\nu_s^{(0)} = \sum_r \nu_{sr}^{(0)}$, where $\nu_{sr}^{(0)}$ are the collision frequencies at equilibrium, i.e. when all the partial temperatures are equal to the mixture temperature T. More specifically, from Eq. (1.198) we have

$$\nu_{sr}^{(0)} = \mathrm{A}(w) n_r K_{sr}^{2/w}\left(\frac{m_s+m_r}{m_s m_r}\right)^{1/2}(2k_BT)^q, \tag{4.67}$$

where $q = 1/2-2/w$ and $\mathrm{A}(w)$ is a numerical constant that can be chosen to optimize the agreement with the Boltzmann equation. In the original formulation of the GK model for Maxwell molecules (Gross and Krook, 1956), as well as in its generalization to $w \neq 4$ (Garzó and López de Haro, 1995), the constant is given from the requirement of reproducing the transfer equations for momentum and energy. This leads to $\mathrm{A}(w) = \frac{8}{3}\sqrt{\pi}\Gamma(3-2/w)w^{2/w}\mathrm{A}^{(1)}(w)$ with $\mathrm{A}^{(1)}(w)$ defined by Eq. (1.118) and given on p. 172 of Chapman and Cowling's textbook (Chapman and Cowling, 1970). For instance, $\mathrm{A}(w=4) = 4\pi\times 0.422$. According to this criterion, ν_{sr} is given by Eq. (1.194) in the case of Maxwell molecules. Here, however, we will take the constant $\mathrm{A}(w)$ to give the Boltzmann Navier–Stokes shear viscosity in the limit of particles mechanically equivalent. This condition implies that $\mathrm{A}(w) = \frac{4}{5}\sqrt{\pi}\Gamma(4-2/w)w^{2/w}\mathrm{A}^{(2)}(w)$, where $\mathrm{A}^{(2)}(w)$ is also tabulated (Chapman and Cowling, 1970). For Maxwell molecules, this second choice yields $\mathrm{A}(w=4) = 3\times 1.3703 = 3\pi\times 0.436$ and then ν_{sr} is given by Eq. (1.221).

Equation (4.66) shows that, in the limit of small shear rates, T_1/T_2 is larger than 1 if $\nu_1^{(0)} < \nu_2^{(0)}$. The latter condition is equivalent to

$$\begin{aligned} &x_1\left[(1+\mu)/2 - (K_{11}/K_{12})^{2/w}\sqrt{(1+\mu)/2}\right] \\ &+x_2\left[(K_{22}/K_{12})^{2/w}\sqrt{\mu(1+\mu)/2} - (1+\mu)/2\right] \;>\; 0, \end{aligned} \tag{4.68}$$

where $\mu \equiv m_1/m_2$. This condition is analogous to the condition $\chi_0 > 0$ in the case of the Boltzmann equation for Maxwell molecules [cf. Eq. (4.48)]. For

Maxwell molecules, the coefficients $\widetilde{A}_{sr}$ defined by Eqs. (4.60) and (4.64) are independent of the temperature ratio and so Eq. (4.62) reduces to a sixth-degree equation for α. Once solved, Eq. (4.63) gives explicitly the temperature ratio. In that case, the condition for $T_1/T_2 > 1$ in the domain of small shear rates does not coincide with the one derived from the Boltzmann equation, Eq. (4.48), but is close to it for not very disparate masses.

In order to present specific results, it is convenient to introduce dimensionless quantities such as μ, x_1, K_{11}/K_{12}, and K_{22}/K_{12}. In addition, we use the cross collision frequency

$$\nu' \equiv \frac{\nu_{12}^{(0)}}{x_2} = \frac{\nu_{21}^{(0)}}{x_1} \tag{4.69}$$

to define the time unit. Note that in the case of Maxwell molecules, the collision frequency ν' is related to the one introduced in the Boltzmann description, Eq. (4.21), by

$$\nu' = \nu \frac{(m_1 + m_2)^2}{4m_1 m_2}. \tag{4.70}$$

While the choice (4.21) seems quite natural in the case of the Boltzmann equation for Maxwell molecules, it is more convenient to take Eq. (4.69) as an effective collision frequency in the case of the GK model for general repulsive interactions. Of course, $\nu' = \nu$ for equal masses. The shear-rate dependence of the temperature ratio T_1/T_2 is shown in Fig. 4.5 for $x_1 = \frac{1}{4}$, $K_{11} = K_{12} = K_{22}$, and two values of the mass ratio. Two interaction potentials are considered: Maxwell molecules ($w = 4$) and hard spheres ($w \to \infty$). The results obtained from the Boltzmann equation for Maxwell molecules are also presented for comparison. We observe that, independently of the interaction considered, the behavior of the temperature ratio is not monotonic. The influence of the interaction potential on T_1/T_2 is rather weak, the breakdown of energy equipartition tending to be less significant as the potential becomes harder. Concerning the comparison with the Boltzmann solution, the qualitative trends predicted by the latter are well captured by the GK model, although, at a quantitative level, the model underestimates the deviation of the temperature ratio from equipartition.

Once the thermostat parameter α and the temperature ratio T_1/T_2 have been determined, all the elements of the pressure tensor are explicitly known. The most important quantity is the viscosity function $F_\eta(a)$ defined by Eq. (2.54) with the collision frequency ν' playing the role of ν, i.e.

$$P_{xy} = -pF_\eta(a)\frac{a}{\nu'}. \tag{4.71}$$

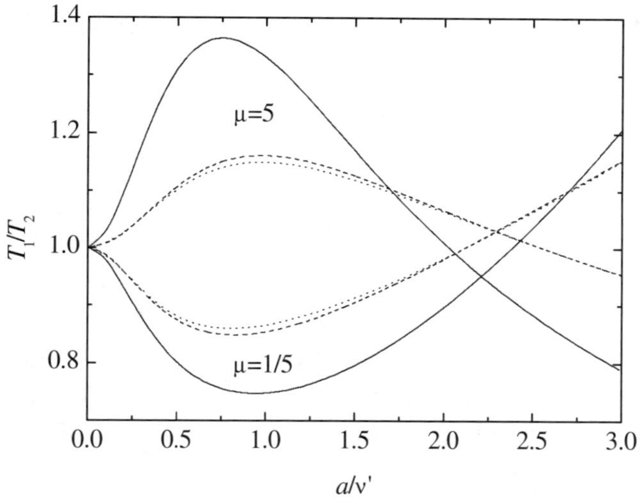

Figure 4.5. Shear-rate dependence of the temperature ratio T_1/T_2 for $x_1 = \frac{1}{4}$, $K_{11} = K_{12} = K_{22}$, and two values of the mass ratio: $\mu = 5$ and $\mu = \frac{1}{5}$. The solid lines refer to the Boltzmann results for Maxwell molecules, while the dashed and dotted lines correspond to the GK model results for Maxwell molecules and hard spheres, respectively.

Its expression is (Marín and Garzó, 1996)

$$\begin{aligned} F_\eta(a) &= \frac{3\alpha(a)}{a^2}\nu' \\ &= \frac{\nu'}{T}\left[x_1\frac{\nu_{11}T_1 + \nu_{12}T_{12}}{(\nu_1 + 2\alpha)^2} + x_2\frac{\nu_{22}T_2 + \nu_{21}T_{21}}{(\nu_2 + 2\alpha)^2}\right]. \end{aligned} \tag{4.72}$$

For vanishing shear rate, the Navier–Stokes shear viscosity coefficient given by the GK model (Goldman and Sirovich, 1967; Garzó et al., 1989) is recovered [cf. Eq. (1.215)], namely

$$F_\eta(0) = \nu'\left(\frac{x_1}{\nu_1^{(0)}} + \frac{x_2}{\nu_2^{(0)}}\right). \tag{4.73}$$

Except in the case of mechanically equivalent particles, the above expression differs from the one obtained from the Boltzmann equation for Maxwell molecules, Eq. (4.52).[2] For a comparison between both Navier–Stokes shear viscosities, see Fig. 1.3. The first correction to the Navier–Stokes value is

$$F_\eta(a^*) \approx F_\eta(0) + F_\eta^{(2)}(a/\nu')^2, \tag{4.74}$$

[2]Note that in units of ν', Eq. (4.52) becomes $F_\eta(0) = [(1+\mu)/2\mu]C/C'$.

where, for Maxwell molecules, the super-Burnett coefficient $F_\eta^{(2)}$ is given by

$$\begin{aligned} F_\eta^{(2)} &= \frac{(m_1+m_2)^2 x_1^2 x_2^2 (\nu_1-\nu_2)^2 \nu'^3}{3m_1 m_2 \nu_1^2 \nu_2^2 x_1 \nu_{12}} \\ &\quad - \frac{2\nu'^3}{3\nu_1^3\nu_2^3}\left[2\left(x_1^2\nu_2^3 + x_2^2\nu_1^3\right) + x_1 x_2 \left(\nu_1+\nu_2\right)\left(\nu_1^2+\nu_2^2\right)\right]. \end{aligned} \tag{4.75}$$

In the particular case of mechanically equivalent particles, we have $\nu_1 = \nu_2 = \nu'$, so we recover the super-Burnett coefficient $F_\eta^{(2)} = -\frac{4}{3}$ obtained in the one-component case [cf. Eq. (2.62)]. Similarly, in the tracer limit $x_2 \to 0$, we get the same result (except for the choice of units), namely $F_\eta^{(2)} = -\frac{4}{3}(\nu'/\nu_1)^3$.

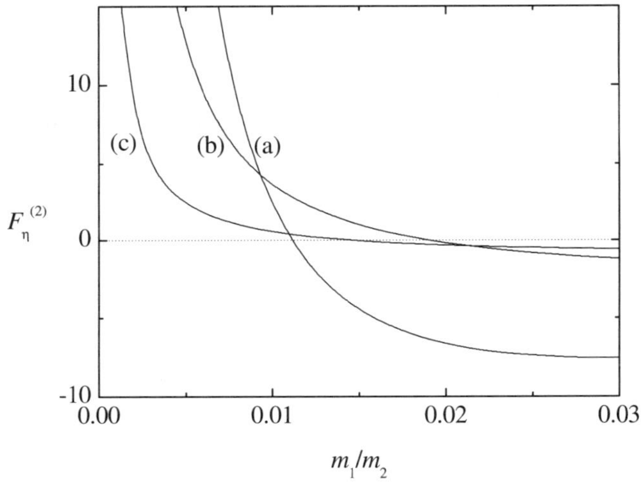

Figure 4.6. Super-Burnett coefficient $F_\eta^{(2)}$ as a function of the mass ratio $\mu = m_1/m_2$ for $K_{11} = K_{12} = K_{22}$ and three values of the mole fraction: (a) $x_1 = 0.25$, (b) $x_1 = 0.5$, and (c) $x_1 = 0.75$.

In the general case, however, the coefficient $F_\eta^{(2)}$ has a complicated dependence on the force constant ratios K_{11}/K_{12} and K_{22}/K_{12}, the mole fraction x_1, and the mass ratio $\mu = m_1/m_2$. Since this coefficient is negative in the single-gas case, one might expect that $F_\eta^{(2)}$ is also negative for a binary mixture, irrespective of the values of the parameters of the mixture. Nevertheless, a careful analysis of Eq. (4.75) shows that there are regions in the parameter space where the first term on the right-hand side of Eq. (4.75) dominates over the second one and $F_\eta^{(2)}$ becomes positive. This happens essentially for very disparate masses, i.e. for $\mu \ll 1$ (where, without loss of generality, we have

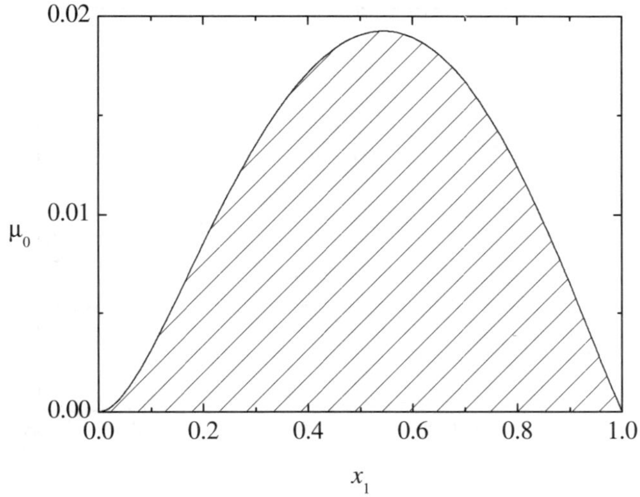

Figure 4.7. Threshold mass ratio $\mu_0(x_1)$ as a function of the mole fraction for $K_{11} = K_{12} = K_{22}$. The shaded region below the curve corresponds to the values of μ and x_1 for which the super-Burnett coefficient $F_\eta^{(2)}$ is positive, so a shear-thickening effect appears for small shear rates.

taken species 1 as the lighter one). To illustrate this change of sign of $F_\eta^{(2)}$, we plot this coefficient as a function of μ for equal force constants and three values of the mole fraction in Fig. 4.6. It can be observed that, at a given value of the mole fraction x_1, the super-Burnett coefficient $F_\eta^{(2)}$ is positive for mass ratios smaller than a certain threshold value $\mu_0(x_1)$. A plot of $\mu_0(x_1)$ is given by Fig. 4.7, which shows that μ_0 has a maximum value $\mu_0 \simeq 0.0193$ at $x_1 \simeq 0.55$, and goes to 0 both for $x_1 \to 0$ and $x_1 \to 1$. The shaded region in Fig. 4.7 represents the values of the mole fraction and the mass ratio (for $K_{11} = K_{12} = K_{22}$) at which $F_\eta^{(2)} > 0$. This means that, for those values, the shear viscosity function starts increasing with the shear rate (*shear-thickening* effect), reaches a maximum and then decreases. Figure 4.6 only shows the branch of μ_0 corresponding to $\mu_0 < 1$. Of course, the symmetry property $\mu_0(x_1) = 1/\mu_0(1-x_1)$ implies the existence of an equivalent branch with $\mu_0 > 1$, so that $F_\eta^{(2)} > 0$ for $\mu > \mu_0(x_1) > 1$.

Once analyzed the small shear-rate behavior of the viscosity function, we turn to explore its full nonlinear dependence. Since we are mainly interested in studying the rheological behavior of the system, we will focus on the ratio $F_\eta(a)/F_\eta(0)$. As an illustration of the shear-thickening effect for very disparate masses, in Fig. 4.8 we plot $F_\eta(a)/F_\eta(0)$ for an equimolar mixture ($x_1 = x_2 = \frac{1}{2}$) with $K_{11} = K_{12} = K_{22}$ and two values of the mass ratio: $\mu = 0.01$ and $\mu = 0.005$. Both cases lie inside the shaded region of Fig. 4.7. We observe

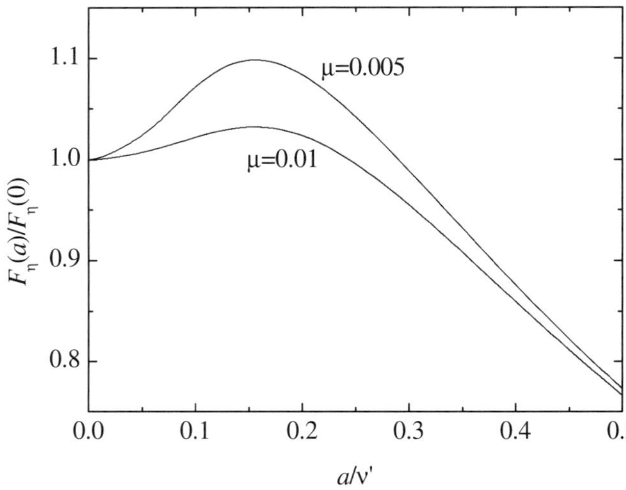

Figure 4.8. Shear-rate dependence of the relative viscosity function $F_\eta(a)/F_\eta(0)$ obtained from the GK model for Maxwell molecules in the case $x_1 = \frac{1}{2}$, $K_{11} = K_{12} = K_{22}$, and two values of the mass ratio: $\mu = 0.01$ and $\mu = 0.005$.

that both curves reach a maximum at about the same shear rate $a/\nu' \simeq 0.15$, the maximum being higher in the most disparate case. After the maximum, the viscosity function decays, crosses its Navier–Stokes value at a certain shear rate ($a^* \simeq 0.24$ for $\mu = 0.01$ and $a^* \simeq 0.29$ for $\mu = 0.005$), and then exhibits a shear-thinning behavior.

Figure 4.9 shows the ratio $F_\eta(a)/F_\eta(0)$ for the same cases as in Fig. 4.5. The two mass ratios considered ($\mu = \frac{1}{5}$ and $\mu = 5$) are not disparate enough to make the shear-thickening effect ($F_\eta^{(2)} > 0$) appear for small shear rates. In these two cases, Fig. 4.9 shows that $F_\eta(a)$ decreases as a increases (shear thinning) whatever the interaction potential considered. The model successfully captures the fact that the inhibition of momentum transport in the case $\mu = 5$ is less noticeable than in the case $\mu = \frac{1}{5}$. In addition, in the latter case the comparison with the exact results shows a better agreement. We also observe that, according to the GK model, the relative viscosity is practically insensitive to the details of the interaction potential. This "universal" character has also been observed in the case of a single component gas under shear flow, as discussed in Section 3.4. The confirmation of this tiny influence of the interaction potential on the transport properties for binary mixtures, as the GK model suggests, would require to solve numerically the Boltzmann equation for non-Maxwell molecules.

From a rheological point of view, normal stresses are also important. They are measured through the viscometric functions defined by Eqs. (2.55) and

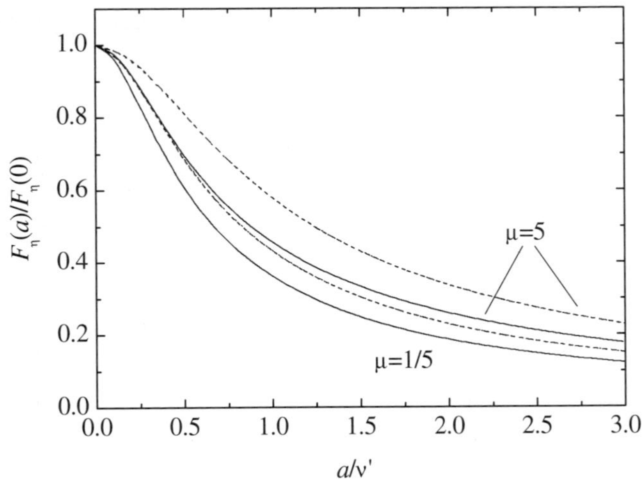

Figure 4.9. Shear-rate dependence of the relative viscosity function $F_\eta(a)/F_\eta(0)$ for $x_1 = \frac{1}{4}$, $K_{11} = K_{12} = K_{22}$, and two values of the mass ratio: $\mu = 5$ and $\mu = \frac{1}{5}$. The solid lines refer to the Boltzmann results for Maxwell molecules, while the dashed and dotted lines correspond to the GK model results for Maxwell molecules and hard spheres, respectively. Note that the dashed and dotted lines are practically indistinguishable.

(2.56) (replacing ν by ν'). As happens in the case of the Boltzmann equation for Maxwell molecules, the GK model leads to $\Psi_2(a) = 0$ regardless of the interaction potential. Based on the results for the single gas (see, for instance, Fig. 3.17), in general $\Psi_2(a) \neq 0$, although its value is expected to be very small. The first viscometric function is given by

$$\Psi_1(a) = -2\frac{\nu'^2}{T}\left[x_1\frac{\nu_{11}T_1 + \nu_{12}T_{12}}{(\nu_1 + 2\alpha)^3} + x_2\frac{\nu_{22}T_2 + \nu_{21}T_{21}}{(\nu_2 + 2\alpha)^3}\right]. \tag{4.76}$$

For vanishing shear rates, one gets the Burnett coefficient

$$\Psi_1(0) = -2\nu'^2\left(\frac{x_1}{\nu_1^{(0)2}} + \frac{x_2}{\nu_2^{(0)2}}\right). \tag{4.77}$$

As expected, this coefficient does not coincide with the one obtained from the Boltzmann equation for Maxwell molecules, Eq. (4.53).[3] The ratio $\Psi_1(a)/\Psi_1(0)$ is plotted in Fig. 4.10 for the same cases as in Figs. 4.5 and 4.9. As in the case of the viscosity function, the influence of the interaction potential on the reduced first viscometric function is very weak. Also, we observe that the GK model

[3]Note that in units of ν', Eq. (4.53) becomes $\Psi_1(0) = -[(1+\mu)^2/2\mu^2]C''/C'^2$.

gives a decay of $\Psi_1(a)/\Psi_1(0)$ smaller than the one predicted by the Boltzmann equation.

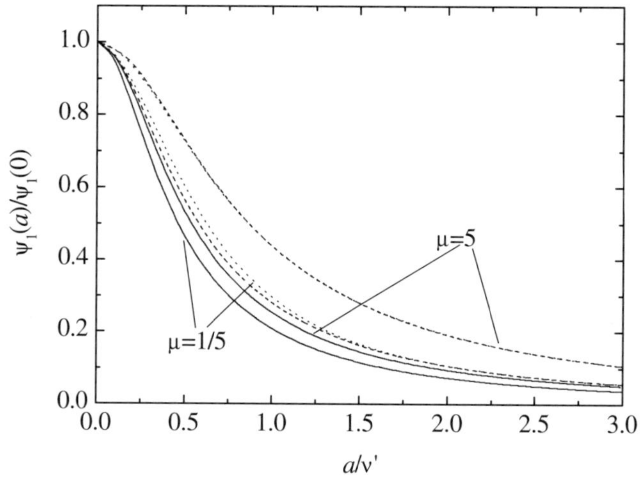

Figure 4.10. Shear-rate dependence of the reduced first viscometric function $\Psi_1(a)/\Psi_1(0)$ for $x_1 = \frac{1}{4}$, $K_{11} = K_{12} = K_{22}$, and two values of the mass ratio: $\mu = 5$ and $\mu = \frac{1}{5}$. The solid lines refer to the Boltzmann results for Maxwell molecules, while the dashed and dotted lines correspond to the GK model results for Maxwell molecules and hard spheres, respectively.

3.2 Velocity distribution functions

The complete description of nonequilibrium states in a mixture requires the knowledge of the velocity distribution functions f_s. Even in the case of Maxwell molecules, an explicit solution to the Boltzmann equation in the USF state is not known and the information about f_s is obtained only indirectly through the knowledge of the first hydrodynamic moments, as derived in Section 4.2. This is one of the main reasons to use a kinetic model since its simplicity allows one to get the distribution functions. In the case of a single component gas under USF, the velocity distribution function obtained from the BGK equation presents a good qualitative agreement with Direct Simulation Monte Carlo (DSMC) simulations, except in the high-velocity region (cf. Fig. 3.1). Now we want to explore whether such an agreement extends to the case of mixtures under USF.

In the steady state, the solution to Eq. (4.58) can be written as

$$\begin{aligned} f_s(\mathbf{V}) &= (\nu_s - 3\alpha + \Lambda)^{-1} \Xi_s(\mathbf{V}) \\ &= \int_0^\infty dt' \, e^{-(\nu_s - 3\alpha)t'} e^{-\Lambda t'} \Xi_s(\mathbf{V}), \end{aligned} \tag{4.78}$$

where $\Xi_s = \sum_r \nu_{sr} f_{sr}$ and Λ is the operator defined by Eq. (3.5). The action of $\exp(-\Lambda t)$ on an arbitrary velocity function is given by Eq. (3.7). To get $f_s(\mathbf{V})$ in a compact form, it is convenient to introduce the dimensionless quantities

$$f_s^*(\boldsymbol{\xi}) = n_s^{-1} \left(\frac{2k_B T}{m_s} \right)^{3/2} f_s(\mathbf{V}), \quad \boldsymbol{\xi} = (m_s/2k_B T)^{1/2} \mathbf{V}. \tag{4.79}$$

From Eq. (4.78) we have

$$\begin{aligned} f_s^*(\boldsymbol{\xi}) &= \pi^{-3/2} \sum_{r=1}^{N} \nu_{sr} (T_{sr}/T)^{-3/2} \int_0^\infty dt'\, e^{-(\nu_s - 3\alpha)t'} \\ &\quad \times \exp\left[-(T_{sr}/T)^{-1} e^{2\alpha t'} (\boldsymbol{\xi} + t' \mathsf{a} \cdot \boldsymbol{\xi})^2 \right]. \end{aligned} \tag{4.80}$$

This expression gives the explicit velocity dependence of the distribution function f_s^*. In the one-component case or in the limit of particles mechanically equivalent, Eq. (4.80) reduces to Eq. (3.8). Equation (4.80) shows that f_s^* is a highly nonlinear function of the shear rate and the parameters of the mixture, both explicitly and through the thermostat parameter α and the temperature ratios $\{T_r/T\}$. In the binary case, those quantities are obtained from the set of equations (4.62) and (4.63). By making use of Eqs. (4.65) and (4.66), the first few terms of the Chapman–Enskog expansion of the distribution function can be easily obtained (Marín and Garzó, 1996).

Using the explicit form of f_s^*, one can easily get the corresponding velocity moments

$$\langle \xi_x^{k_1} \xi_y^{k_2} \xi_z^{k_3} \rangle_s = \int d\boldsymbol{\xi}\, \xi_x^{k_1} \xi_y^{k_2} \xi_z^{k_3} f_s^*(\boldsymbol{\xi}). \tag{4.81}$$

Proceeding in a similar way as in the derivation of Eq. (3.32), one gets

$$\begin{aligned} \langle \xi_x^{k_1} \xi_y^{k_2} \xi_z^{k_3} \rangle_s &= \sum_{r=1}^{N} \nu_{sr} (T_{sr}/T)^{k/2} \\ &\quad \times \sum_{q=0}^{k_1} \frac{k_1!}{(k_1 - q)!} (-a)^q (\nu_s + k\alpha)^{-(q+1)} \langle \xi_x^{k_1 - q} \xi_y^{k_2 + q} \xi_z^{k_3} \rangle_L, \end{aligned} \tag{4.82}$$

where $k \equiv k_1 + k_2 + k_3$ and $\langle \xi_x^{k_1} \xi_y^{k_2} \xi_z^{k_3} \rangle_L$ is given by Eq. (3.31). In particular,

$$\langle \xi^2 \rangle_s = \frac{3}{2(\nu_s + 2\alpha)} \left[1 + \frac{2a^2}{3(\nu_s + 2\alpha)^2} \right] \sum_{r=1}^{N} \nu_{sr} (T_{sr}/T). \tag{4.83}$$

The consistency condition $\langle \xi^2 \rangle_s = \frac{3}{2} T_s/T$ leads to a set of N algebraic coupled equations for the $N-1$ independent temperature ratios and the thermostat parameter α. In the binary case ($N = 2$), Eq. (4.83) reduces to Eqs. (4.62) and (4.63).

The distribution function (4.80) presents a singular behavior not apparent from the velocity moments: f_s^* diverges to infinity at $\boldsymbol{\xi} = \mathbf{0}$ when $\nu_s \leq 3\alpha$. As a consequence, there exists a "condensation" of particles of species s in the state of zero momentum. This singularity was already found in the one-component gas case (cf. Section 3.2) and has a similar physical origin. Since $f_s^*(\boldsymbol{\xi})$ depends on the three components of $\boldsymbol{\xi}$, it is useful to define some marginal distribution functions:

$$f_{s,x}^{(+)}(\xi_x) \equiv \int_0^\infty d\xi_y \int_{-\infty}^\infty d\xi_z f_s^*(\boldsymbol{\xi}), \tag{4.84}$$

$$f_{s,y}^{(+)}(\xi_y) \equiv \int_0^\infty d\xi_x \int_{-\infty}^\infty d\xi_z f_s^*(\boldsymbol{\xi}). \tag{4.85}$$

From Eq. (4.80) it is a simple matter of algebra to get these functions. They are given by

$$\begin{aligned} f_{s,x}^{(+)}(\xi_x) &= \frac{1}{2\sqrt{\pi}} \sum_{r=1}^N \nu_{sr} (T_{sr}/T)^{-1/2} \int_0^\infty dt' \, \frac{e^{-(\nu_s-\alpha)t'}}{(1+a^2t'^2)^{1/2}} \\ & \quad \times \exp\left[-\frac{e^{2\alpha t'}}{(T_{sr}/T)} \frac{\xi_x^2}{1+a^2t'^2} \right] \\ & \quad \times \mathrm{erfc}\left[\frac{e^{\alpha t'}}{(T_{sr}/T)^{1/2}} \frac{at'\xi_x}{(1+a^2t'^2)^{1/2}} \right], \end{aligned} \tag{4.86}$$

$$\begin{aligned} f_{s,y}^{(+)}(\xi_y) &= \frac{1}{2\sqrt{\pi}} \sum_{r=1}^N \nu_{sr} (T_{sr}/T)^{-1/2} \int_0^\infty dt' \, e^{-(\nu_s-\alpha)t'} \\ & \quad \times \exp\left[-\frac{e^{2\alpha t'}}{(T_{sr}/T)} \xi_y^2 \right] \mathrm{erfc}\left[\frac{e^{\alpha t'}}{(T_{sr}/T)^{1/2}} at'\xi_y \right], \end{aligned} \tag{4.87}$$

where $\mathrm{erfc}(x)$ is the complementary error function (Abramowitz and Stegun, 1972). Equations (4.86) and (4.87) are generalizations of Eqs. (3.20) and (3.21), respectively. The marginal distributions $f_{s,x}^{(+)}$ and $f_{s,y}^{(+)}$ diverge at $\xi_x = 0$ and $\xi_y = 0$, respectively, for shear rates such that $\alpha(a) > \nu_s$. This effect is similar to the one discussed in Section 3.2 in the case of a single gas.

For the sake of illustration, let us consider a binary mixture with $x_1 = \frac{1}{6}$, $K_{11} = K_{22} = K_{12}$, and three different values of the mass ratio: $\mu = \frac{1}{10}$, 1, and 10. Figures 4.11 and 4.12 show the ratios

$$R_{2,x}(\xi_x) \equiv \frac{f_{2,x}^{(+)}(\xi_x)}{\frac{1}{2}\pi^{-1/2}e^{-\xi_x^2}}, \quad R_{2,y}(\xi_y) \equiv \frac{f_{2,y}^{(+)}(\xi_y)}{\frac{1}{2}\pi^{-1/2}e^{-\xi_y^2}}, \tag{4.88}$$

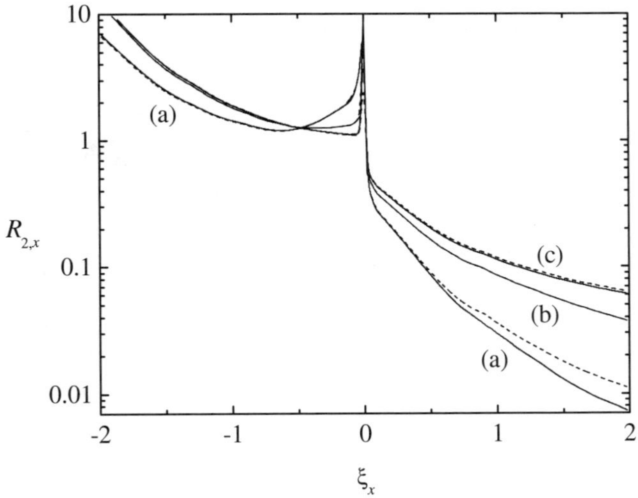

Figure 4.11. Reduced distribution function $R_{2,x}(\xi_x)$ for $x_1 = \frac{1}{6}$, $K_{11} = K_{22} = K_{12}$, $a/\nu' = 8$, and three different values of the mass ratio: (a) $\mu = 0.1$, (b) $\mu = 1$, and (c) $\mu = 10$. The solid lines refer to Maxwell molecules, while the dashed lines refer to hard spheres.

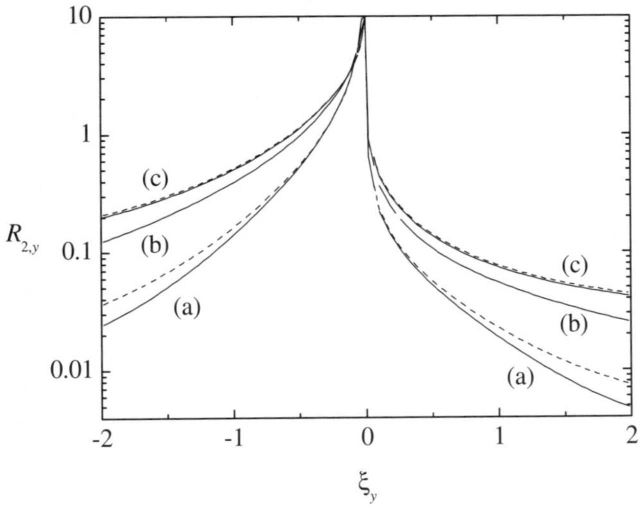

Figure 4.12. Reduced distribution function $R_{2,y}(\xi_y)$ for $x_1 = \frac{1}{6}$, $K_{11} = K_{22} = K_{12}$, $a/\nu' = 8$, and three different values of the mass ratio: (a) $\mu = 0.1$, (b) $\mu = 1$, and (c) $\mu = 10$. The solid lines refer to Maxwell molecules, while the dashed lines refer to hard spheres.

respectively, for $a/\nu' = 8$. The case of equal masses ($\mu = 1$) is equivalent to the solution of the BGK equation for a single gas. Since the shear rate is quite large, the distortion from local equilibrium is significant for all the mass ratios

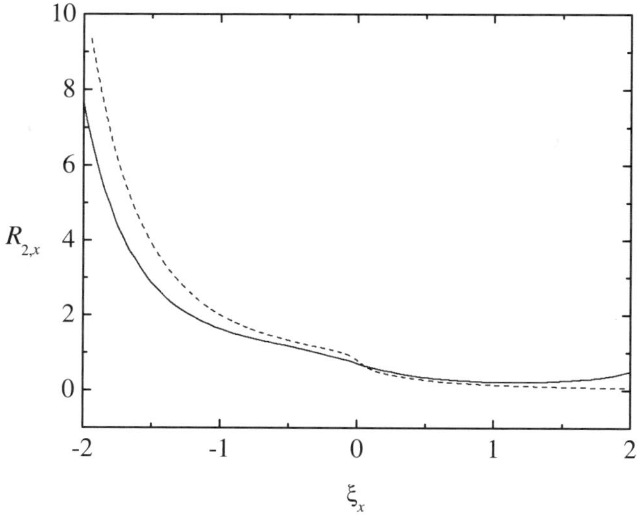

Figure 4.13. Reduced distribution function $R_{2,x}(\xi_x)$ for $x_1 = \frac{1}{4}$, $K_{11} = K_{22} = K_{12}$, $\mu = \frac{1}{5}$ and $a/\nu' = 2.771$, in the case of Maxwell molecules. The solid line corresponds to the numerical solution of the Boltzmann equation via the DSMC method, while the dashed line corresponds to the solution of the GK model.

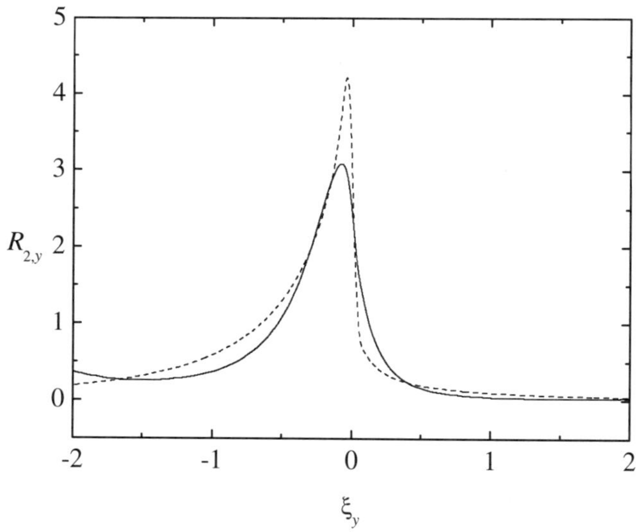

Figure 4.14. Reduced distribution function $R_{2,y}(\xi_y)$ for $x_1 = \frac{1}{4}$, $K_{11} = K_{22} = K_{12}$, $\mu = \frac{1}{5}$ and $a/\nu' = 2.771$, in the case of Maxwell molecules. The solid line corresponds to the numerical solution of the Boltzmann equation via the DSMC method, while the dashed line corresponds to the solution of the GK model.

considered. In fact, the value of a is larger than the threshold value beyond which the marginal distributions $f^*_{2,x}$ and $f^*_{2,y}$ diverge at $\xi_x = 0$ and $\xi_y = 0$. It is interesting to note that the distortion of the solvent distribution becomes larger as the mass of the solute particles decreases. Concerning the influence of the interaction potential, we observe that the distributions for Maxwell molecules and hard spheres are similar for small velocities, while differences are apparent in the region of high velocities.[4] In this region, the population increases as the potential becomes harder, this effect being more significant when the defect particles are lighter than the excess particles. The agreement between the hard sphere and the Maxwell distributions is consistent with the weak influence of the potential on the rheological properties.

Finally, to assess the reliability of the distribution function predicted by the GK model, it is appropriate to compare it with the one obtained from the Boltzmann equation by means of the DSMC method (Garzó and Marín, 1998). In Figs. 4.13 and 4.14 we compare the GK and Boltzmann distributions $R_{2,x}(\xi_x)$ and $R_{2,y}(\xi_y)$ for a binary mixture of Maxwell molecules with $x_1 = \frac{1}{4}$, $K_{11} = K_{22} = K_{12}$, $\mu = \frac{1}{5}$ and $a/\nu' = 2.771$. We see that the GK distributions reproduce fairly well the general qualitative behavior of the Boltzmann distributions. However, there exist discrepancies between the simulation data and the GK predictions, especially in the region of high velocities in the case of $R_{2,x}$ when $\xi_x < 0$ and near the maximum for $R_{2,y}$. In this latter case, the GK distribution predicts well the location of the maximum although not its value. It must be remarked that the shear rate considered in Figs. 4.13 and 4.14 is smaller than the threshold value for the appearance of a divergence at $\xi_x = 0$ or $\xi_y = 0$.

4. Nonequilibrium phase transition in the tracer limit

The study of the properties of tracer particles immersed in a medium is a subject of great interest. For instance, passive tracer particles are used to monitor the characteristics of porous media flows (Bear, 1971; Sahimi, 1993). The so-called Milne problem, which deals with the diffusion of test particles through a background species, has applications in fields such as radiation-transfer problems (Kourganoff, 1963), neutron-transport theory (Duderstadt and Martin, 1979), and rarefied-gas dynamics (Cercignani, 1988). The diffusion of a tagged particle in a concentrated colloidal suspension can also be regarded essentially as a tracer problem (Zhu et al., 1992; Kao et al., 1993).

Since the tracer limit corresponds to a situation in which the mole fraction $x_1 = n_1/n$ of the tracer species is negligible, one expects that the properties of the medium are not affected by the presence of the tracer particles. Let

[4]We recall that the algebraic high-velocity tail obtained from the Boltzmann equation for Maxwell molecules (cf. Chapter 2) is absent in the solution of the BGK model.

us consider, for instance, the relative contribution of the tracer particles to the total energy of the system, E_1/E. At equilibrium, equipartition of energy implies that $E_1/E = x_1$, so that E_1/E goes to zero in the tracer limit. Out of equilibrium, the natural expectation is that $E_1/E \neq x_1$ but $E_1/E \propto x_1$ when $x_1 \ll 1$ and, consequently, the contribution of the tracer species to the total energy is again negligible. In this limit, the breakdown of equipartition would be measured by the proportionality constant. However, the USF state under extreme conditions provides an example of a violation of the above expectation $E_1/E \propto x_1$. In this Section we will show that there exists a *critical* value a_c of the shear rate, such that $\lim_{x_1\to 0} E_1/E = 0$ for $a < a_c$ ("disordered" phase), while $\lim_{x_1\to 0} E_1/E \neq 0$ for $a > a_c$ ("ordered" phase).

In Section 4.2 we studied the time evolution of the second-degree velocity moments, as obtained from the Boltzmann equation for a binary mixture of Maxwell molecules. The evolution of the relevant moments $\mathcal{P}$ is governed by a linear homogeneous set of 6 coupled first-order differential equations with constant coefficients, Eq. (4.17). The general solution is given in terms of the roots of the sixth-degree characteristic equation (4.20). In the tracer limit $x_1 \to 0$, Eq. (4.20) factorizes into two cubic equations, whose smallest roots ℓ_{min} and ℓ'_{min} are given by Eqs. (4.31) and (4.33), respectively. The asymptotic behavior of the temperature of the tracer species, relative to that of the system, is described by Eq. (4.34). In the tracer limit, it is convenient to adopt the collision frequency ν_{22} of the bath particles, given by Eq. (1.221), to define the time unit. It is related to the collision frequency ν defined in Eq. (4.21) by $\nu_{22}/\nu = (1+\mu)\gamma_{22}/2$. In terms of this unit, Eq. (4.34) becomes

$$\frac{T_1}{T} \sim e^{-\Omega t} + \text{const}, \tag{4.89}$$

where

$$\begin{aligned} \frac{\Omega}{\nu_{22}} &= \frac{\nu}{\nu_{22}} \left(\ell'_{\text{min}} - \ell_{\text{min}}\right) \\ &= \frac{2\varpi\mu}{\tau^*} - \frac{G\left(\tau^* a/\nu_{22}\right)}{\tau^*} + G(a/\nu_{22}). \end{aligned} \tag{4.90}$$

Here,

$$\tau^* \equiv \mu(1+\mu)\gamma_{22} = \frac{1}{2}(1+\mu)^2 \nu_{22}/\nu_{12} \tag{4.91}$$

is a measure of the mean free time of the tracer particles, relative to the mean free time of the bath particles. In Eq. (4.90), we recall that $\varpi = 0.645$ and $G(x) \equiv \frac{4}{3}\sinh^2[\frac{1}{6}\cosh^{-1}(1+9x^2)]$. For small shear rates, one has $\Omega > 0$ for any value of the mass ratio μ and the parameter τ^*. More specifically,

$$\frac{\Omega}{\nu_{22}} = \frac{2\varpi\mu}{\tau^*} + \frac{2}{3}\left(1-\tau^*\right)\left(a/\nu_{22}\right)^2 + \mathcal{O}(a^4), \tag{4.92}$$

where use has been made of the behavior $G(x) = \frac{2}{3}x^2 + \mathcal{O}(x^4)$. On the other hand, for asymptotically large shear rates,

$$\frac{\Omega}{\nu_{22}} \approx \left(\frac{2}{3}\right)^{1/3} \left(1 - \tau^{*-1/3}\right) (a/\nu_{22})^{2/3}, \tag{4.93}$$

where we have used the property $G(x) \approx (2/3)^{1/3} x^{2/3}$ for $x \to \infty$. In the special case $\tau^* = 1$, Eq. (4.90) implies that $\Omega/\nu_{22} = 2\varpi\mu$ is independent of the shear rate. In this case, after a transient period of the order of μ/ν_{22}, the temperature ratio T_1/T reaches a stationary value. If $\tau^* < 1$ and the shear rate is small, Eq. (4.92) shows that the transient period is shorter than at equilibrium, while the opposite happens if $\tau^* > 1$ for small shear rates. However, when the shear rate increases, the above trends are reversed, as indicated by Eq. (4.93). In fact, if $\tau^* < 1$, Ω becomes negative for large shear rates and, consequently, the temperature ratio T_1/T monotonically increases with time. Therefore, if $\tau^* < 1$, there must exist a *critical* value of the shear rate, a_c, beyond which $\Omega < 0$ and $T_1/T \to \infty$ (Garzó and López de Haro, 1991). Physically speaking, τ^* is a monotonically increasing function of the mass ratio μ for sensible choices of the force constant ratio. To see this, consider the relationship $\tau^* = (\mu/2)^{1/2}(1 + \mu)^{3/2}(K_{22}/K_{12})^{1/2}$, where we have used Eq. (4.24). In the simple case of equal force constants ($K_{22} = K_{12}$), $\tau^* = (\mu/2)^{1/2}(1 + \mu)^{3/2}$, while in the more realistic case $K_{sr} \propto (m_s m_r)^{1/2}$ (Johnson, 1978), one has $\tau^* = (\mu/4)^{1/4}(1 + \mu)^{3/2}$. In both cases, τ^* is an increasing function of μ. Therefore, the condition $\tau^* < 1$ is equivalent to $\mu < \mu_{\text{th}}$, where the threshold value is $\mu_{\text{th}} = 0.544$ if $K_{22} = K_{12}$ and $\mu_{\text{th}} = 0.443$ if $K_{sr} \propto (m_s m_r)^{1/2}$. For the sake of illustration, in Fig. 4.15 we show the shear-rate dependence of Ω for $\mu = 0.1$ and $\mu = 0.7$ in the case of $K_{sr} \propto (m_s m_r)^{1/2}$. The respective values of τ^* are $\tau^* = 0.459$ and $\tau^* = 1.434$. In both cases, Ω exhibits a non-monotonic behavior, becoming negative for $a > a_c = 7.21\nu_{22}$ in the case $\mu = 0.1$. The dependence of the critical shear rate a_c on the mass ratio $\mu < \mu_{\text{th}}$ is plotted in Fig. 4.16 for the two choices of the force constants previously considered. It is interesting to note that a_c goes to infinity when μ goes to 0 or when it goes to μ_{th}. As a consequence, $a_c(\mu)$ has a non-monotonic behavior and presents a minimum value at a certain value of μ.

According to the above discussion, if either $\mu > \mu_{\text{th}}$ for all shear rates or $\mu < \mu_{\text{th}}$ but $a < a_c(\mu)$, then Ω is positive and, consequently, the temperature ratio T_1/T tends to a stationary value. That stationary value presents a complex dependence on the shear rate, the mass ratio, and the force constant ratio (Garzó and López de Haro, 1993). Its behavior is qualitatively similar to the one observed in the general case (cf. Fig. 4.2). On the other hand, if $\mu < \mu_{\text{th}}$ and $a > a_c(\mu)$, then T_1/T goes to infinity, what represents an extreme breakdown of energy equipartition. This implies that the mean kinetic energy per tracer particle monotonically increases relative to the mean kinetic energy per particle

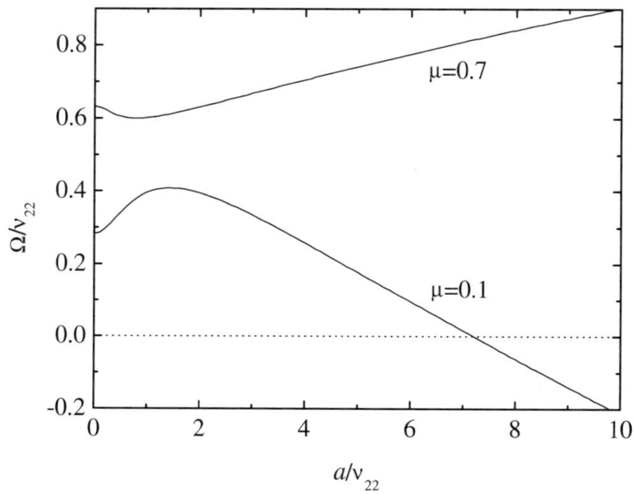

Figure 4.15. Shear-rate dependence of the function Ω for $\mu = 0.1$ and $\mu = 0.7$ in the case $K_{sr} \propto (m_s m_r)^{1/2}$.

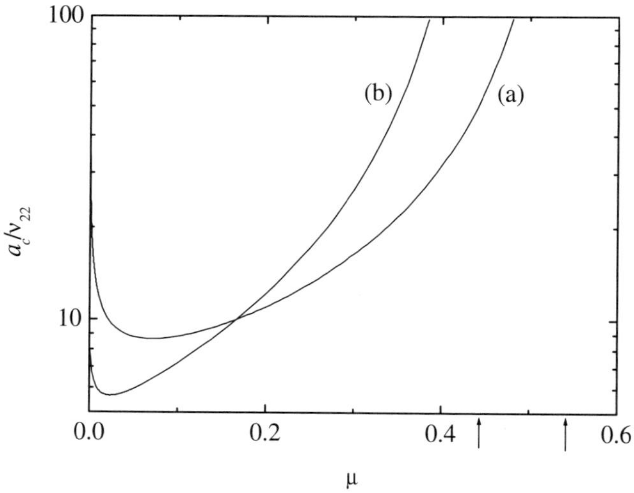

Figure 4.16. Critical shear rate a_c as a function of the mass ratio μ for two choices of the force constants: (a) $K_{12} = K_{22}$ and (b) $K_{sr} \propto (m_s m_r)^{1/2}$. The arrows indicate the location of the corresponding threshold values μ_{th}.

of the bath. The interesting question now is whether the ratio between the total energy of the tracer particles and the total energy of the bath, i.e. $E_1/E = p_1/p = x_1 T_1/T$, reaches a finite value or not. To answer this question one has to start from the general expression of E_1/E for arbitrary mole fractions, as obtained from Eq. (4.44) (Marín et al., 1995). By carefully taking the tracer limit $x_1 \to 0$ (Marín et al., 1996c), one gets

$$\lim_{x_1 \to 0} \frac{E_1}{E} = \begin{cases} F(a/\nu_{22}, \mu, \gamma_{22}) \neq 0 & \text{if } \mu < \mu_{\text{th}} \text{ and } a > a_c(\mu), \\ 0 & \text{otherwise,} \end{cases} \tag{4.94}$$

where $F(a/\nu_{22}, \mu, \gamma_{22})$ is a nonlinear function whose explicit expression can be found elsewhere (Marín et al., 1996c). Therefore, the energy ratio has a nonzero *finite* value in the domain $\mu < \mu_{\text{th}}$ and $a > a_c(\mu)$, which is consistent with the divergence of the temperature ratio in the tracer limit in that domain. Outside that domain, one gets the expected result, namely $E_1/E \to 0$, consistent with the fact that the temperature ratio reaches a stationary value. According to this, a qualitatively different behavior is found in the tracer limit depending on whether the control parameter a is larger or smaller than a_c. As a approaches a_c from above, the function F tends to zero as $F \sim (a - a_c)$ (Marín et al., 1996c). To illustrate the behavior of the energy ratio as the mole fraction x_1 goes to 0, Fig. 4.17 shows $E_1/E - x_1$ versus a for $\mu = 0.1$ and $x_1 = \frac{1}{11}$, $x_1 = 10^{-2}$, and $x_1 = 0$, in the case $K_{11} = K_{12} = K_{22}$. The difference $E_1/E - x_1$ measures the deviation from energy equipartition. For small values of x_1 (say $x_1 \sim 10^{-2}$), that difference is quite small for shear rates $a \lesssim a_c$, but grows rapidly for $a \gtrsim a_c$. This is a precursor of the singular behavior taking place in the tracer limit.

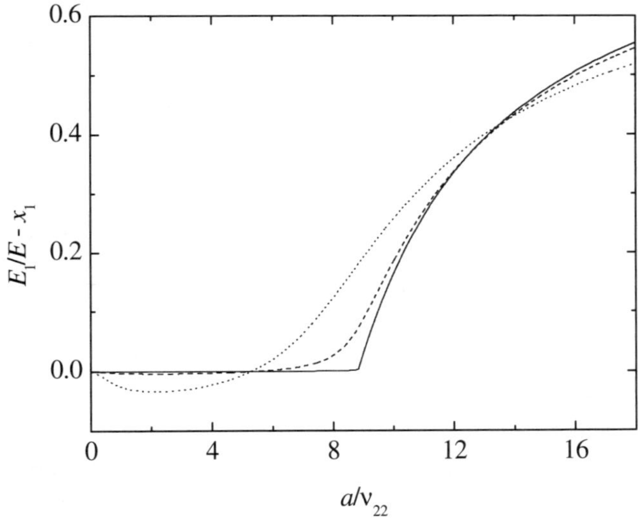

Figure 4.17. Shear-rate dependence of $E_1/E - x_1$ for $\mu = 0.1$, $K_{11} = K_{12} = K_{22}$, and $x_1 = \frac{1}{11}$ (dotted line), $x_1 = 10^{-2}$ (dashed line), and $x_1 = 0$ (solid line).

For given values of the parameters μ and γ_{22}, the existence of a well defined critical shear rate a_c separating different behaviors of the energy ratio E_1/E

in the tracer limit may be interpreted as a *nonequilibrium* phase transition. By borrowing the usual terminology of equilibrium phase transitions (Stanley, 1971), one can identify the energy ratio E_1/E in the tracer limit as an "order" parameter. If $a < a_c(\mu)$, the system evolves in time towards a "disordered" phase, for which $E_1/E = 0$. On the other hand, if $a > a_c(\mu)$, the system tends to an "ordered" phase, where $E_1/E \neq 0$. The phase transition at $a = a_c(\mu)$ only takes place if the mass ratio is smaller than the threshold value μ_{th}, such that $\tau^* < 1$. Figure 4.16 can be then understood as a phase diagram: the points lying above the curve $a_c(\mu)$ represent states belonging to an ordered phase ($E_1/E \neq 0$). The remaining points correspond to a disordered phase ($E_1/E = 0$). In Fig. 4.18 we plot the order parameter E_1/E versus a/a_c for several values of μ in the case $K_{12} = K_{22}$. The fact that the order parameter goes to zero when $a - a_c \to 0^+$ indicates that the transition is of second order. We observe that the order parameter for a given value of $a/a_c > 1$ increases as the mass ratio decreases. For large shear rates, E_1/E tends to the asymptotic value $(1-\tau^*)/(1-\tau^*+\mu)$ (Marín et al., 1996c). Thus, the tracer contribution to the total energy can be even larger than that of the excess component, i.e. $E_1/E > \frac{1}{2}$.

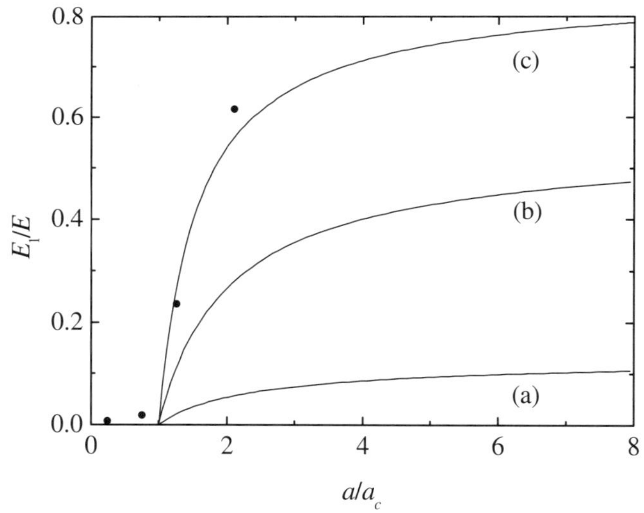

Figure 4.18. "Order parameter" E_1/E versus the shear rate relative to the critical value, a/a_c, for $K_{12} = K_{22}$ and (a) $\mu = 0.5$, (b) $\mu = 0.3$, and (c) $\mu = 0.1$. The circles represent simulation data for $x_1 = 10^{-2}$ and $\mu = 0.1$.

It is tempting to consider an analogy with the *equilibrium* phase transition in a ferromagnetic system. In this sense, the shear rate a plays the role of the (inverse) temperature, the energy ratio E_1/E plays the role of the magnetization, and the mole fraction x_1 plays the role of the external magnetic field.

Consequently, the relationship between E_1/E, a, and x_1 (for a given choice of μ and γ_{22}) may be interpreted as an "equation of state." In general, such a relationship is obtained from Eq. (4.44), where α is the (largest) solution of Eq. (4.39). Near the critical point, i.e. $|a/a_c - 1| \ll 1$ and $x_1 \ll 1$, the sixth-degree equation (4.39) becomes the quadratic equation

$$(\alpha - \alpha_c)^2 - C_1(a - a_c)(\alpha - \alpha_c) + C_2(a - a_c)^2 - C_3 x_1 = 0, \qquad (4.95)$$

where $\alpha_c = (\nu_{22}/2)G(a_c/\nu_{22})$, and C_1, C_2, and C_3 are positive coefficients depending on μ and γ_{22}. The physical root is

$$\alpha - \alpha_c \approx \frac{C_1}{2}(a - a_c) + \frac{1}{2}\left[(C_1^2 - 4C_2)(a - a_c)^2 + 4C_3 x_1\right]^{1/2}. \qquad (4.96)$$

Similarly, Eq. (4.44) in the critical region leads to

$$E_1/E \approx x_1 \left[C_4(\alpha - \alpha_c) - C_5(a - a_c)\right]^{-1}, \qquad (4.97)$$

where C_4 and C_5 are again positive coefficients. The latter can be expressed as $C_5 = (C_4/2)(C_1 + \sqrt{C_1^2 - 4C_2})$. Substituting Eq. (4.96) into Eq. (4.97), one gets

$$E_1/E \approx C(a - a_c) + [C^2(a - a_c)^2 + C' x_1]^{1/2}, \qquad (4.98)$$

where $C \equiv \sqrt{C_1^2 - 4C_2}/2C_3C_4$ and $C' \equiv 1/C_3C_4^2$. Equation (4.98) represents the equation of state in the critical domain. In particular, if now we take the limit $x_1 \to 0$, we get $E_1/E \approx 2C(a - a_c)$ if $a \gtrsim a_c$ and $T_1/T = x_1^{-1}E_1/E \approx (C'/2C)(a_c - a)^{-1}$ if $a \lesssim a_c$. It is remarkable that Eq. (4.98) can be derived by minimizing a Landau-like free energy (Stanley, 1971) given by

$$\Phi_L(a - a_c, x_1; E_1/E) = -x_1(E_1/E) - \frac{C}{C'}(a - a_c)(E_1/E)^2 + \frac{1}{3C'}(E_1/E)^3. \qquad (4.99)$$

In other words, Eq. (4.98) arises from the conditions

$$\frac{\partial \Phi_L}{\partial(E_1/E)} = 0, \qquad \frac{\partial^2 \Phi_L}{\partial(E_1/E)^2} > 0. \qquad (4.100)$$

According to this picture, below (above) the critical shear rate, the ordered (disordered) phase is unstable, the relaxation time going to infinity at criticality. By further exploiting the analogy with equilibrium phase transitions, one can define the following critical exponents (Stanley, 1971):

$$E_1/E \sim x_1^{1/\delta}, \quad a = a_c, \qquad (4.101)$$

$$E_1/E \sim (a - a_c)^{\beta}, \quad x_1 = 0, \quad a > a_c, \qquad (4.102)$$

$$\frac{\partial (E_1/E)}{\partial x_1} \sim |a - a_c|^{-\gamma}, \quad x_1 = 0. \tag{4.103}$$

From Eq. (4.98) one easily gets the values $\delta = 2$, $\beta = 1$, and $\gamma = 1$.

In equilibrium critical phenomena (Stanley, 1971), the appearance of an order parameter is associated with a broken symmetry. The natural question is: which symmetry is broken in our problem? Near equilibrium, E_1/E scales with x_1 (when x_1 is small), so that $x_1^{-1}E_1/E$ is *invariant* under the transformation $x_1 \to \lambda x_1$. We may call "generalized equipartition of energy" to this invariance property. This invariance is broken in the ordered phase, since the mean kinetic energy per tracer particle is very much larger than the one corresponding to the excess particles. In other words, $T_1 \sim T$ in the disordered phase, while $T_1 \sim x_1^{-1}T$ in the ordered phase. Therefore, the phase transition takes place in velocity space. If we define the *scaled* velocity $\widetilde{\boldsymbol{\xi}} = x_1^{1/2}\mathbf{V}/\sqrt{2k_BT/m_1}$ for the tracer particles, then the region of the velocity space explored in the ordered phase corresponds to finite values of $\widetilde{\boldsymbol{\xi}}$, while this region shrinks to $\widetilde{\boldsymbol{\xi}} = \mathbf{0}$ in the disordered phase.

Let us try to gain some insight into the physical origin of the transition studied in this Section. By making $\ell_{\text{min}} = 0$ in Eq. (4.31), we get $\alpha(a) = (\nu_{22}/2)G(a/\nu_{22})$, that can be interpreted as the value of the thermostat parameter needed to get a stationary value for the temperature T_2 of the bath. Of course, this thermostat is also acting on the tracer particles. On the other hand, by making $\ell'_{\text{min}} = 0$ in Eq. (4.33), we get $\alpha'(a) = (\nu_{22}/2)\left[G(\tau^* a/\nu_{22})/\tau^* - 2\varpi\mu/\tau^*\right]$, which is the *minimum* value of the thermostat parameter needed to get a stationary value for the temperature T_1 of the tracer component. The quantity Ω in (4.89) is simply $\Omega = 2(\alpha - \alpha')$. Thus, the tracer particles are subjected to two competing effects. On the one hand, T_1 tends to increase in time due to viscous heating. The strength of this effect is characterized by α', which increases as the collision frequency ν_{12} grows relative to ν_{22} (i.e. as τ^* decreases) and the shear rate increases. On the other hand, the thermostat α tends to control this increase. Therefore, if $\alpha < \alpha'$, the thermostat is not able to control the viscous heating effect and, consequently, T_1/T grows infinitely in time. This situation occurs if $\tau^* < 1$ and the shear rate is large enough (i.e. $a > a_c$). Although the temperature ratio T_1/T goes to infinity when $\alpha < \alpha'$, the *energy* ratio E_1/E reaches a finite value.

Since the nonequilibrium phase transition has been obtained from an algebraic analysis involving a delicate limit, it is convenient to resort to computer simulations to confirm this phenomenon. Figure 4.18 includes DSMC data for E_1/E at $a/a_c = 0.25, 0.76, 1.27$, and 2.11, with a mole fraction $x_1 = 10^{-2}$, a mass ratio $\mu = 0.1$, and equal force constants ($K_{11} = K_{12} = K_{22}$) (Garzó and Marín, 1998). We see that the simulation data, which correspond to quite a small but nonzero mole fraction, clearly support the theoretical predictions.

Thus, for instance, at $a/a_c = 0.76$, the defect particles store less than 2% of the total energy, while the relative contribution to the total energy becomes more than 60% at $a/a_c = 2.11$. This provides a consistent check of the analytical treatment.

An interesting question is whether this transition can also be extended beyond the Maxwell interaction. This question can be answered analytically from the generalized GK model for r^{-w}-potentials. In the context of this model, a detailed analysis (Marín et al., 1997) of the tracer limit allows us again to identify such a transition in velocity space. One important point is that an appropriate unit of time must be taken. In the particular case of Maxwell molecules, the choice between ν_{12}^{-1} (mean free time of the tracer particles) and ν_{22}^{-1} (mean free time of the solvent particles) is not crucial to identify the phase transition since both quantities are independent of the temperature ratio. Nevertheless, for non-Maxwell molecules of the form $\phi(r) \sim r^{-w}$ with $w > 4$, Eq. (1.198) implies that $\nu_{12}/\nu_{22} \sim (1 + \mu^{-1}T_1/T_2)^q$, where $q = 1/2 - 2/w$. Therefore, if there exists a phase transition for non-Maxwell molecules, then the ratio ν_{12}/ν_{22} diverges in the ordered phase. As a matter of fact, taking the tracer limit in the GK solution discussed in Section 4.3, one finds that for any *finite* value of a/ν_{22} the energy ratio E_1/E goes to zero always in the tracer limit, although it can be significant for very large shear rates even for very small (but nonzero) values of the mole fraction x_1 (Marín and Garzó, 1996). However, if we use ν_{12} to nondimensionalize the shear rate, then there exists a *finite* critical value a_c/ν_{12} beyond which the energy ratio takes finite values, provided that the mass ratio is smaller than a certain threshold value μ_{th} (Marín et al., 1997). The dependence of a_c/ν_{12} on μ for several values of the interaction parameter w in the case $K_{12} = K_{22}$ is shown in Fig. 4.19. We observe that the threshold value μ_{th} (at which $a_c/\nu_{12} \to \infty$) increases as the potential becomes harder. Moreover, at a given value of μ, the corresponding critical value a_c/ν_{12} decreases as the hardness of the interaction increases.

The most interesting outcome of the asymptotic analysis from the GK solution (Marín et al., 1997) is that, while the transition is of second order for Maxwell molecules (in agreement with the Boltzmann description), the order parameter E_1/E is discontinuous at the critical point for non-Maxwell molecules, what can be interpreted as a first-order transition. To illustrate this property, in Fig. 4.20 we plot the order parameter E_1/E versus a/a_c for $K_{12} = K_{22}$, $\mu = 0.1$, and $x_1 = 10^{-2}$, 10^{-3}, and 0, in the cases of Maxwell molecules and hard spheres. For hard spheres, it is apparent that in the tracer limit E_1/E goes to 0 when $a \to a_c^-$, while it goes to a finite value when $a \to a_c^+$. It is also interesting to note that, in general, the curves with $x_1 \neq 0$ tend to the asymptotic behavior as $x_1 \to 0$, but this tendency is much slower in the case of hard spheres than for Maxwell molecules.

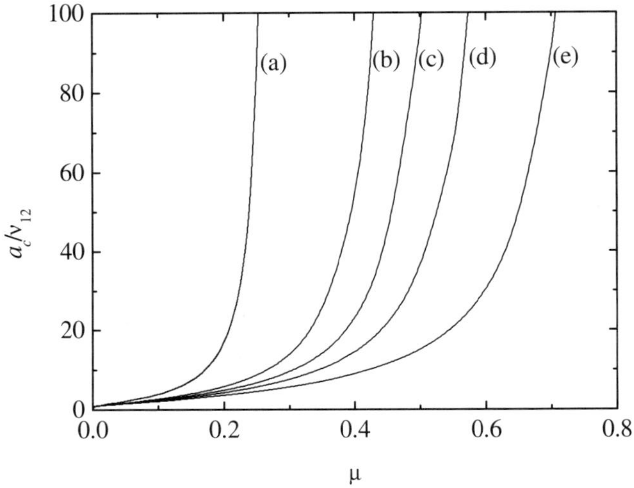

Figure 4.19. Critical shear rate a_c/ν_{12} as a function of the mass ratio μ for $K_{12} = K_{22}$ and (a) $w = 4$ (Maxwell molecules), (b) $w = 6$, (c) $w = 8$, (d) $w = 12$, and (e) $w \to \infty$ (hard spheres).

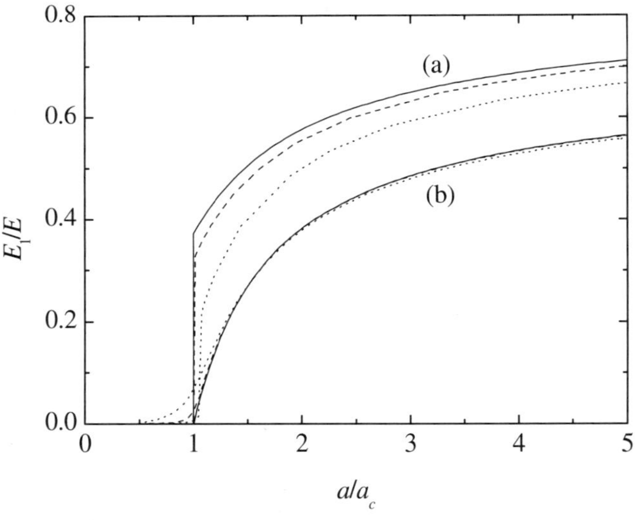

Figure 4.20. Energy ratio E_1/E versus a/a_c for $K_{12} = K_{22}$, $\mu = 0.1$, and $x_1 = 10^{-2}$ (dotted lines), 10^{-3} (dashed lines), and 0 (solid lines), for (a) hard spheres and (b) Maxwell molecules.

5. Generalized diffusion and Dufour coefficients

When a fluid mixture is simultaneously subjected to both weak velocity and concentration gradients, the Curie principle (de Groot and Mazur, 1984) states that the shear field (tensorial quantity) cannot modify vectorial quantities such as mass and heat fluxes, which are generated by a concentration gradient ∇n_s. As a consequence, the mutual diffusion coefficient (which couples the mass current and the concentration gradient) and the Dufour coefficient (which couples the heat flux and the concentration gradient) do not depend on the shear field. Nevertheless, when the shear rate is large, nonlinear effects become important so that the Curie principle does not apply and the above transport coefficients may be modified by the shear flow. If the concentration gradient is weak, one expects that the mass and heat fluxes are still proportional to ∇n_s although the corresponding transport coefficients must be replaced by second-rank tensors with elements that are nonlinear functions of the shear rate. The diffusion process can also be created without a concentration gradient by the action of an external field (e.g. an electric field). The associated transport coefficient is a *mobility* or *conductivity* tensor, which is a nonlinear function of the shear rate (Marín and Garzó, 1999). In this Section we will focus on the diffusion problem induced by a concentration gradient.

In order to gain some insight into this complex nonlinear problem, it is especially useful to consider the situation in which the mixture is subjected to the USF state. As discussed in the previous Sections, this state is well characterized through the knowledge of the first few velocity moments in the context of the Boltzmann equation, as well as at the level of the velocity distribution functions in the case of the GK kinetic model. Let us assume a binary mixture of Maxwell molecules under USF. At a given time we slightly perturb the state by introducing a *weak* density gradient ∇n_s with the constraints $n = n_1 + n_2 = \text{const}$ and $T = \text{const}$. As we will see later, these constraints are consistent with the solution to first order in the concentration gradient. This gradient induces mass and heat fluxes in the system, which are defined by Eqs. (1.138) and (1.141). For zero shear rate, we can apply linear irreversible thermodynamics (de Groot and Mazur, 1984). Thus the mass flux obeys Fick's law [cf. Eq. (1.147)] and the heat flux obeys the Dufour law [cf. Eq. (1.148)]. When the mixture is subjected to USF at finite shear rate, the diffusion and Dufour coefficients become tensorial quantities, so that one can generalize Eqs. (1.147) and (1.148) as

$$\mathbf{j}_s = -\frac{m_1 m_2 n^2}{\rho}\mathsf{D}(a) \cdot \nabla x_s, \tag{4.104}$$

$$\mathbf{J}_q = -k_B \frac{n\rho}{\rho_1 \rho_2}\mathsf{L}(a) \cdot \nabla x_1. \tag{4.105}$$

Our objective in this Section is to determine the diffusion and Dufour tensors in terms of the shear rate and the parameters of the mixture (Marín et al., 2000; Marín and Garzó, 2000).

As said before, let us assume that we perturb the steady USF by introducing a weak concentration gradient ∇x_1. At a kinetic level, the set of Boltzmann equations corresponding to the perturbed USF state is

$$\frac{\partial}{\partial t} f_s - \frac{\partial}{\partial V_i}(a_{ij}V_j + \alpha V_i) f_s + (\mathbf{V} + \mathbf{u}_0) \cdot \nabla f_s = \sum_{r=1}^{N} J_{sr}[f_s, f_r]. \quad (4.106)$$

In this equation we have chosen the thermostat force as proportional to the unperturbed peculiar velocity $\mathbf{V}$. It must be borne in mind that in the perturbed state the true flow velocity $\mathbf{u} = \mathbf{u}_0 + \delta\mathbf{u}$ is, in general, different from the linear velocity profile $\mathbf{u}_0 = \mathsf{a} \cdot \mathbf{r}$. The macroscopic balance equations are obtained by taking moments in Eq. (4.106):

$$\frac{\partial x_s}{\partial t} + \mathbf{u}_0 \cdot \nabla x_s + \nabla \cdot (x_s \delta\mathbf{u}) + \frac{\nabla \cdot \mathbf{j}_s}{m_s n} = 0, \quad (4.107)$$

$$\frac{\partial \delta\mathbf{u}}{\partial t} + \mathsf{a} \cdot \delta\mathbf{u} + \alpha\delta\mathbf{u} + (\mathbf{u}_0 + \delta\mathbf{u}) \cdot \nabla\delta\mathbf{u} + \rho^{-1}\nabla \cdot \mathsf{P} = \mathbf{0}, \quad (4.108)$$

$$aP_{xy} + \nabla \cdot \mathbf{q} + \mathsf{P} : \nabla\delta\mathbf{u} + \alpha(3p + \rho\delta u^2) - \frac{3}{2}k_B T \sum_s \frac{\nabla \cdot \mathbf{j}_s}{m_s} = 0, \quad (4.109)$$

where we have taken into account that n and T are assumed to be constant. We will look for solutions of (4.106) in which all the space and time dependence occurs through a functional dependence on the mole fraction $x_1(\mathbf{r}, t)$. In other words, we look for *normal* solutions of the form $f_s(\mathbf{r}, \mathbf{V}; t) = f_s(x_1; \mathbf{V})$. As in Section 3.6 we apply a generalization of the Chapman–Enskog method, namely (Marín and Garzó, 1998; Marín et al., 2000)

$$f_s = f_s^{(0)} + \varepsilon f_s^{(1)} + \cdots, \quad (4.110)$$

where ε is a formal uniformity parameter of order ∇x_1. The distribution function $f_s^{(0)}$ is the *local* version of the USF distribution, namely the solution to Eq. (4.4) except that now the mole fraction x_1 depends on space and time. Analogously,

$$\delta\mathbf{u} = \varepsilon\mathbf{u}^{(1)} + \cdots, \quad (4.111)$$

$$\mathbf{j}_s = \varepsilon\mathbf{j}_s^{(1)} + \cdots, \quad (4.112)$$

$$\mathsf{P} = \mathsf{P}^{(0)} + \varepsilon\mathsf{P}^{(1)} + \cdots, \quad (4.113)$$

$$\mathbf{q} = \varepsilon\mathbf{q}^{(1)} + \cdots, \quad (4.114)$$

$$\alpha = \alpha^{(0)} + \varepsilon\alpha^{(1)} + \cdots. \quad (4.115)$$

As in the conventional Chapman–Enskog method (Chapman and Cowling, 1970) [cf. Eq. (1.67)], the temporal derivative is expanded in powers of ε as

$$\frac{\partial}{\partial t} = \varepsilon\partial_t^{(1)} + \cdots. \quad (4.116)$$

In particular, from Eq. (4.107), one has

$$\partial_t^{(1)} x_1 = -\mathbf{u}_0 \cdot \nabla x_1. \tag{4.117}$$

In addition, Eqs. (4.108) and (4.109) imply that

$$\mathsf{a} \cdot \mathbf{u}^{(1)} + \alpha^{(0)} \mathbf{u}^{(1)} = -\rho^{-1} \nabla \cdot \mathsf{P}^{(0)} = -\rho^{-1} \frac{\partial \mathsf{P}^{(0)}}{\partial x_1} \cdot \nabla x_1, \tag{4.118}$$

$$a P_{xy}^{(1)} + 3p\alpha^{(1)} = 0. \tag{4.119}$$

The solution to Eq. (4.118) is

$$u_i^{(1)} = -\frac{1}{\rho \alpha^{(0)}} \left(\delta_{ik} - \frac{a_{ik}}{\alpha^{(0)}} \right) \frac{\partial P_{kj}^{(0)}}{\partial x_1} \frac{\partial x_1}{\partial r_j}, \tag{4.120}$$

where use has been made of the identity $(b\mathsf{I} + \mathsf{a})^{-1} = b^{-1}\mathsf{I} - b^{-2}\mathsf{a}$, b being an arbitrary constant.

To first order in ε, the kinetic equation (4.106) for f_1 reduces to

$$\left[\partial_t^{(1)} + (\mathbf{V} + \mathbf{u}_0) \cdot \nabla + \frac{a P_{xy}^{(1)}}{3p} \frac{\partial}{\partial \mathbf{V}} \cdot \mathbf{V} \right] f_1^{(0)}$$
$$- \frac{\partial}{\partial V_i} (a_{ij} V_j + \alpha^{(0)} V_i) f_1^{(1)} = J_{11}^{(1)} + J_{12}^{(1)}, \tag{4.121}$$

where

$$J_{sr}^{(1)} = J_{sr}[f_s^{(0)}, f_r^{(1)}] + J_{sr}[f_s^{(1)}, f_r^{(0)}]. \tag{4.122}$$

Upon writing Eq. (4.121) we have made use of the energy balance equation (4.119). Of course, an equation similar to (4.121) holds for f_2. Since $f_1^{(0)}$ depends on space and time only through x_1, one has

$$(\partial_t^{(1)} + \mathbf{u}_0 \cdot \nabla) f_1^{(0)} = \frac{\partial f_1^{(0)}}{\partial x_1} (\partial_t^{(1)} + \mathbf{u}_0 \cdot \nabla) x_1 = 0, \tag{4.123}$$

where use has been made of the mass balance equation (4.117). Now we focus on the case of Maxwell molecules. If we multiply both sides of Eq. (4.121) by $m_1 \mathbf{V}\mathbf{V}$ and integrate over velocity, the left-hand side becomes a linear combination of the partial pressure tensors $\mathsf{P}_s^{(1)}$. The same happens with the right-hand side, as can be seen from Eq. (1.B.4) to first order in ε. Thus, we get a *homogeneous* set of algebraic equations for $\mathsf{P}_s^{(1)}$, whose solution is just the trivial one, namely $\mathsf{P}_s^{(1)} = 0$. Consequently, Eq. (4.121) simplifies to

$$\mathbf{V} \cdot \nabla f_1^{(0)} - \frac{\partial}{\partial V_i} (a_{ij} V_j + \alpha^{(0)} V_i) f_1^{(1)} = J_{11}^{(1)} + J_{12}^{(1)}. \tag{4.124}$$

5.1 Diffusion tensor

Now we are in conditions of determining the mass flux $\mathbf{j}_1^{(1)}$, which is given by

$$\mathbf{j}_1^{(1)} = m_1 \int d\mathbf{v} \mathbf{V} f_1^{(1)} - \rho_1 \mathbf{u}^{(1)}. \tag{4.125}$$

Multiplying Eq. (4.124) by $\mathbf{V} - \mathbf{u}^{(1)}$ and integrating over velocity one finds that

$$\mathsf{a} \cdot \mathbf{j}_1^{(1)} + \alpha^{(0)} \mathbf{j}_1^{(1)} + \frac{\rho \lambda_{12}}{m_1 m_2} \mathbf{j}_1^{(1)} = \left(\frac{\rho_1}{\rho} \frac{\partial \mathsf{P}^{(0)}}{\partial x_1} - \frac{\partial \mathsf{P}_1^{(0)}}{\partial x_1} \right) \cdot \nabla x_1. \tag{4.126}$$

In this equation we have made use of Eq. (4.118) and Eq. (1.B.1) with $\mathbf{j}_2 = -\mathbf{j}_1$. The solution to Eq. (4.126) can be written in the form of the generalized Fick's law (4.104) with the following expression for the diffusion tensor:

$$\begin{aligned} D_{ij} &= \frac{\rho}{m_1 m_2 n^2} \frac{1}{\alpha^{(0)} + \rho \lambda_{12}/m_1 m_2} \left(\delta_{ik} - \frac{a_{ik}}{\alpha^{(0)} + \rho \lambda_{12}/m_1 m_2} \right) \\ &\quad \times \left(\frac{\partial P_{1,kj}^{(0)}}{\partial x_1} - \frac{\rho_1}{\rho} \frac{\partial P_{kj}^{(0)}}{\partial x_1} \right). \end{aligned} \tag{4.127}$$

This expression gives all the relevant information on the diffusion process in a strongly sheared mixture. The elements D_{ij} present a complex dependence on the shear rate and the parameters of the mixture (masses, concentrations, and force constants) without any constraint on their values. It must be noted that the dependence of the partial pressure tensors $\mathsf{P}_s^{(0)}$ on the mole fraction x_1 is rather intricate, as seen in Section 4.3, so the derivatives in Eq. (4.127) must be carried out with care (Marín and Garzó, 1998). In the absence of shear, we recover the equilibrium mutual diffusion coefficient (Chapman and Cowling, 1970), namely $D_{ij} = D_0 \delta_{ij}$ with D_0 given by Eq. (1.150). According to Eq. (4.127), $D_{xz} = D_{zx} = D_{yz} = D_{zy} = 0$, which is consistent with the symmetry of the problem. In addition, the equality $P_{s,yy}^{(0)} = P_{s,zz}^{(0)}$ for Maxwell molecules implies that $D_{yy} = D_{zz}$. Before illustrating the shear-rate dependence of the elements D_{ij} for different values of the mass ratio and the mole fraction, it is interesting to consider some limit cases. First, if the particles are mechanically equivalent, then $\partial \mathsf{P}_1^{(0)}/\partial x_1 = \mathsf{P}_1^{(0)}/x_1 = \mathsf{P}^{(0)}$ and the total pressure tensor $\mathsf{P}^{(0)}$ is independent of x_1. Thus, Eq. (4.127) becomes

$$D_{ij} = \frac{\rho^{-1}}{\alpha^{(0)} + n\lambda_{12}/m} \left(\delta_{ik} - \frac{a_{ik}}{\alpha^{(0)} + n\lambda_{12}/m} \right) P_{kj}^{(0)}, \tag{4.128}$$

where $n\lambda_{12}/m = 1.29\nu/2$, ν being the effective collision frequency of a single gas defined by Eqs. (1.A.2) or (2.28). Equation (4.128) gives the self-diffusion tensor of tagged particles under USF (Garzó et al., 1990). A generalized Green–Kubo formula expressing the self-diffusion tensor as a time integral of the velocity autocorrelation function was derived by Dufty (1983, 1984) by starting from the Liouville equation. The result is

$$D_{ij} = \lim_{t\to\infty} \int_0^t dt' \, \langle V_i(t) V_j(t') \rangle. \tag{4.129}$$

In the Boltzmann limit, the velocity autocorrelation function of a gas under USF can be exactly evaluated for Maxwell molecules. The result is (Marchetti and Dufty, 1983)

$$\begin{aligned} \langle V_i(t) V_j(t') \rangle &= \rho^{-1} \exp\left[-\left(\alpha^{(0)} + n\lambda_{12}/m \right) (t - t') \right] \\ &\quad \times \left[\delta_{ik} - a_{ik}(t - t') \right] P_{kj}^{(0)}. \end{aligned} \tag{4.130}$$

When this expression is inserted into Eq. (4.129), one recovers Eq. (4.128).

Another interesting case is the tracer limit, namely $x_1 \to 0$ (Garzó and López de Haro, 1991; Garzó and López de Haro, 1993). This implies that $\partial \mathsf{P}_1^{(0)}/\partial x_1 = \mathsf{P}_1^{(0)}/x_1$, while $\mathsf{P}^{(0)} = \mathsf{P}_2^{(0)}$ is independent of x_1. Therefore, Eq. (4.127) yields

$$D_{ij} = \frac{\rho_1^{-1}}{\alpha^{(0)} + n\lambda_{12}/m_1} \left(\delta_{ik} - \frac{a_{ik}}{\alpha^{(0)} + n\lambda_{12}/m_1} \right) P_{1,kj}^{(0)}. \tag{4.131}$$

This case includes the Fokker–Planck limit (Rodríguez et al., 1983), corresponding to the diffusion of a massive tagged particle in a fluid under USF.

Now we turn to the general situation with arbitrary values of μ and x_1. Figures 4.21 and 4.22 show the shear-rate dependence of the yy and xy elements of the reduced diffusion tensor $D_{ij}^* = D_{ij}/D_0$ for $x_1 = \frac{1}{6}$, $K_{11} = K_{22} = K_{12}$, and $\mu = 0.5$ and 2. We have also included the results obtained from the GK model (Marín et al., 2000). The elements D_{yy} and D_{xy} are the relevant ones when the concentration gradient is parallel to the flow velocity gradient, i.e. $\nabla x_1 \| \widehat{\mathbf{y}}$. In the range of shear rates analyzed, deviations with respect to the Navier–Stokes value are important. In the case of D_{yy}^*, we see that this element decreases as the shear rate increases so that the presence of the shear flow inhibits the mass transport along the flow velocity gradient (y-axis). This inhibition becomes more significant when the defect species is heavier than the excess species. The element D_{xy}^* (which is zero in the absence of shear flow) gives the mass flux along the x-direction due to a concentration gradient along the y direction. It can be seen as a measure of the anisotropy generated in the system by the action of the shear field. It is negative, which means that $j_{1,x}$ has the same sign as $\partial x_1/\partial y$. Its dependence on the shear rate is quite similar for the two mass ratios considered: for small shear rates $-D_{xy}^*$ increases with a, while

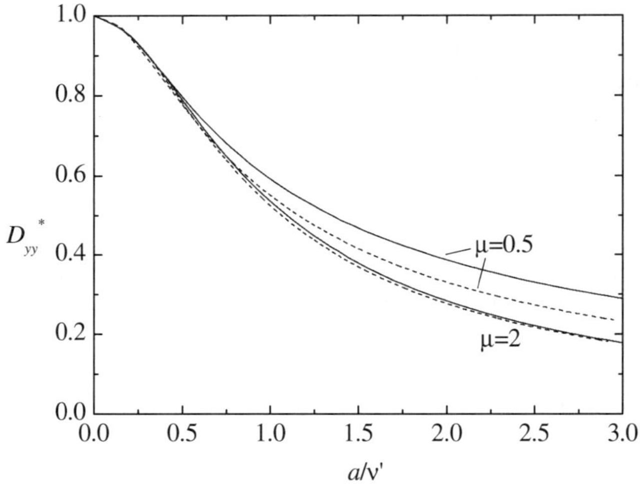

Figure 4.21. Reduced diagonal element D_{yy}^* as a function of the reduced shear rate a/ν' for $K_{11} = K_{22} = K_{12}$, $x_1 = \frac{1}{6}$, and $\mu = 0.5$ and 2. The solid lines correspond to the Boltzmann results, while the dashed lines refer to the GK results.

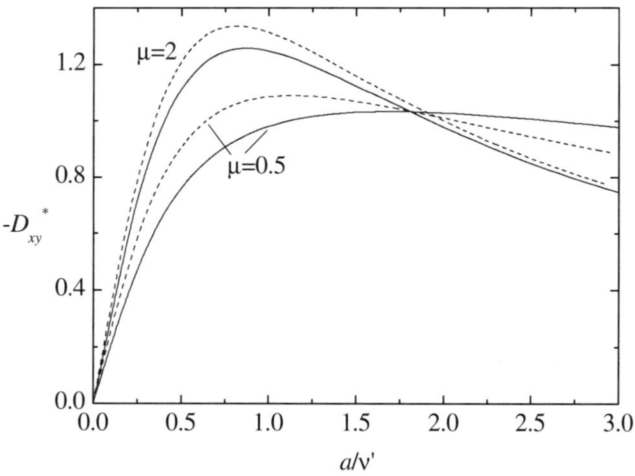

Figure 4.22. Reduced off-diagonal element $-D_{xy}^*$ as a function of the reduced shear rate a/ν' for $K_{11} = K_{22} = K_{12}$, $x_1 = \frac{1}{6}$, and $\mu = 0.5$ and 2. The solid lines correspond to the Boltzmann results, while the dashed lines refer to the GK results.

the opposite happens for large shear rates. It is interesting to remark that the GK predictions present a qualitative good agreement with the Boltzmann ones, especially when the mass of the defect component is larger than that of the excess component. It is interesting to note that the GK and GSB models[5] yield identical results for the mutual diffusion tensor under USF (Garzó and López de Haro, 1992).

The derivation of explicit expressions for the transport properties involved in nonequilibrium states for mixtures may be useful for interpreting experimental or computer simulation results. Although the diffusion process described here has been restricted to Maxwell molecules, it is quite possible that the main trends found for this interaction are representative of what happens for more general interactions. In this context, we recall that the rheological properties derived for a one-component Maxwell gas describe successfully those of a Lennard–Jones gas, as shown in Fig. 3.15. In the case of mixtures, Sarman et al. (1992) performed molecular dynamics simulations in a strongly shearing Lennard–Jones fluid and evaluated the mutual diffusion tensor by means of a Green–Kubo formula (Evans, 1991). Their results show the same qualitative shear-rate dependence of the diffusion tensor as the one presented here. Nevertheless, a quantitative comparison is prevented by the difficulties involved in mapping the Lennard–Jones parameters of the mixture onto the force constants for Maxwell interactions, as well as by the fact that the simulations were carried out for densities beyond the range of applicability of the Boltzmann equation.

It must be borne in mind that the results presented in this Subsection refer to a *thermostatted* gas of Maxwell molecules. The emphasis is in order because, while the distribution function for Maxwell molecules in the unperturbed USF is independent of whether an external force of the form $\mathbf{F} = -m\zeta\mathbf{V}$ exists or not, this is not so in the perturbed state (Garzó et al., 1990). In particular, the self-diffusion tensor is given in the general case by Eq. (4.128) with $\alpha^{(0)}$ replaced by $2\alpha^{(0)} - \zeta$. Thus, the choice $\zeta = \alpha^{(0)}$ corresponds to the particular thermostat force used along this Chapter. However, if the gas is cooled "too rapidly," namely if $\zeta > 2\alpha^{(0)} + n\lambda_{12}/m$, then a hydrodynamic regime is absent and so the diffusion tensor cannot be defined (Garzó et al., 1990).

5.2 Dufour tensor

Now we want to obtain the heat flux $\mathbf{q}^{(1)}$, from which the Dufour tensor defined by Eq. (4.105) can be identified. From Eq. (4.124) it is apparent that to determine $\mathbf{q}^{(1)}$ the fourth-degree moments of the reference distribution function $f_s^{(0)}$ are needed. Since their expressions are not known in the case of a mixture described by the true Boltzmann equation, it is useful to resort to a kinetic model. Here we will consider the GK model for Maxwell molecules, whose solution for USF has been analyzed in Section 4.3. In particular, the generic moments of $f_s^{(0)}$ are given by Eq. (4.82). According to the GK collision term

[5]For a description of the GSB model, see Section 2.7.

(1.182), the perturbed Boltzmann collision operators $J_{sr}^{(1)}$ must be replaced by

$$J_{sr}^{(1)} \to -\nu_{sr}\left[f_s^{(1)} - f_{sr}^{(1)}\right], \tag{4.132}$$

where

$$f_{sr}^{(1)} = \frac{m_s}{k_B T_{sr}} \mathbf{V}\cdot\mathbf{u}_{sr}^{(1)} f_{sr}^{(0)} \tag{4.133}$$

with

$$f_{sr}^{(0)} = n_s\left(\frac{m_s}{2\pi k_B T_{sr}}\right)^{3/2} \exp\left(-\frac{m_s}{2k_B T_{sr}}V^2\right) \tag{4.134}$$

and, according to Eq. (1.193),

$$\mathbf{u}_{sr}^{(1)} = \mathbf{u}^{(1)} + \frac{1}{m_s+m_r}\left(\frac{\mathbf{j}_s^{(1)}}{n_s} + \frac{\mathbf{j}_r^{(1)}}{n_r}\right), \tag{4.135}$$

where we have used the identity $\mathbf{u}_s = \mathbf{u} + \mathbf{j}_s/\rho_s$.

The advantage of using a kinetic model is that it allows us to obtain the velocity distributions $f_s^{(1)}$. The solution to Eq. (4.124) with the replacement (4.132) is

$$\begin{aligned} f_s^{(1)}(\mathbf{V}) &= \left(\nu_s - 3\alpha^{(0)} + \Lambda\right)^{-1} \Xi_s^{(1)}(\mathbf{V}) \\ &= \int_0^\infty dt'\, e^{-(\nu_s - 3\alpha^{(0)})t'} e^{-\Lambda t'} \Xi_s^{(1)}(\mathbf{V}), \end{aligned} \tag{4.136}$$

where

$$\Xi_s^{(1)} = -\mathbf{V}\cdot\nabla f_s^{(0)} + \sum_{r=1}^N \nu_{sr} f_{sr}^{(1)} \tag{4.137}$$

and the operator Λ is defined by Eq. (3.5) with $\alpha \to \alpha^{(0)}$. Equation (4.136) is formally similar to the solution (4.78) of the GK equation for the unperturbed USF. Let us introduce the moments

$$M_{s;k_1,k_2,k_3}^{(1)} = \int d\mathbf{V}\, V_x^{k_1} V_y^{k_2} V_z^{k_3} f_s^{(1)}. \tag{4.138}$$

By following steps similar to those made in the derivation of Eqs. (3.32) and (4.82), one gets

$$M_{s;k_1,k_2,k_3}^{(1)} = \sum_{q=0}^{k_1} \frac{k_1!}{(k_1-q)!} R_{s;k_1-q,k_2+q,k_3}(-a)^q \left(\nu_s + k\alpha^{(0)}\right)^{-(q+1)}, \tag{4.139}$$

where $k \equiv k_1 + k_2 + k_3$ and

$$R_{s;k_1,k_2,k_3} = -\frac{\partial}{\partial x} M^{(0)}_{s;k_1+1,k_2,k_3} - \frac{\partial}{\partial y} M^{(0)}_{s;k_1,k_2+1,k_3} - \frac{\partial}{\partial z} M^{(0)}_{s;k_1,k_2,k_3+1} + 2n_s \sum_{r=1}^{N} \nu_{sr} \left(\frac{2k_B T_{sr}}{m_s} \right)^{(k-1)/2} \left[u^{(1)}_{sr,x} \langle \xi_x^{k_1+1} \xi_y^{k_2} \xi_z^{k_3} \rangle_L + u^{(1)}_{sr,y} \langle \xi_x^{k_1} \xi_y^{k_2+1} \xi_z^{k_3} \rangle_L + u^{(1)}_{sr,z} \langle \xi_x^{k_1} \xi_y^{k_2} \xi_z^{k_3+1} \rangle_L \right]. \quad (4.140)$$

In Eq. (4.140), $M^{(0)}_{s;k_1,k_2,k_3}$ are moments of $f_s^{(0)}$ defined by an equation similar to Eq. (4.138) and $\langle \xi_x^{k_1} \xi_y^{k_2} \xi_z^{k_3} \rangle_L$ are the (reduced) local equilibrium moments given by Eq. (3.31). Note that $R_{s;k_1,k_2,k_3} = 0$ if $k =$ even.

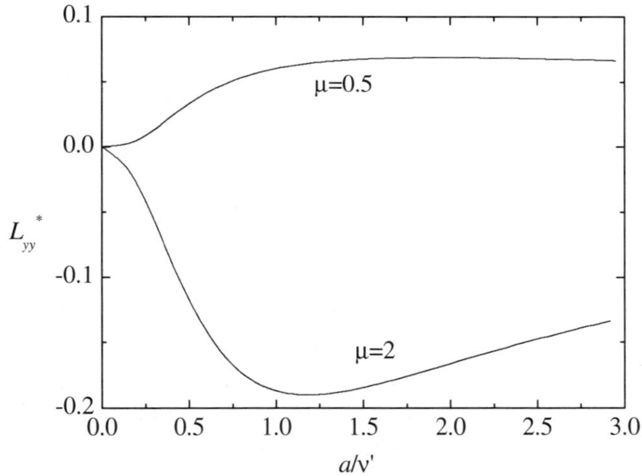

Figure 4.23. Reduced diagonal element L^*_{yy} as a function of the reduced shear rate $a/nu' = a/(n\nu_{12}/n_2)$ for $K_{11} = K_{22} = K_{12}$, $x_1 = \frac{1}{6}$, and $\mu = 0.5$ and 2.

In the particular case $k = 1$, Eq. (4.139) allows us to evaluate the mass fluxes $\mathbf{j}_s^{(1)}$. The associated diffusion tensor is given by Eq. (4.127), except that $\lambda_{12}/m_1 m_2$ must be substituted by $\nu_{12}/n_2(m_1 + m_2)$ and, of course, the unperturbed pressure tensors are different from those obtained in the Boltzmann equation. The next relevant moments are the third-degree moments ($k = 3$). From them, it is a matter of algebra to obtain the heat flux

$$\mathbf{q}^{(1)} = \sum_s \frac{m_s}{2} \int d\mathbf{V}\, V^2 \mathbf{V} f_s(\mathbf{V}) - \frac{3}{2} p \mathbf{u}^{(1)} - \mathsf{P}^{(0)} \cdot \mathbf{u}^{(1)}. \quad (4.141)$$

Once the heat flux is known, the generalized Dufour tensor $\mathsf{L}(a)$ can be identified from Eq. (4.105). Its explicit expression is quite involved and can be found

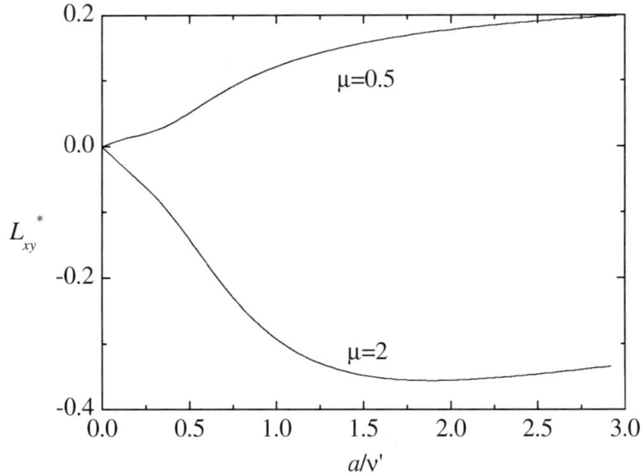

Figure 4.24. Reduced off-diagonal element L_{xy}^* as a function of the reduced shear rate $a/\nu' = a/(n\nu_{12}/n_2)$ for $K_{11} = K_{22} = K_{12}$, $x_1 = \frac{1}{6}$, and $\mu = 0.5$ and 2.

elsewhere (Marín et al., 2000). In the limit of zero shear rate, the tensor L vanishes, which is consistent with the known result $L_0 = 0$ for binary Maxwell molecules in equilibrium (Chapman and Cowling, 1970) [cf. the paragraph below Eq. (1.150)]. The tensor $\mathsf{L}(a)$ has an obvious symmetry property. According to Eq. (4.105), $\mathsf{L}(a) \leftrightarrow -\mathsf{L}(a)$ when the labels 1 and 2 are exchanged. This implies that, in the particular case of particles mechanically equivalent, $\mathsf{L}(a) = 0$. Due to the geometry of the problem, the elements L_{xz}, L_{yz}, L_{zx}, and L_{zy} vanish, similarly to the case of the diffusion tensor. However, in contrast to the property $D_{yy} = D_{zz}$, now we have three different diagonal elements (i.e. $L_{xx} \neq L_{yy} \neq L_{zz}$). Figures 4.23 and 4.24 show the shear-rate dependence of the reduced elements L_{yy}^* and L_{xy}^*, where

$$L_{ij}^* = L_{ij} \frac{\nu'}{nk_B T^2}. \tag{4.142}$$

We observe that the elements of the Dufour tensor are non-monotonic functions of the shear rate, reaching either a maximum or a minimum for a certain value of a. Depending on the values of the mass ratio $\mu = m_1/m_2$ and/or the mole fraction $x_1 = n_1/n$, the elements L_{ij} may be either positive or negative. More specifically, if $x_1 < \frac{1}{2}$ (i.e. species 1 is the defect component), then L_{yy} and L_{xy} are positive for $\mu < 1$, while they are negative for $\mu > 1$. In contrast to what happens for the diffusion tensor, the Dufour tensor in the GK and the GSB models differs (Garzó and López de Haro, 1992).

Chapter 5

PLANAR COUETTE FLOW IN A SINGLE GAS

1. Introduction

In the three preceding chapters we have considered the properties of the uniform shear flow state (USF) far from equilibrium. As said in Section 2.1, this represents a prototype situation to unveil non-Newtonian features in a gas, obtain exact results, and test approximate theories. In addition, the USF is relatively easy to implement in simulations by means of (Lees–Edwards) periodic boundary conditions. As a counterpart, this state is an idealized version of a realistic shear flow problem since the gas is formally unbounded, the flow velocity profile is assumed to be strictly linear, and the density and temperature are spatially uniform. The price to be paid is that the temperature increases with time (due to viscous heating effects), unless an (artificial) thermostat force is applied.

On the other hand, the planar Couette flow describes a gas enclosed in a slab by two parallel plates in relative motion (see Fig. 5.1). The walls act as thermal baths, removing the increase of energy due to the shearing work. As a consequence, the gas reaches a steady state with non-uniform density, temperature, and shear rate. This gives rise to the presence of a nonzero heat flux describing energy transport coupled to momentum transport. Furthermore, the rheological properties are different in the planar Couette flow from those of the USF, as will be seen in this Chapter.

The aim of this chapter is to obtain the elements of the pressure tensor and the components of the heat flux for a dilute gas subjected to the Couette flow with arbitrary values of the velocity and temperature gradients. These fluxes define five generalized transport coefficients: a viscosity function, two viscometric functions, a thermal conductivity function, and a cross transport coefficient measuring the heat flux orthogonal to the thermal gradient. They are nonlinear

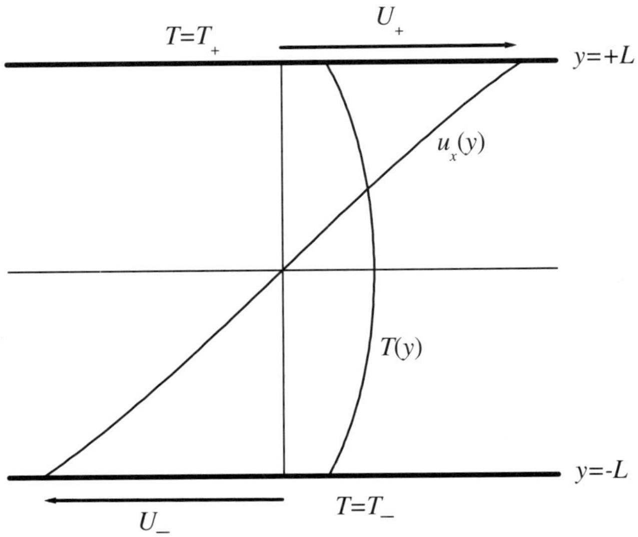

Figure 5.1. Sketch of the planar Couette flow. The gas is enclosed between two infinite parallel walls located at $y = \pm L$, moving along the x-direction with velocities $U_\pm$, and kept at temperatures $T_\pm$. The situation depicted in the figure corresponds to $U_- = -U_+$ and $T_- = T_+$.

functions of the (reduced) shear rate, but are independent of the thermal gradient. After providing a hydrodynamic description of the Couette flow beyond the Navier–Stokes regime, we show that the Boltzmann equation for Maxwell molecules admits a normal solution consistent with a uniform pressure, and linear velocity and quadratic temperature profiles with respect to a scaled spatial variable. In order to determine the explicit shear-rate dependence of the five generalized transport coefficients, as well as the full velocity distribution function, the BGK model of the Boltzmann equation is solved . This solution provides a nice example of the asymptotic character of the Chapman–Enskog expansion. A comparison with the results corresponding to the USF shows that, as expected, the non-Newtonian behavior is different in both shearing problems. In particular, the shear-thinning effect is more pronounced in the Couette flow than in the USF. As in the USF problem, the explicit knowledge of the distribution function allows us to get the nonequilibrium entropy. This quantity is evaluated as a series expansion in powers of the shear rate and the thermal gradient, as well as numerically for finite values of those gradients.

The problem can also be solved by using other model kinetic equations (cf. Section 1.6) and Grad's moment method. The different theoretical predictions are compared with molecular dynamics simulations and numerical simulations of the Boltzmann equation by the Direct Simulation Monte Carlo (DSMC) method. The results indicate that the kinetic models agree quite well with the simulation data, even for large values of the shear rate and the thermal gradient. On the other hand, while Grad's method is qualitatively good at the level of the shear and normal stresses, it gives wrong qualitative predictions for the heat flux.

2. Hydrodynamic description

The specific problem considered in this chapter is steady planar Couette flow (cf. Fig. 5.1). The gas is enclosed between two parallel plates at $y = \pm L$ in relative motion along the x-axis and maintained, in general, at different temperatures. The resulting flow velocity is directed along the x-axis and, from symmetry, it is expected that the hydrodynamic fields vary only in the y-direction. The pressure tensor and the heat flux characterize a macroscopic state with combined heat and momentum transport. Under the above conditions, the mass balance equation Eq. (1.38) is identically satisfied, while the balance equations for momentum, Eq. (1.39), and energy, Eq. (1.44) yield (in the absence of an external force)

$$\frac{\partial P_{xy}}{\partial y} = 0, \tag{5.1}$$

$$\frac{\partial P_{yy}}{\partial y} = 0, \tag{5.2}$$

$$P_{xy}\frac{\partial u_x}{\partial y} + \frac{\partial q_y}{\partial y} = 0. \tag{5.3}$$

The stationarity of the state implies that the shear stress P_{xy} and the normal stress P_{yy} are uniform across the system. The presence of q_y in Eq. (5.3) indicates that a thermal gradient $\partial T/\partial y$ is induced by the velocity gradient, even if both plates are kept at the same temperature ($T_+ = T_-$). This is a crucial difference with respect to the USF, where the viscous heating term $P_{xy}\partial u_x/\partial y$ is not balanced (since $\mathbf{q} = \mathbf{0}$) and produces a monotonic increase of temperature, unless a thermostat is introduced.

There are three independent hydrodynamic fields in the problem, namely the pressure, temperature, and the x-component of the flow velocity (the other two components vanish). The balance equations (5.1)–(5.3) are exact but do not constitute a closed set, unless the dependence of the pressure tensor and the heat flux on the hydrodynamic fields is known. In order to gain some insight into the general hydrodynamic profiles, it is instructive first to assume that the hydrodynamic gradients are small. Thus, the fluxes P and $\mathbf{q}$ are described by the Navier–Stokes constitutive equations (1.106) and (1.108), which in this problem yield

$$P_{xx} = P_{yy} = P_{zz}, \quad P_{xy} = -\eta_0\frac{\partial u_x}{\partial y}, \tag{5.4}$$

$$q_x = 0, \quad q_y = -\kappa_0 \frac{\partial T}{\partial y}, \tag{5.5}$$

where η_0 and κ_0 are the Navier–Stokes shear viscosity and thermal conductivity coefficients, respectively. As a consequence of the absence of normal stress differences in the Navier–Stokes description, the hydrostatic pressure $p = (P_{xx} + P_{yy} + P_{zz})/3$ is a constant,

$$p = \text{const}, \tag{5.6}$$

on account of the balance equation (5.2). One still needs to know the spatial dependence of the transport coefficients to integrate the Navier–Stokes hydrodynamic equations. Since in a dilute gas the ratio η_0/κ_0 is a constant [see, for instance, Eqs. (1.111) and (1.112)], it follows from Eqs. (5.1)–(5.5) that the flow velocity profile is quasi-linear,

$$\eta_0 \frac{\partial u_x}{\partial y} = \text{const} \tag{5.7}$$

and the temperature is quasi-parabolic,

$$\left(\eta_0 \frac{\partial}{\partial y}\right)^2 T = -\frac{\eta_0}{\kappa_0} \left(\eta_0 \frac{\partial u_x}{\partial y}\right)^2 = \text{const}. \tag{5.8}$$

Note that the profile of u_x is not strictly linear, due to the space dependence of η_0 through the temperature. Analogously, the temperature profile is not strictly quadratic. In fact, the specific form of both profiles depends on the interaction potential under consideration. For instance, $\eta_0, \kappa_0 \propto T$ for Maxwell molecules [cf. Eqs. (1.111) and (1.112)] and $\eta_0, \kappa_0 \propto T^{1/2}$ for hard spheres [cf. Eq. (1.119)]. On From Eqs. (5.7) and (5.8) it is easy to derive an interesting result, namely that if the temperature T is seen as a function of u_x rather than as a function of the coordinate space y, then one has

$$\frac{\partial^2 T}{\partial u_x^2} = -\frac{\eta_0}{\kappa_0}, \tag{5.9}$$

regardless of the interaction model. This can be seen as a sort of nonequilibrium "equation of state," according to which the temperature is a quadratic function of the flow velocity. Moreover, the "curvature" of the profile is practically universal, given the weak influence of the interaction potential on the Prandtl number $\text{Pr} \equiv 5k_B\eta_0/2m\kappa_0 \simeq \frac{2}{3}$.

As done in Chapters 2 and 3, it is convenient to define a (local) effective collision frequency as $\nu = \nu_\eta = p/\eta_0$. In the case of Maxwell molecules, it is given by Eq. (2.28). Associated with ν we can define a (local) mean free path

$\ell = v_0/\nu$, where $v_0 = \sqrt{2k_BT/m}$ is the thermal velocity. In the Couette flow problem, the (local) shear rate

$$a = \frac{\partial u_x}{\partial y} \tag{5.10}$$

allows one to define a hydrodynamic length $h = v_0/a$. This quantity represents the vertical distance over which the flow velocity changes significantly. The ratio between both length scales,

$$a^* = \frac{\ell}{h} = \frac{1}{\nu}\frac{\partial u_x}{\partial y}, \tag{5.11}$$

defines a reduced shear rate that plays the role of a Knudsen number in this problem. As in the USF case, it is the primary control parameter of the steady Couette flow measuring the departure of the system from the equilibrium state. From Eqs. (5.6) and (5.7), it follows that a^* is uniform in the Navier–Stokes approximation. This means that, while both ℓ and h depend on the position due to the spatial dependence of the temperature and the shear rate, their ratio is uniform across the system. The Navier–Stokes description (5.4)–(5.9) is expected to apply when $L \gg h \gg \ell$. The first condition ($L \gg h$) guarantees that the gas reaches a hydrodynamic regime in which all space dependence of the heat and momentum fluxes occurs via a *functional* dependence on the hydrodynamic fields, so that the macroscopic balance equations become a closed set of hydrodynamic equations. This hydrodynamic description holds in the bulk region, i.e. for points sufficiently far from the boundaries. The second condition ($h \gg \ell$ or, equivalently, $a^* \ll 1$) means that such functional relations have the linear forms given by Eqs. (5.4) and (5.5). It is important to note that, even if the reduced shear rate a^* is not small, a hydrodynamic description can still be valid in the bulk region if $L \gg \max\{h, \ell\}$. In that case, the constitutive equations are not expected to be linear. For instance,

$$P_{xy} = -\eta[p, T, \mathbf{u}]\frac{\partial u_x}{\partial y}, \tag{5.12}$$

$$q_y = -\kappa[p, T, \mathbf{u}]\frac{\partial T}{\partial y}, \tag{5.13}$$

where $\eta[p, T, \mathbf{u}]$ and $\kappa[p, T, \mathbf{u}]$ are *generalized* shear viscosity and thermal conductivity, respectively. The square brackets indicate that they are functionals of p, T, and $\mathbf{u}$ or, equivalently, functions of these variables and their gradients.

Now we assume that a hydrodynamic description (not restricted to small a^*) exists consistent with profiles with the same form as the Navier–Stokes ones, Eqs. (5.6)–(5.8). More specifically, we look for special solutions to the macroscopic balance equations characterized by

$$p = \text{const}, \quad a^* = \text{const}, \quad \left(\frac{1}{\nu}\frac{\partial}{\partial y}\right)^2 T = \text{const}. \tag{5.14}$$

Equation (5.12) can be written as $P_{xy} = -pa^*(\eta/\eta_0)$. Since the shear stress P_{xy} is spatially constant and a^* has been assumed to be constant, it follows that the ratio η/η_0 is constant as well. This suggests that the nonlinear shear viscosity η has the form

$$\eta = \eta_0 F_\eta(a^*). \tag{5.15}$$

The function $F_\eta(a^*)$ is the viscosity function that includes all nonlinear rheological effects. It plays the same role as the viscosity function defined by Eq. (2.54) in the USF. Now we consider the energy balance equation (5.3). It yields

$$\left(\frac{1}{\nu}\frac{\partial}{\partial y}\right)\left(\frac{\kappa}{\kappa_0}\frac{1}{\nu}\frac{\partial}{\partial y}\right) T = -\frac{\eta_0}{\kappa_0} a^{*2} F_\eta(a^*) = \text{const}. \tag{5.16}$$

Comparison with the last equality in (5.14) suggests that $\kappa/\kappa_0 = \text{const}$, so the nonlinear thermal conductivity has the form

$$\kappa = \kappa_0 F_\kappa(a^*). \tag{5.17}$$

This yields a generalization of Fourier's law with the thermal conductivity modified by the nonlinear rheological factor $F_\kappa(a^*)$. Henceforth, we will refer to $F_\kappa(a^*)$ as the thermal conductivity function. According to Eqs. (5.15) and (5.17), the generalized shear viscosity and thermal conductivity are nonlinear functions of the velocity gradient, but they do not depend on the thermal gradient. This is a consequence of the simplicity of the assumed hydrodynamic profiles (5.14). Inserting Eq. (5.17) into Eq. (5.16) we get

$$\left(\frac{1}{\nu}\frac{\partial}{\partial y}\right)^2 T = -\frac{2m}{k_B}\text{Pr}\gamma(a^*), \tag{5.18}$$

where

$$\gamma(a^*) = \frac{a^{*2}}{5}\frac{F_\eta(a^*)}{F_\kappa(a^*)}. \tag{5.19}$$

Equation (5.18) is a generalization of Eq. (5.8). In general, the thermal "curvature" parameter $\gamma(a^*)$ is a nonlinear function of the reduced shear rate that behaves as $\gamma \simeq a^{*2}/5$ in the Navier–Stokes regime.

Apart from the generalized constitutive equations (5.12) and (5.13), complemented by Eqs. (5.15) and (5.17), one needs additional constitutive equations for the diagonal elements of the pressure tensor and for the component of the heat flux parallel to the flow velocity. As in the USF problem [cf. Eqs. (2.55) and

(2.56)], the normal stress differences are specified in terms of the dimensionless viscometric functions

$$\Psi_1(a^*) = \frac{P_{yy} - P_{xx}}{pa^{*2}}, \tag{5.20}$$

$$\Psi_2(a^*) = \frac{P_{zz} - P_{yy}}{pa^{*2}}. \tag{5.21}$$

These equations imply that, not only P_{yy} is a constant, as required by the balance equation (5.2), but also P_{xx} and P_{zz} are constant. The anisotropy of the Couette flow induces a heat flux component normal to the temperature gradient but parallel to the flow direction. This effect can be characterized through a scalar function $\Phi(a^*)$ by

$$q_x = -\kappa_0 a^* \Phi(a^*) \frac{\partial T}{\partial y}. \tag{5.22}$$

This cross contribution to the heat flux has no analogue at Navier–Stokes order, being of first order in both the shear rate and the temperature gradient. We will coin the term *visco-thermal function* to refer to the cross transport coefficient $\Phi(a^*)$.

The existence of normal stress differences and an x-component of the heat flux are effects of Burnett order. According to the general structure of the Burnett constitutive equations, they are not *apparently* compatible with the simplicity of Eqs. (5.20)–(5.22). However, this discrepancy disappears *exactly* in the particular case of Maxwell molecules and *approximately* for any other interaction potential. Let us analyze this point more carefully. Taking into account the geometry of the problem and particularizing to $\nabla p = \mathbf{0}$, the Burnett equations for the normal stress differences become (Chapman and Cowling, 1970)

$$P_{yy} - P_{xx} = \frac{\eta_0^2}{\rho T}\varpi_3 \frac{\partial^2 T}{\partial y^2} + \frac{\eta_0^2}{\rho T^2}\varpi_5 \left(\frac{\partial T}{\partial y}\right)^2 - \frac{\eta_0^2}{p}\varpi_2 \left(\frac{\partial u_x}{\partial y}\right)^2, \tag{5.23}$$

$$P_{zz} - P_{yy} = -\frac{\eta_0^2}{\rho T}\varpi_3 \frac{\partial^2 T}{\partial y^2} - \frac{\eta_0^2}{\rho T^2}\varpi_5 \left(\frac{\partial T}{\partial y}\right)^2 - \frac{\eta_0^2}{p}\left(\frac{1}{4}\varpi_6 - \varpi_2\right) \left(\frac{\partial u_x}{\partial y}\right)^2, \tag{5.24}$$

where ϖ_i are pure numbers whose values depend, in general, on the interaction model (Chapman and Cowling, 1970). Now we make use of the exact relation (for $\nabla p = \mathbf{0}$)

$$\frac{\partial^2 T}{\partial y^2} = \nu^2 \left(\frac{1}{\nu}\frac{\partial}{\partial y}\right)^2 T - \frac{1}{T}\frac{\partial \ln \eta_0}{\partial \ln T}\left(\frac{\partial T}{\partial y}\right)^2. \tag{5.25}$$

In the case of Maxwell molecules, one has $\varpi_3(\partial \ln \eta_0/\partial \ln T) - \varpi_5 = 0$. This relation also holds for any interaction potential in the first Sonine approximation.

Consequently, Eqs. (5.23) and (5.24) reduce to

$$P_{yy} - P_{xx} = -\left(\frac{2}{5}\text{Pr}\,\varpi_3 + \varpi_2\right) pa^{*2}, \tag{5.26}$$

$$P_{zz} - P_{yy} = \left(\frac{2}{5}\text{Pr}\,\varpi_3 - \frac{1}{4}\varpi_6 + \varpi_2\right) pa^{*2}, \tag{5.27}$$

where use has been made of Eq. (5.8) and the definition of the reduced shear rate, Eq. (5.11). We see that the Burnett equations (5.26) and (5.27) are consistent with the assumption of a uniform pressure tensor. In addition, the above equations yield the values of the viscometric functions in the limit of small shear rates:

$$\Psi_1(0) = -\left(\frac{2}{5}\text{Pr}\varpi_3 + \varpi_2\right) = -\frac{14}{5}, \tag{5.28}$$

$$\Psi_2(0) = \left(\frac{2}{5}\text{Pr}\varpi_3 - \frac{1}{4}\varpi_6 + \varpi_2\right) = \frac{4}{5}, \tag{5.29}$$

where we have considered the values $\varpi_2 = 2$, $\varpi_3 = 3$, and $\varpi_6 = 8$ for Maxwell molecules. Those values are also valid for non-Maxwell molecules in the first Sonine approximation.

Let us consider now the x-component of the heat flux to Burnett order. It is given by

$$q_x = \frac{\eta_0^2}{\rho T}\frac{3}{2}\theta_5 \frac{\partial T}{\partial y}\frac{\partial u_x}{\partial y} + \frac{\eta_0^2}{\rho}\frac{1}{2}\theta_4 \frac{\partial^2 u_x}{\partial y^2}, \tag{5.30}$$

where again θ_4 and θ_5 are pure numbers (Chapman and Cowling, 1970). Since $a^* = \text{const}$ and $p = \text{const}$, the second term can be easily cast into the same form as the first term, namely

$$\frac{\partial^2 u_x}{\partial y^2} = -\frac{\nu}{T}a^* \frac{\partial \ln \eta_0}{\partial \ln T}\frac{\partial T}{\partial y}. \tag{5.31}$$

Therefore,

$$q_x = \kappa_0 \left(3\theta_5 - \theta_4 \frac{\partial \ln \eta_0}{\partial \ln T}\right) \frac{\text{Pr}}{5} a^* \frac{\partial T}{\partial y}. \tag{5.32}$$

This Burnett contribution to the heat flux is consistent with Eq. (5.22). In addition, the cross transport coefficient $\Phi(0)$ can be identified as

$$\Phi(0) = -\left(3\theta_5 - \theta_4 \frac{\partial \ln \eta_0}{\partial \ln T}\right) \frac{\text{Pr}}{5} = -\frac{7}{2}, \tag{5.33}$$

where we have taken into account that for Maxwell molecules one has $\theta_4 = 3$ and $\theta_5 = \frac{35}{4} + \partial \ln \eta_0 / \partial \ln T$ (Chapman and Cowling, 1970). Moreover, those values are a good approximation for non-Maxwell molecules.

The fact that the hydrodynamic profiles (5.14) and the constitutive equations (5.15), (5.17), (5.20), (5.21), and (5.22) are self-consistent to Burnett order in the case of Maxwell molecules does not imply, in principle, that this consistency extends to higher hydrodynamic orders. Nevertheless, the general consistency of the assumed profiles and constitutive equations is supported by an exact *normal* solution of the Boltzmann equation for Maxwell molecules (Makashev and Nosik, 1981; Nosik, 1983), although the explicit shear-rate dependence of the five generalized transport coefficients is only known to super-Burnett order (Tij and Santos, 1995). For general interactions, the above hydrodynamic description is a good approximation, as supported by exact solutions of kinetic models (Brey et al., 1987; Dufty, 1990; Garzó and López de Haro, 1994; Garzó and López de Haro, 1997; Montanero and Garzó, 2000), as well as by computer simulations (Risso and Cordero, 1997; Montanero et al., 2000). In addition, the kinetic model solutions provide explicit expressions for the generalized transport coefficients.

To put the constitutive equations for the planar Couette flow, Eqs. (5.15), (5.17), and (5.20)–(5.22), in a proper context, some comments are in order. The Chapman–Enskog expansion treats the hydrodynamic fields as unknown and independent. However, if the fields are known, many different gradients can be combined to look like a Navier–Stokes or Burnett coefficient. For instance, this is the case with Eq. (5.22), where an infinite number of terms have been expressed in terms of the thermal gradient and then summed to give the Burnett-like coefficient $\Phi(a^*)$. This will always be possible if the profiles are simple, as happens with the assumed profiles (5.14).

Before analyzing the problem from a kinetic theory point of view, it is interesting to write the y-dependence of the hydrodynamic fields. First, let us express the temperature in terms of the flow velocity: $T(y) \to T(u_x)$. From Eq. (5.18), one has

$$\frac{\partial^2 T}{\partial u_x^2} = -\frac{2m}{k_B}\mathrm{Pr}\frac{\gamma(a^*)}{a^{*2}}, \tag{5.34}$$

which generalizes Eq. (5.9) outside the Navier–Stokes regime. Equation (5.34) implies that the temperature is simply a quadratic function of the flow velocity,

$$T = T_0\left[1 - c^{-2}(u_x - u_0)^2\right], \quad c^{-2} \equiv \frac{m}{k_B T_0}\mathrm{Pr}\frac{\gamma(a^*)}{a^{*2}}, \tag{5.35}$$

where u_0 is an arbitrary constant and T_0 is the temperature at the point where $\partial T/\partial u_x = 0$, i.e. T_0 is the maximum temperature. The characteristic speed c is of the order of the sound velocity $c_s = \sqrt{5k_B T_0/3m}$, so $|u_x - u_0|/c$ represents a sort of Mach number. The explicit y-dependence of u_x can be easily obtained from (5.35) and the condition $a^* = \text{const}$ in Eq. (5.11). For instance, in the case of Maxwell molecules, where $\eta_0 \propto T$,

$$\left[1 - c^{-2}(u_x - u_0)^2\right] \frac{\partial u_x}{\partial y} = a^* \frac{p}{\eta_0(T_0)}, \tag{5.36}$$

whose (implicit) solution is

$$y = y_0 + \frac{\eta_0(T_0)}{pa^*} u_x \left(1 - \frac{1}{3} u_x^2/c^2\right). \tag{5.37}$$

Here, we have called y_0 to the point where the temperature is equal to T_0 and, without loss of generality, we have taken $u_x(y_0) \equiv u_0 = 0$. Analogously, in the case of hard spheres, where $\eta_0 \propto T^{1/2}$, one has

$$y = y_0 + \frac{\eta_0(T_0)}{2pa^*} c \left[\sin^{-1}(u_x/c) + u_x/c\left(1 - u_x^2/c^2\right)^{1/2}\right]. \tag{5.38}$$

In the Couette flow geometry with equal wall temperatures, $y_0 = 0$ represents the middle point, where the temperature is maximum. Note that Eq. (5.35) is only meaningful for $u_x^2 \leq c^2$ (where we have again taken $\mathbf{u}_0 = \mathbf{0}$). In other words, the flow velocity, relative to the thermal velocity $v_0(T_0) = \sqrt{2k_B T_0/m}$ at $y = y_0$ has an upper bound given by $|u_x|/v_0(T_0) \leq a^*/\sqrt{2\mathrm{Pr}\gamma(a^*)}$. According to Eqs. (5.36) and (5.37), this implies that $|y - y_0|/\ell_0 \leq b/\sqrt{2\mathrm{Pr}\gamma(a^*)}$, where $\ell_0 = v_0(T_0)/\nu(T_0)$ is the mean free path at $y = y_0$ and b is a pure number ($b = \frac{2}{3}$ for Maxwell molecules and $b = \pi/4$ for hard spheres).[1]

The velocity profiles (5.37) and (5.38) and the temperature profile (5.35) are shown in Fig. 5.2 for the representative case $a^* = 1$, in which case $\gamma \simeq 0.3$ (Montanero et al., 2000), with $y_0 = 0$. Up to $y/\ell_0 \approx 0.5$, the velocity profile is practically linear and the temperature profile is practically parabolic. Beyond that point, as the temperature keeps decreasing, the shear rate $a = \nu(T)a^*$ increases due to the temperature dependence of the collision frequency [$\nu(T) \propto pT^{-1}$ for Maxwell molecules and $\nu(T) \propto pT^{-1/2}$ for hard spheres] and, consequently, the temperature drops more rapidly than parabolically. The deviations of u_x and T from a linear and a parabolic profile, respectively, is more pronounced for Maxwell molecules than for hard spheres since the temperature-dependence of the collision frequency is more significant in the former case than in the latter.

3. The Boltzmann equation for the planar Couette flow

In the planar Couette flow problem, the system is driven out of equilibrium by the combined action of the velocity and thermal gradients along the direction

[1] In the more general case of repulsive potentials of the form $\phi(r) \propto r^{-w}$, one has $\eta_0(T) \propto T^{1/2+2/w}$ and then $b = \sqrt{\pi}\Gamma(3/2 + 2/w)/2\Gamma(2 + 2/w)$.

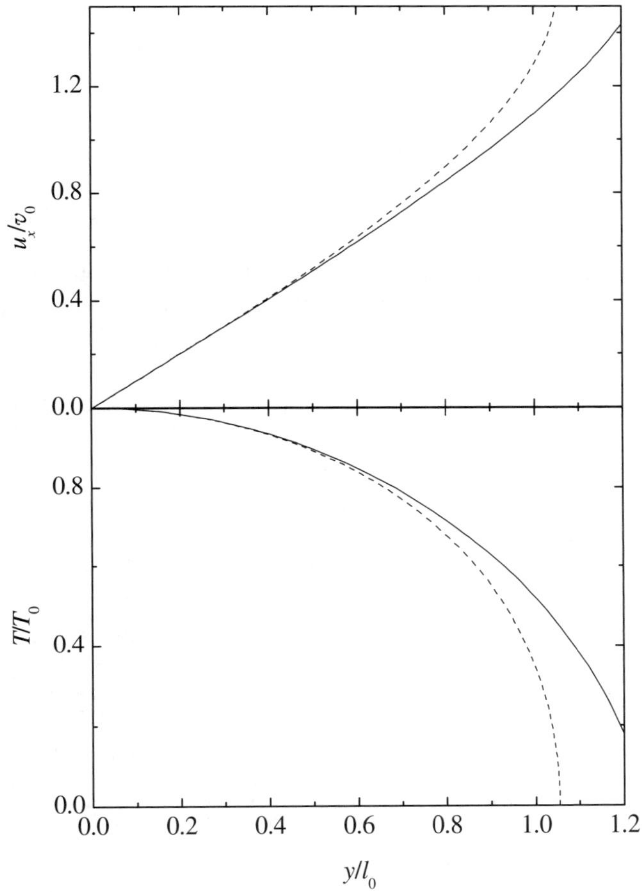

Figure 5.2. Velocity and temperature profiles for hard spheres (solid lines) and Maxwell molecules (dashed lines) in the case $a^* = 1$. The magnitudes T_0, v_0, and ℓ_0 are the temperature, thermal velocity, and mean free path, respectively, at the layer $y_0 = 0$ where the temperature has a maximum.

normal to the plates. After a transient period, the gas is expected to reach a steady state and the corresponding Boltzmann equation reads

$$v_y \frac{\partial}{\partial y} f = J[f, f], \tag{5.39}$$

where, due to the geometry of the problem, the spatial dependence of the distribution function is restricted to the y-direction. In general, this equation must be solved subjected to specific boundary conditions. They can be expressed in terms of the kernels $K_{\pm}(\mathbf{v}, \mathbf{v}')$ defined as follows. When a particle with velocity $\mathbf{v}'$ hits the wall at $y = L$, the probability of being reemitted with a velocity $\mathbf{v}$

within the range $d\mathbf{v}$ is $K_+(\mathbf{v},\mathbf{v}')d\mathbf{v}$; the kernel $K_-(\mathbf{v},\mathbf{v}')$ represents the same but at $y=-L$. The boundary conditions are then (Dorfman and van Beijeren, 1977)

$$\Theta(\pm v_y)|v_y|f(y=\mp L,\mathbf{v}) = \Theta(\pm v_y)\int d\mathbf{v}'\,|v_y'|K_\mp(\mathbf{v},\mathbf{v}') \times\Theta(\mp v_y')f(y=\mp L,\mathbf{v}'), \tag{5.40}$$

where $\Theta(x)$ is Heaviside's step function. In the case of boundary conditions of complete accommodation with the walls, so that $K_\pm(\mathbf{v},\mathbf{v}')=K_\pm(\mathbf{v})$ does not depend on the incoming velocity $\mathbf{v}'$, the kernels can be written as

$$K_\mp(\mathbf{v}) = A_\mp^{-1}\Theta(\pm v_y)|v_y|\varphi_\mp(\mathbf{v}),\quad A_\mp=\int d\mathbf{v}\,\Theta(\pm v_y)|v_y|\varphi_\mp(\mathbf{v}), \tag{5.41}$$

where $\varphi_\mp(\mathbf{v})$ represents the probability distribution of a fictitious gas in contact with the system at $y=\mp L$. Equation (5.41) can then be interpreted as meaning that when a particle hits a wall, it is instantaneously absorbed and replaced by a particle leaving the fictitious bath. Of course, any choice of $\varphi_\mp(\mathbf{v})$ must be consistent with the imposed wall velocities and temperatures, i.e.

$$U_\mp = \int d\mathbf{v}\,v_x\varphi_\mp(\mathbf{v}), \tag{5.42}$$

$$k_BT_\mp = \frac{1}{3}m\int d\mathbf{v}\,(\mathbf{v}-\mathbf{U}_\mp)^2\varphi_\mp(\mathbf{v}). \tag{5.43}$$

Inserting (5.41) into (5.40), one has

$$\Theta(\pm v_y)f(y=\mp L,\mathbf{v}) = \Theta(\pm v_y)n_\mp\varphi_\mp(\mathbf{v}), \tag{5.44}$$

where

$$n_\mp = \frac{\int d\mathbf{v}'\,\Theta(\mp v_y')|v_y'|f(y=\mp L,\mathbf{v}')}{\int d\mathbf{v}'\,\Theta(\pm v_y')|v_y'|\varphi_\mp(\mathbf{v}')}. \tag{5.45}$$

The boundary conditions (5.44) are usually referred to as *diffuse* (or stochastic) boundary conditions. The simplest and most common choice is that of a Maxwell-Boltzmann (MB) distribution (Tehver et al., 1998):

$$\varphi_\mp^{\mathrm{MB}}(\mathbf{v}) = \left(\frac{m}{2\pi k_BT_\mp}\right)^{3/2}\exp\left[-\frac{m(\mathbf{v}-\mathbf{U}_\mp)^2}{2k_BT_\mp}\right]. \tag{5.46}$$

Under these conditions, the system is understood to be enclosed between two independent baths *in equilibrium*. While the conditions (5.46) are adequate for analyzing boundary effects, (Cercignani, 1990; Wadsworth, 1993; Sone, 2002) they are not very efficient when one is interested in obtaining the transport properties in the bulk region. In order to inhibit the influence of boundary effects, it is more convenient to imagine that the two fictitious baths are in *nonequilibrium* states resembling the state of the actual gas near the walls (Montanero et al., 1994; Montanero et al., 2000).

The general solution to the stationary Boltzmann equation (5.39) with the boundary conditions (5.44) can be split into two parts (Sone, 1991; Sone, 2000; Sone, 2002):

$$f(y, \mathbf{v}) = f_H(y, \mathbf{v}) + f_B(y, \mathbf{v}). \tag{5.47}$$

Here, f_H represents the *hydrodynamic*, Hilbert-class, or *normal* contribution to the distribution function. This means that f_H depends on space only through a functional dependence on the hydrodynamic fields,i.e.

$$f_H(y) = f_H[p, T, u_x]. \tag{5.48}$$

The contribution $f_B(y, \mathbf{v})$ represents the boundary-layer correction to f_H, so that $f = f_H + f_B$ verifies the specified boundary conditions. The correction f_B is appreciably different from zero only in a thin layer (the so-called or Knudsen layer), adjacent to the plates, of thickness of the order of the mean free path. Consequently, if the separation $2L$ between the plates is much larger than the characteristic mean free paths, there exists a well defined bulk region where the boundary correction vanishes and the distribution function is fully given by its hydrodynamic part. In the boundary layers the hydrodynamic profiles are much less smooth than in the bulk domain. The values of the flow velocity near the walls are different from the velocity of the plates, i.e. $u_x(y = \pm L) \neq U_\pm$. This is usually referred to as velocity slip. Besides, the extrapolation of the velocity profile in the bulk to the boundaries, $u_{x,H}(y = \pm L)$ is also different from both the actual values $u_x(y = \pm L)$ and the wall velocities $U_\pm$. Of course, an analogous temperature jump effect takes place with the temperature profile. The boundary contribution f_B for small Knudsen numbers has been widely analyzed elsewhere (Cercignani, 1990; Sone, 1997; Sone, 2002).

In the remaining of this Section we will focus on the hydrodynamic part f_H, with special emphasis on the corresponding hydrodynamic contributions to the momentum and heat fluxes. As seen in Section 5.2, the hydrodynamic equations to Burnett order for Maxwell molecules support a solution consistent with the hydrodynamic profiles (5.14). Makashev and Nosik (1981) have proven that this consistency extends to arbitrary values of the reduced shear rate a^* and the thermal gradient. Here, we outline a proof along the lines followed by Tij and Santos (1995).

It is convenient to work with the hierarchy of moment equations associated with Eq. (5.39). To do so, let us introduce the moments

$$M_{k_1,k_2,k_3}(y) = \int d\mathbf{v}\, V_x^{k_1} V_y^{k_2} V_z^{k_3} f(y, \mathbf{v}), \tag{5.49}$$

where $\mathbf{V} = \mathbf{v} - \mathbf{u}(y)$ is the peculiar velocity. From Eq. (5.39) one gets

$$\frac{\partial}{\partial y} M_{k_1,k_2+1,k_3} + \frac{\partial u_x}{\partial y} k_1 M_{k_1-1,k_2+1,k_3} = J_{k_1,k_2,k_3}, \tag{5.50}$$

where

$$J_{k_1,k_2,k_3}(y) = \int d\mathbf{v}\, V_x^{k_1} V_y^{k_2} V_z^{k_3} J[f,f]. \tag{5.51}$$

In the sequel, we will use the roman boldface $\mathbf{k}$ to denote the triad $\{k_1, k_2, k_3\}$ and the italic lightface k to denote the sum $k_1+k_2+k_3$. Thus, $M_{\mathbf{k}} \equiv M_{k_1,k_2,k_3}$ is a moment of degree $k \equiv k_1+k_2+k_3$. In the special case of Maxwell molecules, $J_{\mathbf{k}}$ can be expressed as a bilinear combination of moments of degree equal to or smaller than k (Truesdell and Muncaster, 1980; McLennan, 1989):

$$J_{\mathbf{k}} = \sum_{\mathbf{k}',\mathbf{k}''}{}' \widehat{J}_{\mathbf{k}\mathbf{k}'\mathbf{k}''} M_{\mathbf{k}'} M_{\mathbf{k}''}, \tag{5.52}$$

where the prime in the summation denotes the constraint $k' + k'' = k$. The coefficients $\widehat{J}_{\mathbf{k}\mathbf{k}'\mathbf{k}''}$ in Eq. (5.52) are closely related to the coefficients $J_{\mathbf{k}\mathbf{k}'\mathbf{k}''}$ defined by Eq. (2.103). While $\widehat{J}_{\mathbf{k}\mathbf{k}'\mathbf{k}''}$ are associated with the Cartesian moments (5.49), the coefficients $J_{\mathbf{k}\mathbf{k}'\mathbf{k}''}$ are associated with the (dimensionless) moments defined from the eigenfunctions (1.86). Both sets of coefficients can be expressed as linear combinations of the eigenvalues $\lambda_{r\ell}^0$ of the linearized collision operator [cf. Eq. (1.90)]. The expressions for $J_{\mathbf{k}}$ through order $k = 4$ can be easily obtained from those in Appendix 1.A.

Let us introduce the scaled space variable

$$s(y) = \int_{y_0}^{y} dy'\, \nu(y'), \tag{5.53}$$

where, for Maxwell molecules, the (local) effective collision frequency $\nu = p/\eta_0$ is given by Eq. (2.28). Thus, $\partial u_x/\partial s = a^*$ represents the ratio between the local mean free path ℓ and the hydrodynamic length h, as shown in Eq. (5.11). Now we make the assumption, to be confirmed by consistency, that a *hydrodynamic* solution to Eq. (5.39) or, equivalently, to Eq. (5.50) exists compatible with Eq. (5.14). This implies that the pressure is constant, the flow velocity is a linear function of s, and the temperature is a quadratic function of s, i.e.

$$u_x(s) = a^* s, \tag{5.54}$$

$$T(s) = T_0 - \mathrm{Pr}\frac{m}{k_B}\gamma(a^*)s^2. \tag{5.55}$$

Here we have taken into account that $y = y_0$ is the plane where the temperature $T(y_0) = T_0$ is a maximum, and that, without loss of generality, $u_x(y_0) = u_0 =$

0. The relationship between the scaled variable s and the true space variable y is given by Eq. (5.37) with $u_x = a^* s$. The thermal curvature parameter γ is a function of the shear rate and is related to the viscosity function F_η and the thermal conductivity function F_κ by Eq. (5.19). So far, the space dependence of the hydrodynamic fields is established but the coefficients $F_\eta(a^*)$ and $F_\kappa(a^*)$ remain unknown.

In addition to the shear rate $a = \partial u_x/\partial y$, the thermal gradient $\partial T/\partial y$ is the second nonequilibrium parameter in the problem. In the particular case of zero shear rate (i.e. both plates moving with the same velocity), the thermal gradient is different from zero provided that the plate temperatures are different. This corresponds to the so-called Fourier's flow (Reif, 1965), for which an exact normal solution of the Boltzmann equation for Maxwell molecules has been found (Asmolov et al., 1979; Nosik, 1981; Nosik, 1983; Santos and Garzó, 1995b). When $a \neq 0$, even if the two plates are kept at the same temperature, the stationarity of the state requires the presence of a nonzero thermal gradient, as the energy balance equation (5.3) indicates. In dimensionless form, the strength of the thermal gradient can be represented by the parameter

$$\epsilon = \left(\frac{2k_B T}{m}\right)^{1/2} \frac{\partial}{\partial s} \ln T. \tag{5.56}$$

This parameter can be interpreted as the ratio $\epsilon = \ell/h_T$ between the (local) mean free path $\ell = v_0/\nu$ and the (local) hydrodynamic length $h_T = (\partial \ln T/\partial y)^{-1}$ associated with the thermal gradient. Thus ϵ measures the relative variation of temperature along the y-direction over a distance equal to the local mean free path, while the reduced shear rate a^* measures the variation of flow velocity relative to the thermal velocity over the same distance. Both dimensionless parameters are independent and describe how far the system is from equilibrium at each point. On the other hand, while a^* is constant across the system, ϵ is a local function as Eq. (5.55) shows. More explicitly,

$$\frac{\partial \epsilon}{\partial s} = -\left(\frac{m}{2k_B T}\right)^{1/2} \left[\frac{\epsilon^2}{2} + 4\mathrm{Pr}\,\gamma(a^*)\right]. \tag{5.57}$$

According to Eq. (5.48), the hydrodynamic (or normal) solution to Eq. (5.39) depends on space only through the hydrodynamic fields and their gradients. This implies that, assuming the consistency of (5.14), f_H adopts the form

$$f_H(y, \mathbf{v}) = \frac{p}{k_B T(y)} \left[\frac{2k_B T(y)}{m}\right]^{-3/2} f^*\left(\boldsymbol{\xi}(y); a^*, \epsilon(y)\right), \tag{5.58}$$

where

$$\boldsymbol{\xi}(y) = \left[\frac{2k_B T(y)}{m}\right]^{-1/2} [\mathbf{v} - \mathbf{u}(y)] \tag{5.59}$$

is the peculiar velocity relative to the thermal velocity. The function $f^*(\boldsymbol{\xi})$ is the probability distribution function when the velocities are reduced with respect to the thermal velocity $v_0 = (2k_BT/m)^{1/2}$. It depends parametrically on the reduced shear rate a^* and the reduced thermal gradient ϵ. Note that $f^*(\boldsymbol{\xi})$ still depends on space through $\epsilon(y)$. For small gradients, f^* is given by the Navier–Stokes approximation (1.116). Particularized to the Couette flow geometry, one has

$$f^*\left(\boldsymbol{\xi}(y); a^*, \epsilon(y)\right) \to \pi^{-3/2} e^{-\xi^2}\left[1 - 2a^*\xi_x\xi_y - \mathrm{Pr}^{-1}\epsilon\left(\xi^2 - \frac{5}{2}\right)\xi_y\right]. \tag{5.60}$$

Our purpose is to prove that the function (5.58) is indeed a consistent solution to Eq. (5.39), without the need of obtaining it explicitly. Equation (5.58) implies that the velocity moments defined by Eq. (5.49) have the form

$$M_{\mathbf{k}}(y) = n(y)\left[\frac{2k_BT(y)}{m}\right]^{k/2} M^*_{\mathbf{k}}\left(a^*, \epsilon(y)\right), \tag{5.61}$$

where

$$M^*_{k_1,k_2,k_3}(a^*, \epsilon) = \int d\boldsymbol{\xi}\, \xi_x^{k_1}\xi_y^{k_2}\xi_z^{k_3} f^*(\boldsymbol{\xi}; a^*, \epsilon). \tag{5.62}$$

According to the symmetry properties of the problem, the reduced distribution function f^* must be invariant under the transformations

$$\xi_z \to -\xi_z, \tag{5.63}$$

$$(\xi_x, \xi_y, \epsilon) \to (-\xi_x, -\xi_y, -\epsilon), \tag{5.64}$$

$$(\xi_x, a^*) \to (-\xi_x, -a^*). \tag{5.65}$$

As a consequence, the reduced moments $M^*_{k_1,k_2,k_3}(a^*, \epsilon)$ vanish if $k_3 = \text{odd}$ and have the parity properties

$$M^*_{k_1,k_2,k_3}(a^*, \epsilon) = (-1)^{k_1+k_2} M^*_{k_1,k_2,k_3}(a^*, -\epsilon), \tag{5.66}$$

$$M^*_{k_1,k_2,k_3}(a^*, \epsilon) = (-1)^{k_1} M^*_{k_1,k_2,k_3}(-a^*, \epsilon). \tag{5.67}$$

Making use of Eqs. (5.56) and (5.57), we have

$$\frac{\partial}{\partial s} M_{\mathbf{k}} = n\left(\frac{2k_BT}{m}\right)^{(k-1)/2}\left[\frac{k-2}{2}\epsilon M^*_{\mathbf{k}} - \left(\frac{\epsilon^2}{2} + \frac{8}{3}\gamma\right)\frac{\partial}{\partial\epsilon}M^*_{\mathbf{k}}\right], \tag{5.68}$$

where the numerical value of the Prandtl number for Maxwell molecules, i.e. $\mathrm{Pr} = \frac{2}{3}$, has already been taken. Then Eq. (5.50) becomes

$$\begin{aligned}\frac{k-1}{2}\epsilon M^*_{k_1,k_2+1,k_3} &- \left(\frac{\epsilon^2}{2} + \frac{8}{3}\gamma\right)\frac{\partial}{\partial\epsilon}M^*_{k_1,k_2+1,k_3} + k_1 a^* M^*_{k_1-1,k_2+1,k_3} \\ &= \sum_{\mathbf{k}',\mathbf{k}''}{}' \widehat{J}^*_{\mathbf{k}\mathbf{k}'\mathbf{k}''} M^*_{\mathbf{k}'} M^*_{\mathbf{k}''},\end{aligned} \tag{5.69}$$

where $\widehat{J}^*_{\mathbf{kk'k''}} \equiv (n/\nu)\widehat{J}_{\mathbf{kk'k''}}$ The hierarchy (5.69) is equivalent to the hierarchy (5.50) (for Maxwell molecules), provided that the profiles (5.14) are self-consistent. By definition of the moments (5.62),

$$M^*_{000} = 1, \tag{5.70}$$

$$M^*_{100} = M^*_{010} = M^*_{001} = 0, \tag{5.71}$$

$$M^*_{200} + M^*_{020} + M^*_{002} = \frac{3}{2}. \tag{5.72}$$

Insertion into Eq. (5.69) yields

$$\frac{\partial}{\partial \epsilon} M^*_{110} = \frac{\partial}{\partial \epsilon} M^*_{020} = 0, \tag{5.73}$$

$$\left[\frac{\epsilon}{2} - \left(\frac{\epsilon^2}{2} + \frac{8}{3}\gamma\right)\frac{\partial}{\partial \epsilon}\right](M^*_{210} + M^*_{030} + M^*_{012}) = -2a^* M^*_{110}. \tag{5.74}$$

Equations (5.73) and (5.74) express in dimensionless form the conservation of momentum and energy and are equivalent to Eqs. (5.1)–(5.3), once the consistency of a hydrodynamic solution with the profiles (5.14) is assumed. Equation (5.73) shows that M^*_{110} does not depend on ϵ. This is consistent with the generalized Newton's law, Eqs. (5.12) and (5.15), where the generalized shear viscosity is independent of the thermal gradient. The general mathematical solution of Eq. (5.74) is

$$M^*_{210} + M^*_{030} + M^*_{012} = \frac{3}{4}\frac{a^*}{\gamma(a^*)} M^*_{110}\epsilon + C(a^*)\left[\frac{\epsilon^2}{2} + \frac{8}{3}\gamma(a^*)\right]^{1/2}, \tag{5.75}$$

where $C(a^*)$ is an integration constant. According to the symmetry property (5.66), the left-hand side of Eq. (5.75) must be an odd function of ϵ. Consequently, $C = 0$ and the physically meaningful solution is

$$M^*_{210} + M^*_{030} + M^*_{012} = \frac{3}{4}\frac{a^*}{\gamma(a^*)} M^*_{110}\epsilon. \tag{5.76}$$

The left-hand side is proportional to the component q_y of the heat flux. The fact that it depends linearly on ϵ indicates that the generalized Fourier's law, Eqs. (5.13) and (5.17), holds for *arbitrary* thermal gradients.

In order to complete the proof of the consistency of Eq. (5.14), we need to show that the hierarchy (5.69), seen as an infinite set of first-order differential equations with respect to ϵ, admits a solution. We have already seen that the second-degree moments M^*_{110} and M^*_{020} are independent of ϵ, while the combination of third-degree moments $M^*_{210} + M^*_{030} + M^*_{012}$ is linear in ϵ. This suggests to guess that, in general, the moments of degree $k \geq 2$ are *polynomials* in ϵ of degree $k-2$, the coefficients being functions of the shear rate a^*.[2] This guess is consistent with Eq. (5.69): the first two terms on the left-hand side are polynomials of degree k, while the third one and the right-hand side are polynomials of degree $k-2$. More specifically, the moments of degree $k \geq 2$ have the structure

$$M^*_{\mathbf{k}}(a^*, \epsilon) = \sum_{\ell=0}^{k-2} \mu^{(\ell)}_{\mathbf{k}}(a^*)\epsilon^{\ell}. \tag{5.77}$$

The symmetry property (5.66) implies that $\mu^{(\ell)}_{\mathbf{k}} = 0$ if $k_1 + k_2 + \ell =$ odd, while Eq. (5.67) yields $\mu^{(\ell)}_{\mathbf{k}}(-a^*) = (-1)^{k_1}\mu^{(\ell)}_{\mathbf{k}}(a^*)$. By equating the coefficients of the same degree in Eq. (5.69), one has

$$\begin{aligned} \frac{k-\ell}{2}\mu^{(\ell-1)}_{k_1,k_2+1,k_3} \quad &- \quad \frac{8}{3}(\ell+1)\gamma\mu^{(\ell+1)}_{k_1,k_2+1,k_3} + k_1 a^* \mu^{(\ell)}_{k_1-1,k_2+1,k_3} \\ &= \quad \sum_{\mathbf{k}',\mathbf{k}''}{}' \widehat{J}^*_{\mathbf{k}\mathbf{k}'\mathbf{k}''} \sum_{\ell'=0}^{k'} \mu^{(\ell')}_{\mathbf{k}'} \mu^{(\ell-\ell')}_{\mathbf{k}''} \end{aligned} \tag{5.78}$$

with the convention that $\mu^{(\ell)}_{\mathbf{k}} = 0$ if $\ell < 0$ or $\ell > k-2$, except $\mu^{(\ell)}_{000} = \delta_{\ell 0}$, on account of Eq. (5.70).

Thus, we have shown that the Boltzmann equation (5.39) (for Maxwell molecules) admits a hydrodynamic solution with the *space* dependence expressed by Eq. (5.14). This solution is characterized by reduced moments of the form (5.77). On the other hand, the numerical coefficients $\gamma(a^*)$ and $\mu^{(\ell)}_{\mathbf{k}}(a^*)$ are still unknown. They obey the algebraic hierarchy (5.78), which cannot be solved in a recursive way for arbitrary shear rates, since the coefficients of degree k and order ℓ are coupled to those of degree $k+1$ and order $\ell+1$. However, the hierarchy can be solved step by step if one performs a perturbation expansion in powers of a^* (Tij and Santos, 1995). From the results up to order a^{*3}, one has the following expressions for the generalized transport coefficients defined in Section 5.2 [cf. Eqs. (5.15), (5.17), (5.20)–(5.22)]:

$$F_\eta(a^*) = 1 - \frac{149}{45}a^{*2} + \mathcal{O}(a^{*4}), \tag{5.79}$$

$$F_\kappa(a^*) = 1 - 7.260a^{*2} + \mathcal{O}(a^{*4}), \tag{5.80}$$

$$\Psi_1(a^*) = -\frac{14}{5} + \mathcal{O}(a^{*2}), \tag{5.81}$$

$$\Psi_2(a^*) = \frac{4}{5} + \mathcal{O}(a^{*2}), \tag{5.82}$$

[2] Of course, the only moment of degree 0 is a constant, namely $M^*_{000} = 1$.

$$\Phi(a^*) = -\frac{7}{2} + \mathcal{O}(a^{*2}). \tag{5.83}$$

Equations (5.81)–(5.83) coincide with those obtained from the Burnett constitutive equations, Eqs. (5.28), (5.29), and (5.33). On the other hand, Eqs. (5.79) and (5.80) include super-Burnett order terms. From (5.19) we get the thermal curvature parameter γ as

$$\gamma(a^*) = \frac{a^{*2}}{5}\left[1 + 3.948a^{*2} + \mathcal{O}(a^{*4})\right]. \tag{5.84}$$

This parameter grows with the shear rate more rapidly than $\frac{1}{5}a^{*2}$ because the decrease of $F_\kappa(a^*)$ is more pronounced than that of $F_\eta(a^*)$. As we will see in the next Section, the BGK model predicts that $\gamma \sim a^{*2}$ for large shear rates, so that the ratio $\gamma(a^*)/a^{*2}$ has an upper bound.

4. BGK kinetic model description

Although the Boltzmann equation for Maxwell molecules admits a hydrodynamic solution for planar Couette flow characterized by the hydrodynamic profiles (5.14), it is not possible to get closed expressions for the generalized transport coefficients for *arbitrary* values of the shear rate. This contrasts with the related case of the USF, where the relevant rheological properties were explicitly obtained for Maxwell molecules. This is basically due to the fact that in the latter state the convective term $\mathbf{v}\cdot\nabla$ is absent in the Lagrangian frame [cf. Eq. (2.6)], while it cannot be eliminated in the Couette flow due to the spatial dependence of the temperature. Consequently, if one wants to determine the nonlinear transport coefficients one has to resort to approximate schemes or to computer simulations. From the theoretical point of view, the BGK kinetic model represents an excellent compromise between simplicity and reliability, as seen in the USF problem. As will be shown in this Section, the use of the BGK model in the Couette flow problem not only allows us to get the nonlinear transport coefficients (Brey et al., 1987; Garzó and López de Haro, 1997), but also the explicit form of the velocity distribution function (Kim et al., 1989a; Dufty, 1990; Garzó and López de Haro, 1997).

According to the BGK model, the Boltzmann collision operator for an arbitrary interaction potential is replaced by a single relaxation-time term, so Eq. (5.39) becomes

$$v_y \frac{\partial}{\partial y} f = -\nu(f - f_L), \tag{5.85}$$

where ν is the effective collision frequency (whose temperature dependence models the influence of the interaction potential) and f_L is the local equilibrium distribution function. In general, Eq. (5.85) must be supplemented by the appropriate boundary conditions. The case of diffuse boundary conditions is given by Eq. (5.44). One of the advantages of the BGK equation is that it can be formally solved in terms of the hydrodynamic fields. Assuming that the latter fields were known, then f_L would be completely specified and so Eq. (5.85) could be seen as a linear, inhomogeneous, first-order partial differential equation. Its solution can be expressed as the sum of two terms:

$$f(y, \mathbf{v}) = f_{B'}(y, \mathbf{v}) + f_{H'}(y, \mathbf{v}). \tag{5.86}$$

Here, $f_{B'}$ is the solution to the homogeneous equation (obtained by setting $f_L \to 0$) that verifies the complete boundary conditions, while $f_{H'}$ is the solution to the complete equation that verifies homogeneous boundary conditions (namely, half-distributions vanishing at the boundaries). More specifically, $f_{B'}$ is given by

$$\begin{aligned} f_{B'}(s, \mathbf{v}) &= \Theta(v_y) n_- \varphi_-(\mathbf{v}) e^{-(s-s_-)/v_y} \\ &\quad + \Theta(-v_y) n_+ \varphi_+(\mathbf{v}) e^{-(s-s_+)/v_y}, \end{aligned} \tag{5.87}$$

where we have introduced the scaled space variable $s(y)$ defined by Eq. (5.53) and we have called $s_\pm \equiv s(y = \pm L)$. In contrast, $f_{H'}$ is

$$f_{H'}(s, \mathbf{v}) = \int_0^\infty dt\, e^{-t} e^{-t v_y \partial_s} \Theta(s - s_-) \Theta(s_+ - s) f_L(s, \mathbf{v}), \tag{5.88}$$

where $\partial_s \equiv \partial/\partial s$. The action of the shift operator $e^{-t v_y \partial_s}$ is

$$e^{-t v_y \partial_s} X(s) = X(s - t v_y). \tag{5.89}$$

At the boundary $s = s_\pm$, the function $f_{H'}$ is

$$f_{H'}(s_\pm, \mathbf{v}) = \int_0^\infty dt\, e^{-t} \Theta(\pm t v_y) \Theta(s_+ - s_- \mp t v_y) f_L(s_\pm - t v_y, \mathbf{v}). \tag{5.90}$$

Thus $\Theta(\mp v_y) f_{H'}(s_\pm, \mathbf{v}) = 0$, i.e. $f_{H'}$ verifies homogeneous boundary conditions. Equation (5.86) is a *formal* solution to the BGK equation (5.85) since the hydrodynamic fields are not known "a priori." Nonetheless, taking the first few velocity moments in Eq. (5.86), one gets a closed set of nonlinear integral equations for the fields $p(s)$, $u_x(s)$, and $T(s)$ (Kim et al., 1989a; Dufty, 1990). The behavior of the hydrodynamic profiles in the boundary layer adjacent to the walls at $y = \pm L$ has been widely studied (Sone and Yamamoto, 1970; Sone and Onishi, 1979; Sone et al., 1990; Sone, 1991).

4.1 Generalized transport coefficients

As in Section 5.3, we are interested in the *hydrodynamic* solution f_H of the kinetic equation (5.85), where all the spatial dependence occurs through the hydrodynamic fields. Strictly speaking, the contribution $f_{H'}$, Eq. (5.88), is not fully hydrodynamic because of the explicit s-dependence of the Heaviside functions. To determine the hydrodynamic solution we formally extend the domain $s_- \leq s \leq s_+$ to a larger domain $-s_0 \leq s \leq s_0$ with the only constraint that the temperature must be positive definite. This is schematically illustrated in Fig. 5.3. The solid line represents the actual temperature profile for the system bounded between $s = s_-$ ($y = -L$) and $s = s_+$ ($y = L$). The temperatures of the fluid adjacent to the walls are different from the imposed wall temperatures $T_\pm$ (temperature jump effect). The dashed line represents the *hydrodynamic* temperature profile (i.e. the one associated with the hydrodynamic solution), which coincides with the actual one in the bulk domain, but deviates from the latter in both boundary layers (which are strongly exaggerated in Fig. 5.3). When this hydrodynamic profile is extrapolated outside the system, it comprises the region $-s_0 \leq s \leq s_0$, where $s = \pm s_0$ are the points where the (extrapolated) hydrodynamic temperature vanishes. Therefore, the hydrodynamic distribution

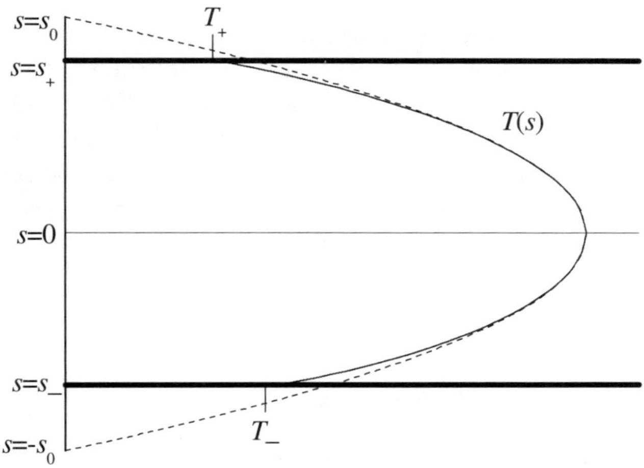

Figure 5.3. Sketch of the extended domain $-s_0 \leq s \leq s_0$ to define the hydrodynamic temperature (dashed line). The solid line represents the actual temperature for the system bounded between $s = s_-$ and $s = s_+$. The wall temperatures are T_- and T_+, respectively. The case depicted in the figure corresponds to $T_- > T_+$.

f_H has the form (5.88), except that the constraint $s_- \leq s \leq s_+$ is replaced by $-s_0 \leq s \leq s_0$, i.e.

$$f_H(s, \mathbf{v}) = \int_0^\infty dt\, e^{-t} e^{-tv_y \partial_s} \Theta\left(T(s)\right) f_L(s, \mathbf{v}). \tag{5.91}$$

This distribution gives the entire solution to the kinetic equation (5.85) corresponding to *idealized* boundary conditions of infinitely cold walls ($T_\pm \to 0$), in which case $\varphi_\pm(\mathbf{v}) \to 0$ and, consequently, $f_{B'}(s, \mathbf{v}) \to 0$.

It is apparent that Eq. (5.91) does not determine completely the velocity distribution function unless the hydrodynamic profiles entering through the local equilibrium distribution are known. Suggested by the results obtained in Section 5.3 from the Boltzmann equation, we guess that f_H is characterized by a constant pressure, a linear velocity profile, and a parabolic temperature profile. More specifically, the two latter are given by Eqs. (5.54) and (5.55) with the BGK value $\mathrm{Pr} = 1$. In order to prove the self-consistency of this guess, one needs to prove that

$$k_B T(s) \int d\mathbf{v}\, f_H(s, \mathbf{v}) = p = \text{const}, \tag{5.92}$$

$$\frac{k_B T(s)}{p} \int d\mathbf{v}\, v_x f_H(s, \mathbf{v}) = u_x(s) = a^* s, \tag{5.93}$$

$$\frac{m}{3} \int d\mathbf{v}\, [\mathbf{v} - \mathbf{u}(s)]^2 f_H(s, \mathbf{v}) = p = \text{const}. \tag{5.94}$$

These conditions are proved in Appendix 5.A from the evaluation of the generating function for the velocity moments of f_H, confirming that Eq. (5.91) is an exact solution to the BGK kinetic equation. While Eqs. (5.92) and (5.93) are verified irrespective of the relationship between the reduced shear rate a^* and the thermal curvature parameter γ, Eq. (5.94) provides the functional dependence of $\gamma(a^*)$ for arbitrary shear rate. The details of the analysis is given in Appendix 5.A, leading to the implicit equation

$$a^{*2} = \gamma \frac{3F_1(\gamma) + 2F_2(\gamma)}{F_1(\gamma)}. \tag{5.95}$$

Here, the functions $F_r(x)$ are defined as

$$F_r(x) = \left(\frac{\partial}{\partial x} x\right)^r F_0(x), \tag{5.96}$$

where

$$F_0(x) = \frac{2}{x} \int_0^\infty dt\, t e^{-t^2/2} K_0(2x^{-1/4} t^{1/2}), \tag{5.97}$$

$K_0(x)$ being the zeroth-order modified Bessel function (Abramowitz and Stegun, 1972; Gradshteyn and Ryzhik, 1980). In the limits of small and large shear rates, the thermal curvature parameter $\gamma(a^*)$ behaves as (cf. Appendix 5.A)

$$\gamma(a^*) \approx \frac{1}{5} a^{*2} \left(1 + \frac{72}{25} a^{*2}\right), \quad a^* \ll 1, \tag{5.98}$$

$$\gamma(a^*) \approx \frac{1}{3} a^{*2}, \quad a^* \gg 1. \tag{5.99}$$

Equation (5.98) gives the super-Burnett correction to the Navier–Stokes prediction $\gamma = a^{*2}/5$. This correction is about 27% smaller than the one given by the Boltzmann equation for Maxwell molecules [cf. Eq. (5.84)]. Since the expansion of γ in powers of the shear rate is only asymptotic in the BGK description, as shown in Appendix 5.A, it is quite possible that the same happens in the Boltzmann description. The nonlinear shear-rate dependence of γ, as given by Eq. (5.95) is shown in Fig. 5.4, which also includes the Navier–Stokes and super-Burnett order predictions. We observe that the Navier–Stokes expression is in general closer to the full nonlinear curve than the super-Burnett expression. This is a consequence of the asymptotic character of the expansion in powers of a^*. Since this is an *alternating* series, the error of a truncated series is bounded by the absolute value of the first discarded term (Gradshteyn and Ryzhik, 1980). Therefore, $|5\gamma/a^{*2} - 1| \leq \frac{72}{25}a^{*2}$, while $|5\gamma/a^{*2} - 1 - \frac{72}{25}a^{*2}| \leq \frac{84672}{625}a^{*4}$ [cf. Eq. (5.A.48)]; the second bound becomes larger than the first bound if $a^* > \frac{5\sqrt{6}}{84} \simeq 0.15$.

It may be noted that, although the independent nonequilibrium parameter is the shear rate a^*, from a practical point of view it is more convenient to take γ as the independent variable, so that Eq. (5.95) can be seen as giving explicitly a^{*2} in terms of γ. As a consequence, all the transport coefficients of the problem will be expressed in terms of γ only.

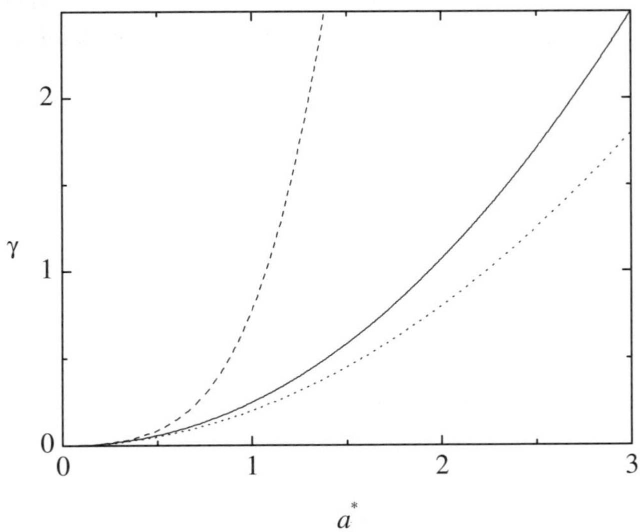

Figure 5.4. Nonlinear shear-rate dependence of the thermal curvature parameter γ in the BGK model (solid line). The Navier–Stokes prediction $\gamma = \frac{1}{5}a^{*2}$ (dotted line) and the super-Burnett prediction $\gamma = \frac{1}{5}a^{*2}\left(1 + \frac{72}{25}a^{*2}\right)$ (dashed line) are also included.

The explicit expressions of the pressure tensor and the heat flux are derived in Appendix 5.B. From them, one can identify the generalized transport coefficients defined by Eqs. (5.15), (5.17), and (5.20)–(5.22). The viscosity function $F_\eta(a^*)$ is

$$F_\eta(a^*) = F_0(\gamma). \tag{5.100}$$

The generalized thermal conductivity coefficient $F_\kappa(a^*)$ is directly related to $F_\eta(a^*)$ and $\gamma(a^*)$ through Eq. (5.19), so that

$$F_\kappa(a^*) = \frac{1}{5}F_0(\gamma)\frac{3F_1(\gamma) + 2F_2(\gamma)}{F_1(\gamma)}. \tag{5.101}$$

The viscometric functions $\Psi_{1,2}(a^*)$ characterizing normal stress differences are

$$\Psi_1(a^*) = -2F_1(\gamma)\frac{3F_1(\gamma) + 4F_2(\gamma)}{3F_1(\gamma) + 2F_2(\gamma)}, \tag{5.102}$$

$$\Psi_2(a^*) = 4F_1(\gamma)\frac{F_2(\gamma)}{3F_1(\gamma) + 2F_2(\gamma)}. \tag{5.103}$$

Finally, the expression for the visco-thermal function $\Phi(a^*)$ is

$$\Phi(a^*) = \frac{F_0(\gamma) - 1}{10\gamma} - F_1(\gamma) - \frac{4}{5}F_2(\gamma)\left[1 - \frac{F_2(\gamma)}{F_1(\gamma)}\right]. \tag{5.104}$$

Using the asymptotic forms of $F_r(x)$ for small x, Eq. (5.A.45), one can derive the first few terms in the (asymptotic) expansion of the generalized transport coefficients in powers of the shear rate:

$$F_\eta(a^*) \approx 1 - \frac{18}{5}a^{*2}, \tag{5.105}$$

$$F_\kappa(a^*) \approx 1 - \frac{162}{25}a^{*2}, \tag{5.106}$$

$$\Psi_1(a^*) \approx -\frac{14}{5}\left(1 - \frac{1476}{175}a^{*2}\right), \tag{5.107}$$

$$\Psi_2(a^*) \approx \frac{4}{5}\left(1 - \frac{288}{25}a^{*2}\right), \tag{5.108}$$

$$\Phi(a^*) \approx -\frac{14}{5}\left(1 - \frac{468}{35}a^{*2}\right). \tag{5.109}$$

Equations (5.107) and (5.108) extend the Burnett-order values given by Eqs. (5.28) and (5.29). Note that the numerical value of the other Burnett coefficient, namely $\Phi(0)$, differs from the value (5.33) obtained in the Boltzmann equation. In addition, the numerical values of the super-Burnett coefficients in F_η and F_κ

are also different from those obtained in the Boltzmann equation, as shown in Eqs. (5.79) and (5.80). In the opposite limit of large shear rates, and using the dominant behaviors of $F_r(x)$ for large x, one gets

$$F_\eta(a^*) \approx 3a^{*-2} \ln a^*, \tag{5.110}$$

$$F_\kappa(a^*) \approx \frac{9}{5} a^{*-2} \ln a^*, \tag{5.111}$$

$$\Psi_1(a^*) \approx -3a^{*-2}, \tag{5.112}$$

$$\Psi_2(a^*) \approx \sqrt{\frac{3\pi}{2}} a^{*-3} \ln a^*, \tag{5.113}$$

$$\Phi(a^*) \approx -\frac{9}{5} a^{*-2}. \tag{5.114}$$

The five generalized transport coefficients are plotted in Figs. 5.5 and 5.6 in the range $0 \leq a^* \leq 3$. It is apparent that these coefficients exhibit a highly nonlinear dependence on the shear rate, monotonically decreasing (in magnitude) as the shear rate increases. This means that the shear stress and the heat flux component parallel to the thermal gradient are inhibited with respect to their Navier–Stokes values, the normal stress differences grow more slowly than a^{*2}, and the heat flux component orthogonal to the thermal gradient grows more slowly than a^*.

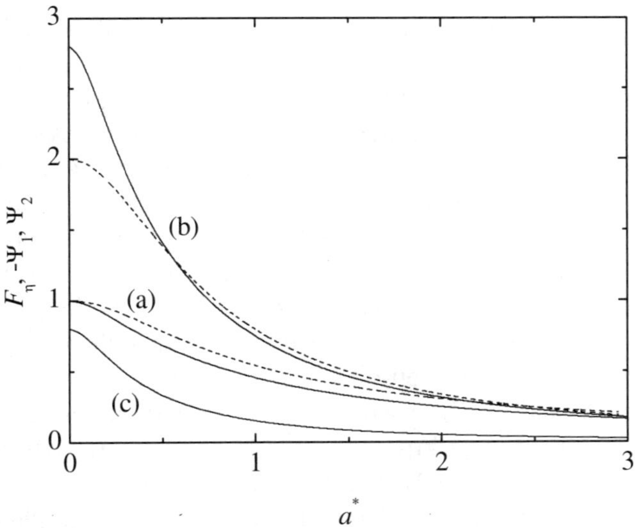

Figure 5.5. Plot of (a) the viscosity function $F_\eta(a^*)$, (b) the first viscometric function $-\Psi_1(a^*)$, and (c) the second viscometric function $\Psi_2(a^*)$. The solid lines correspond to the Couette flow results, while the dashed lines refer to the USF. Note that $\Psi_2 = 0$ in the USF.

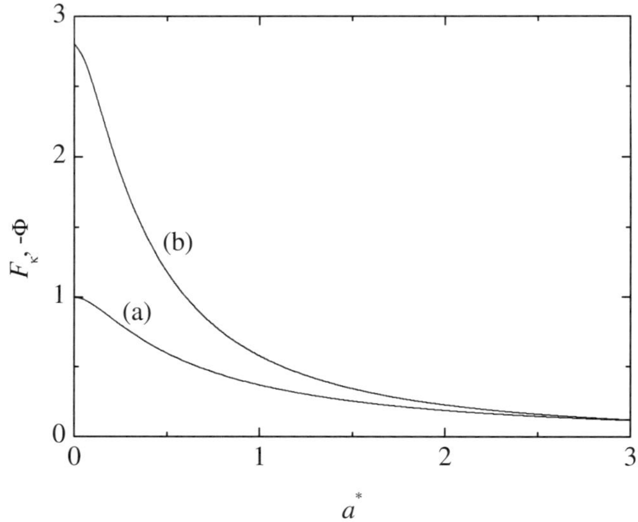

Figure 5.6. Shear-rate dependence of (a) the thermal conductivity function $F_\kappa(a^*)$ and (b) the visco-thermal function $-\Phi(a^*)$.

We have emphasized in Section 5.1 that the Couette flow problem is intrinsically different from the USF problem (Santos and Brey, 1991b; Santos et al., 1992). Both have in common that the pressure and the *reduced* shear rate a^* are uniform. However, while the temperature is uniform but unsteady (unless a thermostat force is introduced) in the USF, it is steady but non-uniform in the Couette flow. An obvious consequence is that the heat flux is absent in the USF, unless a weak thermal gradient perturbation is imposed, as has been analyzed in Sections 3.6 and 3.7. A subtler effect appears in the rheological quantities. Since they do not depend explicitly on the thermal gradient $\partial T/\partial s$ in the Couette flow, one might be tempted to think that they have the same shear-rate dependence in both problems (Liem et al., 1992). However, this is not so because the rheological properties of the Couette flow include terms associated with $\partial^2 T/\partial s^2 \propto \gamma(a^*)$, which are absent in the case of the USF. This is exemplified by the Burnett-order calculation of the normal stress differences, Eqs. (5.26) and (5.27), where the terms proportional to ϖ_3 come from the second derivative $\partial_s^2 T$, as Eqs. (5.23) and (5.24) show. If we formally make $\varpi_3 \to 0$ in Eqs. (5.26) and (5.27), we recover the Burnett values $\Psi_1(0) = -2$ and $\Psi_2(0) = 0$ of the USF case. Notwithstanding this, one might expect that the main trends of the nonlinear properties of the pressure tensor are similar in both flows, at least at a qualitative level (Santos et al., 1992). In order to assess this point, we have also included the curves representing $F_\eta(a^*)$ and $\Psi_1(a^*)$ of the USF problem in Fig. 5.5. It is apparent that the viscosity function is quite similar in both states, the shear-thinning effect being more pronounced in the Couette flow. With respect to the first viscometric function (which measures the normal stress difference in the shearing flow plane), we observe that it decays more rapidly in the Couette flow than in the USF. Of course, since the second viscometric function vanishes in the USF, it behaves in a way different from that of the Couette flow.

Although the five generalized transport coefficients provide the complete information on the momentum and heat transport in the Couette flow problem, it is also instructive to consider other physical quantities directly related to the above transport coefficients. This helps to gain insight into the nonlinear features of the Couette flow. Let us consider first the relative shear stress $-P_{xy}/p = a^* F_\eta(a^*)$. Figure 5.7 shows that $-P_{xy}/p$ starts growing almost linearly with a^*, reaches a maximum $-P_{xy}/p \simeq 0.50$ around $a^* = 2.03$, and then decays slowly. This non-monotonic behavior of $-P_{xy}/p$ is a consequence of a strong shear-thinning effect making $F_\eta(a^*)$ decay more rapidly than $1/a^*$ for large shear rates, as shown by Eq. (5.110). The non-monotonic behavior of the shear stress is also present in the USF, although in that case the maximum is higher and occurs at a larger shear rate. We refer the reader to the final paragraph of Section 2.3 for a discussion on this unexpected behavior of the shear stress.

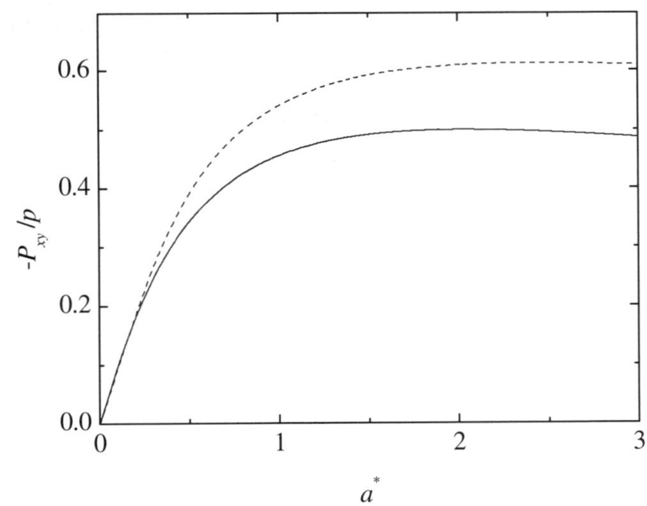

Figure 5.7. Relative shear stress $-P_{xy}/p$ as a function of the shear rate a^* in the Couette flow (solid line) and in the USF (dashed line).

Another interesting quantity related to the pressure tensor is the "friction" coefficient $-P_{xy}/P_{yy}$, measuring the tangential force relative to the normal force on a plane parallel to the plates. In the Navier–Stokes domain the friction coefficient is simply $-P_{xy}/P_{yy} \approx a^*$. Therefore, non-Newtonian behavior of this quantity is measured by the friction function $F_\mu(a^*) = -(P_{xy}/P_{yy})/a^*$, which is given by

$$F_\mu(a^*) = \frac{F_\eta(a^*)}{1 - [\Psi_2(a^*) - \Psi_1(a^*)]\,a^{*2}/3} = \frac{F_0(\gamma)}{1 - 2\gamma\,[F_1(\gamma) + 2F_2(\gamma)]}. \tag{5.115}$$

Its asymptotic behaviors for small and large a^* are, respectively,

$$F_\mu(a^*) \approx 1 - \frac{12}{5}a^{*2}, \tag{5.116}$$

$$F_\mu(a^*) \approx \sqrt{\frac{6}{\pi}}a^{*-1}\ln a^*. \tag{5.117}$$

The friction coefficient $-P_{xy}/P_{yy}$ is plotted in Fig. 5.8. In contrast to the reduced shear stress $-P_{xy}/p$, the ratio $-P_{xy}/P_{yy}$ is a monotonically increasing function of the shear rate, both in the Couette flow and in the USF, the increase being larger in the latter case. It is interesting to note that the friction coefficient can be even larger than 1 for shear rates beyond a certain value ($a^* \simeq 2.09$ in the Couette flow, $a^* \simeq 1.67$ in the USF). For shear rates larger than this value, one has $P_{zz} \leq P_{yy} < |P_{xy}| < p < P_{xx}$.

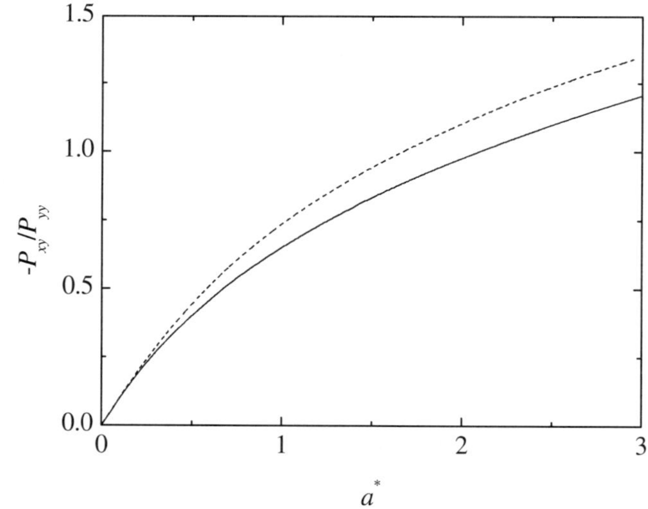

Figure 5.8. Friction coefficient $-P_{xy}/P_{yy}$ as a function of the shear rate a^* in the Couette flow (solid line) and in the USF (dashed line).

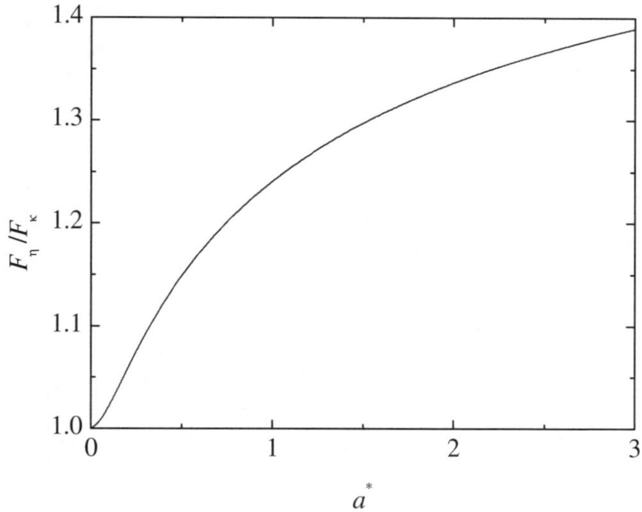

Figure 5.9. Shear-rate dependence of ratio between the viscosity function and the thermal conductivity function, $F_\eta(a^*)/F_\kappa(a^*)$.

As said above, the main qualitative difference between the Couette flow and the USF is perhaps the existence of a non-vanishing heat flux in the former. In other words, the momentum and heat fluxes are inextricably coupled in the Couette problem. A way of characterizing the strength of the coupling is through the ratio $F_\eta(a^*)/F_\kappa(a^*) = 5\gamma(a^*)/a^{*2}$, which plays the role of a *generalized* Prandtl number. This quantity is plotted in Fig. 5.9. We observe that the generalized Prandtl number monotonically increases from its Navier–Stokes value to a constant asymptotic value $F_\eta(\infty)/F_\kappa(\infty) = \frac{5}{3}$ [cf. Eqs. (5.110) and (5.111)]. Thus, although both the momentum and the heat fluxes are inhibited (with respect to their Navier–Stokes values) by the shear flow, this effect is less significant in the case of the momentum transport.

An interesting non-Newtonian effect is the existence of a component of the heat flux orthogonal to the thermal gradient, i.e. $q_x \neq 0$. For small shear rates, the magnitude of q_x is much smaller than that of q_y. However, as the shear rate increases, the ratio $-q_x/q_y$ monotonically increases without a bound, behaving as $-q_x/q_y \approx a^*/\ln a^*$ for large shear rates [cf. Eqs. (5.111) and (5.114)]. The ratio $-q_x/q_y = -a^*\Phi(a^*)/F_\kappa(a^*)$ is shown in Fig. 5.10. It is noteworthy that for a not extremely large shear rate (namely $a^* \simeq 0.5$), the component $|q_x|$ starts to dominate over the component $|q_y|$. This is partially explained by the fact that, while $|q_y|$ decreases with increasing shear rate, $|q_x|$ starts growing, reaches a maximum at $a^* \simeq 0.66$ and then decreases, although more slowly than $|q_y|$. The fact that $P_{xx} > |P_{xy}| > P_{yy}$ and $|q_x| > |q_y|$ for sufficiently large

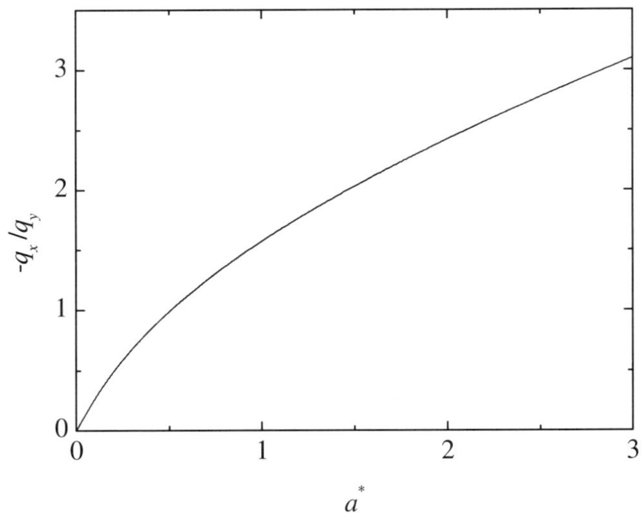

Figure 5.10. Ratio between the two components of the heat flux, $-q_x/q_y$, as a function of the shear rate.

shear rates indicates a tendency of the particles to move with peculiar velocities almost parallel to the flow direction.

4.2 Velocity distribution function

Once the consistency of the assumed hydrodynamic profiles has been proved, we are in conditions of writing the velocity distribution function from Eq. (5.91) in an explicit way. Since (5.91) qualifies as a normal solution, it is convenient to scale the velocities with respect to the (local) thermal velocity and define a reduced distribution function f^* and a reduced velocity $\boldsymbol{\xi}$, according to Eqs. (5.58) and (5.59). The reduced distribution f^* depends parametrically on the reduced shear rate a^* and the reduced thermal gradient ϵ defined by Eq. (5.56). From Eq. (5.91) it is easy to get

$$f^*(\boldsymbol{\xi}) = \pi^{-3/2} \int_0^\infty dt\, e^{-t} \left[\vartheta(t\xi_y)\right]^{-5/2} \Theta\left(\vartheta(t\xi_y)\right) e^{a^* t\xi_y \partial/\partial \xi_x} e^{-\xi^2/\vartheta(t\xi_y)}, \tag{5.118}$$

where the shift operator $e^{a^* t\xi_y \partial/\partial \xi_x}$ acts as

$$e^{a^* t\xi_y \partial/\partial \xi_x} X(\xi_x, \xi_y, \xi_z) = X(\xi_x + a^* t\xi_y, \xi_y, \xi_z) \tag{5.119}$$

and

$$\vartheta(u) \equiv \frac{T(s - v_0(s)u)}{T(s)}, \tag{5.120}$$

$v_0(s) \equiv [2k_BT(s)/m)]^{1/2}$ being the thermal velocity. Since $T(s)$ is a quadratic function, one has

$$T(s-z) = T(s) - \frac{\partial T}{\partial s} z - \frac{m}{k_B}\gamma z^2. \tag{5.121}$$

Therefore,

$$\vartheta(u) = 1 - \epsilon u - 2\gamma u^2. \tag{5.122}$$

The symmetry of the problem (5.64) allows us to focus on the half-distribution $f_+^*(\boldsymbol{\xi}; a^*, \epsilon) = \Theta(\xi_y) f^*(\boldsymbol{\xi}; a^*, \epsilon)$, so that the other half-distribution can be obtained as $f_-^*(\boldsymbol{\xi}; a^*, \epsilon) = f_+^*(-\boldsymbol{\xi}; a^*, -\epsilon)$. Thus, making the change $u = t\xi_y$ in Eq. (5.118), one gets

$$f_\pm^*(\boldsymbol{\xi}) = \frac{\pi^{-3/2}}{|\xi_y|} \int_0^{\beta_\pm} du\, e^{-u/|\xi_y|} \left[\vartheta(\pm u)\right]^{-5/2} e^{\pm a^* u \partial/\partial \xi_x} e^{-\xi^2/\vartheta(\pm u)}, \tag{5.123}$$

where

$$\beta_\pm \equiv \frac{\sqrt{\epsilon^2 + 8\gamma} \mp \epsilon}{4\gamma}. \tag{5.124}$$

It is interesting to note from Eq. (5.118) that the distribution function at $\xi_y = 0$ is the same as that of local equilibrium, namely

$$f(\xi_x, \xi_y = 0, \xi_z) = \pi^{-3/2} \exp(-\xi_x^2 - \xi_z^2). \tag{5.125}$$

This means that the population of particles moving on a plane parallel to the plates is not affected by shearing motion. Paradoxically, this property does not hold in the USF case, as shown by Eq. (3.8) or Fig. 3.1.

Equation (5.123) gives the explicit form of the velocity distribution as a function of the reduced peculiar velocity $\boldsymbol{\xi}$, the reduced shear rate a^* and the reduced thermal gradient ϵ. It is apparent the highly nonlinear dependence on the nonequilibrium control parameters a^* and ϵ. If we particularize to $\epsilon = 0$, we obtain the velocity distribution function at the specific point ($s = 0$) where the temperature has a maximum. This is the point where the Couette flow state has the same symmetry properties as the USF, although both distribution functions are clearly different. On the other hand, if we set $a^* = 0$ with $\epsilon \neq 0$, the Couette flow reduces to the planar Fourier flow and Eq. (5.118) becomes the corresponding exact solution of the BGK equation (Santos et al., 1989; Kim et al., 1989b). In that case, the linear Fourier's law is exactly verified for arbitrary values of the thermal gradient (Santos et al., 1986b), a result that also holds in the Boltzmann equation for Maxwell molecules (Asmolov et al., 1979; Nosik, 1981; Nosik, 1983; Santos and Garzó, 1995b).

In principle, it is possible to get the expansion of f^* in powers of a^* and ϵ (Chapman–Enskog expansion) from Eq. (5.123). However, from a practical point of view it is more convenient to get those terms from the kinetic equation itself, Eq. (5.85). Let us rewrite that equation as

$$V_y\left(\partial_s - a^*\frac{\partial}{\partial V_x}\right)f = -f + f_L, \tag{5.126}$$

where here the operator ∂_s is taken at fixed $\mathbf{V}$ (not at fixed $\mathbf{v}$) and use has been made of the relation (5.A.1). Taking into account the relationship (5.58) between f and f^*, one has

$$\partial_s f = -\frac{5}{2}f\partial_s \ln T + \frac{p}{k_BT}v_0^{-3}\left(\frac{\partial f^*}{\partial \epsilon}\frac{\partial \epsilon}{\partial s} + \frac{\partial f^*}{\partial \boldsymbol{\xi}}\cdot\frac{\partial \boldsymbol{\xi}}{\partial s}\right). \tag{5.127}$$

Moreover, $v_0\partial_s\boldsymbol{\xi} = -\frac{1}{2}\epsilon\boldsymbol{\xi}$ and, from Eq. (5.57), $v_0\partial_s\epsilon = -\frac{1}{2}\epsilon^2 - 4\gamma$. Consequently, Eq. (5.126) becomes

$$\xi_y\left[a^*\frac{\partial}{\partial \xi_x} + \epsilon\left(1 + \frac{1}{2}\frac{\partial}{\partial \boldsymbol{\xi}}\cdot\boldsymbol{\xi}\right) + \frac{1}{2}\left(\epsilon^2 + 8\gamma\right)\frac{\partial}{\partial \epsilon}\right]f^* = f^* - \varphi_0, \tag{5.128}$$

where $\varphi_0(\boldsymbol{\xi}) \equiv f_L^*(\boldsymbol{\xi}) = \pi^{-3/2}e^{-\xi^2}$. Of course, the solution to Eq. (5.128) is given by (5.123). Now we expand f^* in the form

$$f^*(\boldsymbol{\xi}) = \varphi_0(\boldsymbol{\xi})\left[1 + \sum_{k=1}^{\infty}\Phi^{(k)}(\boldsymbol{\xi})\right], \tag{5.129}$$

where the contribution $\Phi^{(k)}$ is of order k in a^* and ϵ, i.e. it contains terms of the form $a^{*k_1}\epsilon^{k_2}$ with $k_1 + k_2 = k$. Inserting the expansion (5.129) into Eq. (5.127) and equating terms of the same order, we get

$$\begin{aligned}\Phi^{(k)} &= a^*\xi_y\left(-2\xi_x + \frac{\partial}{\partial \xi_x}\right)\Phi^{(k-1)} + \epsilon\xi_y\left(1 - \xi^2 + \frac{1}{2}\frac{\partial}{\partial \boldsymbol{\xi}}\cdot\boldsymbol{\xi}\right)\Phi^{(k-1)} \\ &\quad + \frac{1}{2}\xi_y\epsilon^2\frac{\partial}{\partial \epsilon}\Phi^{(k-1)} + 4\xi_y\sum_{\ell=0}^{k-1}\gamma_{k-\ell+1}a^{*k-\ell+1}\frac{\partial}{\partial \epsilon}\Phi^{(\ell)}.\end{aligned} \tag{5.130}$$

Here, γ_k are the corresponding coefficients in the expansion of $\gamma(a^*)$ in powers of a^* [cf. Eq. (5.A.48)]. Note that $\gamma_k = 0$ when $k = 0$ or $k =$ odd. The first few terms in the expansion (5.129) are

$$\Phi^{(1)} = -\xi_y\left[2a^*\xi_x + \left(\xi^2 - \frac{5}{2}\right)\epsilon\right], \tag{5.131}$$

$$\Phi^{(2)} = \xi_y^2\left[a^{*2}\left(4\xi_x^2 - \frac{4}{5}\xi^2\right) + 4a^*\epsilon\xi_x\left(\xi^2 - \frac{7}{2}\right) + \epsilon^2\left(\frac{35}{4} - 7\xi^2 + \xi^4\right)\right], \tag{5.132}$$

$$\Phi^{(3)} = -\xi_y^3 \left[\frac{8}{5} a^{*3} \xi_x \left(3 - 3\xi^2 + 5\xi_x^2\right) - \frac{6}{5} a^{*2} \epsilon \left(2\xi^4 - 9\xi^2 + 45\xi_x^2 - 10\xi^2\xi_x^2\right) + \frac{3}{2} a^* \epsilon^2 \xi_x \left(63 - 36\xi^2 + 4\xi^4\right) + \frac{1}{8} \epsilon^3 \left(8\xi^6 - 108\xi^4 + 378\xi^2 - 315\right) \right]. \tag{5.133}$$

Of course, the first few terms in the expansion of the momentum and heat fluxes in powers of the shear rate, Eqs. (5.105)–(5.109), coincide with those obtained from Eq. (5.129). However, although the expansion (5.129) involves all the powers in ϵ, only the terms of order ϵ^0 contribute to the pressure tensor, while only the terms of order ϵ contribute to the heat flux. This is so because the terms of order ϵ and higher are orthogonal to $\xi_i\xi_j$, while those of order ϵ^2 and higher are orthogonal to $\xi^2\xi_i$. In general, it can be proved that the velocity moments of degree k are polynomials in ϵ of degree $k-2$, with coefficients that are nonlinear functions of a^*. Comparison of Eqs. (5.131)–(5.133) with those corresponding to the USF, Eqs. (3.11)–(3.13), shows that, even if $\epsilon = 0$, they differ beyond the Navier–Stokes order.

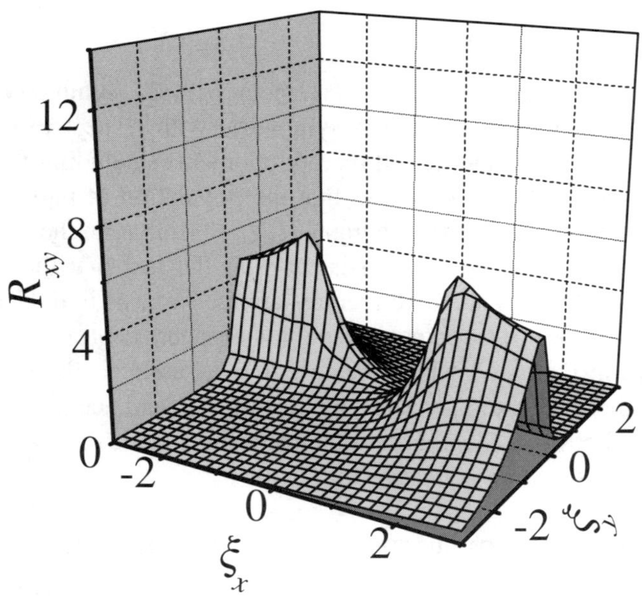

Figure 5.11. Surface plot of the ratio R_{xy} for $a^* = 1$ and $\epsilon = 0$.

To illustrate the distortion of the velocity distribution function from local equilibrium, it is useful to consider the marginal distribution function f_{xy}^* defined by Eq. (3.14). Figure 5.11 shows the ratio $R_{xy}(\xi_x, \xi_y)$ defined by Eq.

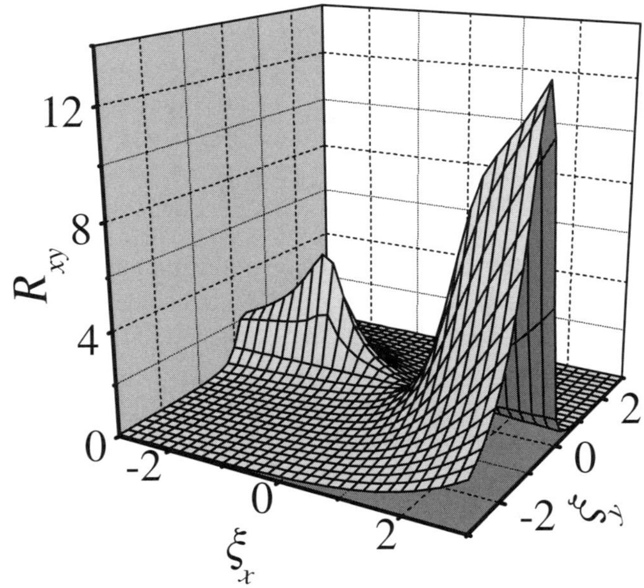

Figure 5.12. Surface plot of the ratio R_{xy} for $a^* = 1$ and $\epsilon = 0.2$.

(3.28), for $a^* = 1$ and $\epsilon = 0$. In agreement with the symmetry properties (5.63) and (5.64), the distribution is symmetric with respect to reflection of the velocity. As a consequence, all the odd moments (including the heat flux) vanish. In addition, Fig. 5.11 shows that the population of particles is much larger in the second and fourth quadrants ($\xi_x\xi_y < 0$) than in the first and third quadrants ($\xi_x\xi_y > 0$). This is consistent with the fact that the shear stress P_{xy} is negative. These qualitative features are the same as in the USF flow, as illustrated in Fig. 3.2. The above reflection symmetry is broken when $\epsilon \neq 0$. This is shown in Fig. 5.12, where the case $a^* = 1$ and $\epsilon = 0.2$ is considered. Since $\epsilon > 0$, there are more particles with positive ξ_x and negative ξ_y than with negative ξ_x and positive ξ_y, what results in a positive value of q_x and a negative value of q_y. The situation is reversed if $\epsilon < 0$.

5. Nonequilibrium entropy of the Couette flow

As done in Section 3.5 for the USF, we can also get the nonequilibrium entropy of a dilute gas under Couette flow by using the exact form of the velocity distribution function in the BGK model, Eq. (5.123). Again, the computation of the triple velocity integral (3.64) needs to be carried out numerically (cf. Appendix 3.A). Apart from the numerical evaluation, one can get analytical information in the form of a series expansion. In order to get the entropy through order k in the gradients, one needs to know the expansion of the velocity distribution function through order $k-1$. This expansion can be recursively obtained from Eq. (5.130). For simplicity, and to compare with the USF results, we focus on the point where $\epsilon = 0$, i.e. where the local thermal gradient and heat flux vanish. In this point, the expansion of the reduced excess entropy through eighth order is

$$s_{\text{ex}}^*(a^*) = -\frac{1}{2}a^{*2} + \frac{177}{100}a^{*4} - \frac{40\,041}{1250}a^{*6} + \frac{43\,975\,989}{25\,000}a^{*8} + \mathcal{O}(a^{*10}), \quad (5.134)$$

where use has been made of Eqs. (5.131)–(5.133) (with $\epsilon = 0$), plus the more intricate expressions for $\Phi^{(k)}$ with $k = 4$–7. The first term coincides with that of the USF [cf. Eq. (3.65)] and comes from the Navier–Stokes approximation. The second term is larger than the one found in the USF, thus showing that the entropy in the Couette flow is *larger* than the entropy in the USF, despite the fact that the Couette flow problem can be considered as being more complex than the USF problem, since in the former $P_{zz} \neq P_{yy}$ and the shear-thinning effect is more pronounced than in the latter (cf. Fig. 5.5). This shows that, as already noted in Section 3.5, an intuitive interpretation of the *nonequilibrium* entropy in terms of apparently "more" or "less ordered" states can be misleading. The magnitude of the coefficients in the expansion (5.134) increases dramatically with the order, suggesting that the series is only asymptotic. This is not surprising since we know that the series expansions of the pressure tensor and of the heat flux, Eqs. (5.105)–(5.109) are asymptotic. The series (5.134) diverges so rapidly that the use of a Padé approximant is not practical.

The above analytical results can be complemented by using information theory to get an upper bound of the entropy from the exact knowledge of the pressure tensor, as given by Eq. (2.133). Taking into account the explicit form for the elements of the pressure tensor in the Couette flow, Eqs. (5.B.12)–(5.B.15), one gets

$$\begin{aligned} s_{\text{ex,IT}}^* &= \frac{1}{2}\ln\Big\{(1-2\gamma F_1)\Big[1 - a^{*2}F_0^2 + 2\gamma F_1 \\ &\qquad -8\gamma^2 (F_1+F_2)(F_1+2F_2)\Big]\Big\}, \end{aligned} \quad (5.135)$$

where the functions $F_r(x)$ are evaluated at $x = \gamma(a^*)$. The first few terms in the series expansion of $s_{\text{ex,IT}}^*$ are

$$s_{\text{ex,IT}}^* = -\frac{1}{2}a^{*2} + \frac{251}{100}a^{*4} - \frac{188\,959}{3750}a^{*6} + \mathcal{O}(a^{*8}). \quad (5.136)$$

In Fig. 5.13 we plot s_{ex}^* versus the shear rate a^* as obtained by numerically performing the integral in Eq. (3.64) with f^* given by Eq. (5.123), as well as the information-theory upper bound $s_{\text{ex,IT}}^*$. For comparison, the numerical data for the USF entropy (cf. Section 3.5) are also included. We see that information theory gives quite a good estimate of the actual entropy in the Couette flow even for shear rates as large as $a^* \simeq 0.5$. As said before when comparing the series expansion of the entropy for USF and the Couette flow, s_{ex}^* is smaller in the USF than in the Couette flow. This fact is captured by information theory, which gives an estimate based only on the knowledge of the determinant of the pressure tensor, Eq. (2.133). For any value of the reduced shear rate, this determinant is smaller in the USF than in the Couette flow.

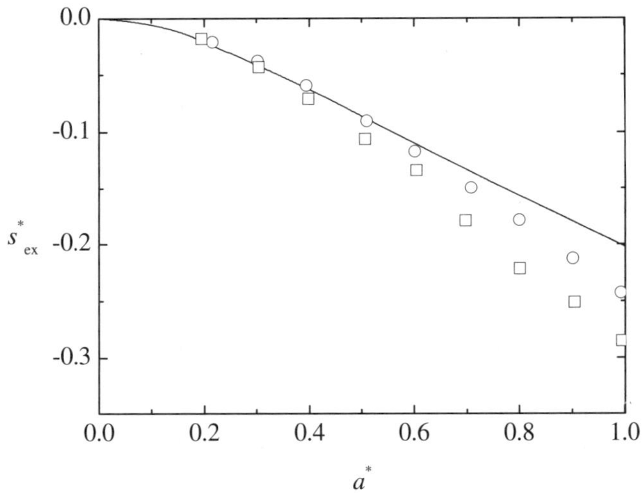

Figure 5.13. Shear-rate dependence of the reduced excess entropy, according to the numerical evaluation of Eq. (3.64) by using the distribution function Eq. (5.123) (circles) and information theory (solid line). The squares represent the numerical values corresponding to USF.

6. Other kinetic theories

One of the main drawbacks of the BGK kinetic model is that it does not give the correct value of the Prandtl number (Pr $\simeq \frac{2}{3}$). This shortcoming is not relevant in those nonequilibrium problems where either only momentum transport (e.g. in the USF) or only heat conduction (e.g. in the Fourier flow) are present in the system. In those cases, the BGK collision frequency can be adjusted to reproduce either the Boltzmann shear viscosity or the Boltzmann thermal conductivity. However, when both momentum and heat transport coexist, as happens in the Couette flow, the BGK model cannot reproduce simultaneously the Boltzmann transport coefficients.

As discussed in Section 1.6, the above limitation of the BGK model is corrected by the ellipsoidal statistical (ES) model (Holway, 1966; Cercignani, 1988; Andries et al., 2000) and the Liu model (Liu, 1990). In both models, the

Boltzmann collision operator is replaced by Eq. (1.172), the reference function f_R being given by Eq. (1.173) in the ES model and by Eq. (1.181) in the Liu model. The solution of both kinetic models in the planar Couette flow follows mathematical steps similar to those made in Section 5.4 for the BGK model. Thus, the hydrodynamic solution has the same structure as in Eq. (5.91), namely

$$f(s,\mathbf{v}) = \int_0^\infty dt\, e^{-t} e^{-(\nu/\zeta)tv_y\partial_s} \Theta\left(T(s)\right) f_R(s,\mathbf{v}), \tag{5.137}$$

where the space variable s is again defined by Eq. (5.53). Assuming that the pressure is constant and the hydrodynamic profiles are given by Eqs. (5.54) and (5.55), the spatial dependence of f_R is explicit. It turns out that, as in the BGK case, the consistency conditions (5.92) and (5.93) are automatically satisfied.

In the case of the Liu model, the consistency condition (5.94) gives an implicit equation for $\gamma(a^*)$. The result is (Garzó and López de Haro, 1994)

$$a^{*2} = 2\gamma \frac{3F_1(\overline{\gamma}) + 2F_2(\overline{\gamma})}{3F_1(\overline{\gamma}) - F_0(\overline{\gamma})}, \tag{5.138}$$

where $\overline{\gamma} \equiv \frac{2}{3}\gamma(\nu/\zeta)^2$ and we have already taken $\mathrm{Pr} = \frac{2}{3}$. Moreover, the pressure tensor and the heat flux can be evaluated from Eq. (5.137) by the same method as in Appendix 5.A. The viscosity function is

$$F_\eta(a^*) = F_0(\overline{\gamma}) - 2\overline{\gamma}\left[F_2(\overline{\gamma}) + 2F_3(\overline{\gamma})\right]. \tag{5.139}$$

The thermal conductivity function $F_\kappa(a^*)$ is easily obtained from $\gamma(a^*)$ and $F_\eta(a^*)$ through Eq. (5.19). The two viscometric functions are (Garzó and López de Haro, 1994)

$$\Psi_1(a^*) = -\frac{\nu}{\zeta} \frac{\left[3F_1(\overline{\gamma}) + 4F_2(\overline{\gamma})\right]\left[3F_1(\overline{\gamma}) - F_0(\overline{\gamma})\right]}{3F_1(\overline{\gamma}) + 2F_2(\overline{\gamma})}, \tag{5.140}$$

$$\Psi_2(a^*) = \frac{\nu}{\zeta} \frac{2F_2(\overline{\gamma})\left[3F_1(\overline{\gamma}) - F_0(\overline{\gamma})\right]}{3F_1(\overline{\gamma}) + 2F_2(\overline{\gamma})}. \tag{5.141}$$

As in the BGK case, the Chapman–Enskog expansion of the rheological properties is only asymptotic . The first few terms are

$$\gamma(a^*) \approx \frac{1}{5}a^{*2}\left[1 + \frac{18}{25}\left(\frac{\nu}{\zeta}\right)^2 a^{*2}\right], \tag{5.142}$$

$$F_\eta(a^*) \approx 1 - \frac{16}{5}\left(\frac{\nu}{\zeta}\right)^2 a^{*2}, \tag{5.143}$$

$$F_\kappa(a^*) \approx 1 - \frac{98}{25}\left(\frac{\nu}{\zeta}\right)^2 a^{*2}, \tag{5.144}$$

$$\Psi_1(a^*) \approx -\frac{14}{5}\frac{\nu}{\zeta}\left[1 - \frac{1194}{175}\left(\frac{\nu}{\zeta}\right)^2 a^{*2}\right], \tag{5.145}$$

$$\Psi_2(a^*) \approx \frac{4}{5}\frac{\nu}{\zeta}\left[1 - \frac{222}{25}\left(\frac{\nu}{\zeta}\right)^2 a^{*2}\right]. \tag{5.146}$$

The predictions of the Liu model are meaningless in the limit of asymptotically large shear rates. This is clearly seen in the expression (5.139) for F_η, which becomes negative for $\overline{\gamma} \gtrsim 5.3$. According to Eq. (5.138), this unphysical behavior corresponds to $a^* \gtrsim 10\zeta/\nu$. This is an artifact of the Liu model, whose reference function (1.181) is not positive definite, so that the positiveness of its solution is not guaranteed. This drawback also appears in earlier extensions of the BGK equation proposed by Gross and Jackson (Gross and Jackson, 1959).

Now we consider the solution of the ES model in the Couette flow. Since f_R involves the pressure tensor [cf. Eq. (1.173)], to get the shear-rate dependence of γ, we need to complement the consistency condition (5.94) with similar conditions for the remaining three independent elements of the pressure tensor. This leads to a coupled set of four equations (Garzó and López de Haro, 1997; Montanero and Garzó, 1998), whose solution can be expressed in terms of an auxiliary parameter β defined as the solution of the implicit equation

$$a^{*2} = \frac{4\beta}{9}\frac{[2\beta(F_1+2F_2)-3]^2[3F_1+2F_2-2\beta F_1(F_1+2F_2)]}{F_0^2[2\beta(F_1+F_2)-3]+F_1[2\beta(F_1+2F_2)-3]^2}, \tag{5.147}$$

where now the functions F_r are evaluated at $x = \beta$. The relationship between the thermal curvature parameter γ and the parameter β is

$$\gamma(a^*) = \frac{2}{9}\beta[3 - 2\beta(F_1 + 2F_2)]. \tag{5.148}$$

The generalized transport coefficients associated with the pressure tensor are

$$F_\eta(a^*) = \frac{9F_0}{[2\beta(F_1+2F_2)-3]^2}, \tag{5.149}$$

$$\Psi_1(a^*) = -\frac{12\beta}{a^{*2}}\frac{3F_1+4F_2-2\beta F_1(F_1+2F_2)}{(3-2\beta F_1)[3-2\beta F_1(F_1+2F_2]}, \tag{5.150}$$

$$\Psi_2(a^*) = \frac{24\beta}{a^{*2}}\frac{F_2}{(3-2\beta F_1)[3-2\beta F_1(F_1+2F_2]}. \tag{5.151}$$

The generalized thermal conductivity can be obtained again from Eq. (5.19). The expression for the visco-thermal function $\Phi(a^*)$ is rather involved:

$$\begin{aligned}\Phi(a^*) &= \frac{1}{40}H_1\{a^{*2}[H_1^4F_0^3(F_1+2F_2)-9H_1^3F_0^2(F_2+2F_3)\\&\quad+54H_1^2F_0(F_2+4(F_3+F_4))-108H_1(F_2+5F_3\\&\quad+4(2F_4+F_5))]-H_3(H_1F_0F_1-6F_2)+4H_1^2F_0(F_1+2F_2)\\&\quad+4H_1[H_2F_0F_1-6(F_2+2F_3)]-24H_2F_2\},\end{aligned} \tag{5.152}$$

where we have called

$$H_1 \equiv \frac{3}{3-2\beta(F_1+2F_2)}, \tag{5.153}$$

$$H_2 \equiv \frac{3}{3-2\beta F_1}, \tag{5.154}$$

$$H_3 \equiv 3a^{*2}H_1^3F_0^2+4H_2\{2\beta[H_1(F_1+4F_2)+3F_1]-3H_1\}. \tag{5.155}$$

For small shear rates, one has

$$\gamma(a^*) \approx \frac{1}{5}a^{*2}\left(1+\frac{92}{25}a^{*2}\right), \tag{5.156}$$

$$F_\eta(a^*) \approx 1-\frac{21}{5}a^{*2}, \tag{5.157}$$

$$F_\kappa(a^*) \approx 1-\frac{197}{25}a^{*2}, \tag{5.158}$$

$$\Psi_1(a^*) \approx -\frac{14}{5}\left(1-\frac{2126}{175}a^{*2}\right), \tag{5.159}$$

$$\Psi_2(a^*) \approx \frac{4}{5}\left(1-\frac{413}{25}a^{*2}\right). \tag{5.160}$$

In the above equations for the ES model we have taken Pr $= \frac{2}{3}$. The expressions for arbitrary values of the Prandtl number are given by Garzó and López de Haro (1997) and by Montanero and Garzó (1998). Of course, they reduce to the BGK results if one formally sets Pr $= 1$.

As said at the end of Section 1.3, an alternative route to the use of kinetic models consists of finding approximate solutions of the Boltzmann equation by means of Grad's moment method (Grad, 1949; Grad, 1963). In the thirteen-moment method, the hierarchy of moment equations is truncated after the first 13 equations (i.e. the rate of change equations for n, $\mathbf{u}$, T, P, and $\mathbf{q}$) by estimating the higher moments and the collisional moments with the approximate velocity distribution function

$$f \to f_L \left\{ 1 + \frac{m}{n(k_BT)^2} \left[\left(\frac{mV^2}{5k_BT} - 1 \right) \mathbf{V} \cdot \mathbf{q} + \frac{1}{2} \Pi_{ij} V_i V_j \right] \right\}, \qquad (5.161)$$

where f_L is the local equilibrium distribution defined by Eq. (1.59) and Π_{ij} is the irreversible part of the pressure tensor, Eq. (1.43). In general, Eq. (5.161) yields a closed set of 13 nonlinear partial differential equations. According to the geometry of the planar Couette flow, $u_y = u_z = P_{xz} = P_{yz} = q_z = 0$, so the problem involves 8 independent moments only. Taking into account the stationarity of the state, neglecting nonlinear terms in the fluxes, and assuming that the reduced shear rate a^* defined by Eq. (5.11) is constant, one gets a closed set of algebraic equations whose solution gives the following expressions for the generalized transport coefficients (Risso and Cordero, 1997; Risso and Cordero, 1998):

$$F_\eta(a^*) = \frac{2}{1 + \frac{72}{25}a^{*2} + \Delta(a^*)}, \qquad (5.162)$$

$$F_\kappa(a^*) = \frac{4}{1 - \frac{216}{25}a^{*2} + 3\Delta(a^*)}, \qquad (5.163)$$

$$\Phi(a^*) = -7 \frac{1 - \frac{36}{125}a^{*2}}{1 + \frac{6}{5}a^{*2} + (1 - \frac{63}{25}a^{*2})\Delta(a^*)}, \qquad (5.164)$$

where $\Delta(a^*) \equiv \sqrt{1 + \frac{116}{25}a^{*2} - \frac{864}{625}a^{*4}}$. The viscometric functions are not evaluated by Risso and Cordero (1997, 1998), but the friction function $F_\mu(a^*) = -(P_{xy}/P_{yy})/a^*$ is provided. It is given by

$$F_\mu(a^*) = \frac{2}{1 + \frac{12}{25}a^{*2} + \Delta(a^*)}. \qquad (5.165)$$

Note that $F_\eta(a^*)$ and $F_\mu(a^*)$ become meaningless when $\Delta(a^*)$ ceases to be a real number, namely for $a^{*2} \geq 25(\sqrt{1057} + 29)/432 \simeq 3.56$. On the other hand, $F_\kappa(a^*)$ and $\Phi(a^*)$ already take unphysical negative values for $a^2 \geq 50/63 \simeq 0.79$. For small shear rates, the above transport coefficients behave as

$$F_\eta(a^*) \approx 1 - \frac{13}{5}a^{*2}, \qquad (5.166)$$

$$F_\kappa(a^*) \approx 1 + \frac{21}{50}a^{*2}, \qquad (5.167)$$

$$\Phi(a^*) \approx -\frac{7}{2}\left(1 - \frac{197}{250}a^{*2}\right), \qquad (5.168)$$

Table 5.1. Values of the first few numerical coefficients appearing in the expansion of the generalized transport coefficients in powers of the shear rate. The second column gives the hydrodynamic order of each coefficient: Burnett (B), super-Burnett (s-B), or super-super-Burnett (s-s-B). The Liu-1 model refers to the results obtained from the Liu model with the choice $\zeta/\nu = 1$, while the Liu-2 model refers to the choice $\zeta/\nu = \frac{2}{3}$. The unknown coefficients are indicated by long dashes.

Coefficient	Order	Boltzmann	BGK	Liu-1	Liu-2	ES	Grad
$F_\eta^{(2)}$	s-B	-3.311	-3.6	-3.2	-7.2	-4.2	-2.6
$F_\kappa^{(2)}$	s-B	-7.260	-6.48	-3.92	-8.82	-7.88	0.42
$\Psi_1^{(0)}$	B	-2.8	-2.8	-2.8	-4.2	-2.8	-2.8
$\Psi_2^{(0)}$	B	0.8	0.8	0.8	1.2	0.8	0.8
$\Psi_1^{(2)}$	s-s-B	—	23.616	19.104	64.476	34.016	—
$\Psi_2^{(2)}$	s-s-B	—	-9.216	-7.104	-23.976	-13.216	—
$\Phi^{(0)}$	B	-3.5	-2.8	-2.333	-3.5	-3.5	-3.5
$\Phi^{(2)}$	s-s-B	—	37.44	—	—	—	2.758

$$F_\mu(a^*) \approx 1 - \frac{7}{5}a^{*2}. \tag{5.169}$$

Before comparing the predictions given by the different kinetic theories against Monte Carlo simulations, let us compare the first few numerical coefficients appearing in the expansion of the generalized transport coefficients in powers of the shear rate. Let us use the notation

$$F_\eta(a^*) = 1 + F_\eta^{(2)} a^{*2} + \cdots, \tag{5.170}$$

$$\Psi_1(a^*) = \Psi_1^{(0)} + \Psi_1^{(2)} a^{*2} + \cdots, \tag{5.171}$$

and so on. Note that, for instance, $F_\eta^{(2)}$ is a super-Burnett coefficient, while $\Psi_1^{(0)}$ is a Burnett coefficient and $\Psi_1^{(2)}$ is a super-super-Burnett coefficient. In Table 5.1 we show the values of some of those coefficients, according to the Boltzmann equation for Maxwell molecules, the BGK kinetic model, the Liu model, the ES model, and Grad's method. The Burnett coefficients $\Psi_1^{(0)}$ and $\Psi_2^{(0)}$ obtained from the Boltzmann equation are reproduced by all the approximations except by the Liu model with the choice $\zeta/\nu = \frac{2}{3}$ (Liu-2 model). On the other hand, the Burnett coefficient $\Phi^{(0)}$ is reproduced by all the approximations except by the BGK model and the Liu model with the choice $\zeta/\nu = 1$ (Liu-1 model). In the case of the super-Burnett coefficients $F_\eta^{(2)}$ and $F_\kappa^{(2)}$, they are reasonably well estimated by the BGK and the ES models. The Liu-1 model gives a good estimate for $F_\eta^{(2)}$ but a poor one for $F_\kappa^{(2)}$, while the opposite happens for the Liu-2 model. Grad's method gives an estimate for $F_\kappa^{(2)}$ with the wrong sign. As the hydrodynamic order increases, the relative deviations between the Liu-1 and the Liu-2 models grow. This fact, along with the lack of positive definiteness of the distribution function suggests to discard the predictions of the Liu model in the comparison with computer simulations carried out in Section 5.7. In conclusion, Table 5.1 suggests that the best general agreement with the Boltzmann results is provided by the ES model, followed by the BGK model.

7. Comparison with computer simulations

Table 5.1 is not sufficient to assess the reliability of the different kinetic theories in the nonlinear regime because the exact solution of the Boltzmann equation for Maxwell molecules is not known, except for the super-Burnett coefficients (Tij and Santos, 1995). Even if the solution for Maxwell molecules were known, it is also important to investigate whether the results for a system of Maxwell molecules extend to other interaction potentials. This extension holds when the Boltzmann equation is replaced by kinetic models or solved by Grad's method where, in terms of a conveniently scaled space variable, all the results are independent of the interaction law. Moreover, in all the kinetic theories discussed in the previous Sections, we have assumed the existence of a *normal* or hydrodynamic solution, even far from equilibrium. In order to confirm such a hydrodynamic description, as well as to test the theoretical predictions for the generalized transport coefficients, one has to resort to computer simulations.

One of the first molecular dynamics simulations of the planar Couette flow was carried out by Trozzi and Ciccotti (1984), who considered a dense Lennard–Jones fluid near the triple point and applied stochastic or diffuse boundary conditions [cf. Eqs. (5.44) and (5.46)]. They were able to identify a bulk region with a linear velocity profile and a parabolic temperature profile. In addition, Trozzi and Ciccotti found a shear-thinning effect, i.e. the nonlinear shear viscosity decreases as the shear rate increases. However, a comparison with the kinetic theory predictions cannot be performed due to the high densities considered in their simulations.

The planar Couette flow for a *dilute* gas of hard disks has been studied by means of molecular dynamics (Risso and Cordero, 1997). The density considered (about 1% of the total area is covered by the disks) is small enough to allow for a comparison with the kinetic theories specialized to a two-dimensional system (Montanero and Garzó, 1998). The simulations indicate that there is a bulk region where the pressure and the reduced shear rate a^* are constant. This confirms that, in appropriate reduced units, the properties of the system are practically "universal", i.e. independent of the interaction potential. As an example of nonlinear transport, Fig. 5.14 compares the simulation values of the visco-thermal function $\Phi(a^*)$ with the theoretical predictions given by the BGK model, the ES model, and Grad's method. We observe that the ES model reproduces pretty well the simulation data. The BGK model follows the main trends but is not quantitatively accurate since it does not give the correct Burnett value $\Phi^{(0)} = \Phi(a^* = 0) = -\frac{9}{2}$ for a two-dimensional gas. Although Grad's method gives correctly the latter value, its predictions worsen significantly as the shear rate increases. With respect to the viscosity function $F_\eta(a^*)$, the three theories present a quantitative good agreement with the simulation results for the range of shear rates considered (Montanero and Garzó, 1998). Nevertheless, in the case of the thermal conductivity function $F_\kappa(a^*)$, Grad's method predicts an increasing function, while the BGK and ES models agree quantitatively well with the decreasing dependence observed in the simulations.

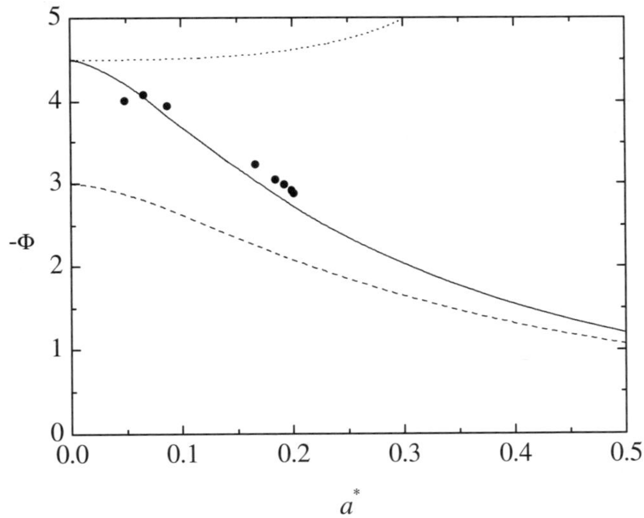

Figure 5.14. Visco-thermal function $-\Phi(a^*)$ versus the reduced shear rate a^* for a two-dimensional gas, as obtained from the BGK model (dashed line), the ES model (solid line), and Grad's method (dotted line). The circles correspond to molecular dynamics simulations of a hard-disk gas.

Given the difficulties associated with molecular dynamics simulations to achieve large shear rates in a dilute gas, the above comparison is restricted to a range of shear rates ($a^* \lesssim 0.2$) for which non-Newtonian effects are hardly significant. For instance, the shear viscosity has only decreased around 10% with respect to its Navier–Stokes value for the largest value of the shear rate considered in the molecular dynamics simulations. In order to overcome such difficulties and extend the range of values of a^*, one may use the Direct Simulation Monte Carlo (DSMC) method (Bird, 1994; Alexander and Garcia, 1997; Oran et al., 1998) to numerically solve the Boltzmann equation. Since we are interested in describing transport properties in the bulk region, i.e. free from finite-size effects, it is appropriate to consider boundary conditions of the type (5.44) but more efficient than the conventional ones. In the conventional boundary conditions (Risso and Cordero, 1997; Tehver et al., 1998), the gas is assumed to be enclosed between two baths *in equilibrium* in relative motion and, in general, at different temperatures. These boundary conditions are represented by the probability distribution (5.46). Under these conditions, a particle leaving the system is formally replaced by a particle coming from the bath, so the incoming velocity is sampled from a local equilibrium distribution. As a consequence, there exists a mismatch between the velocity distribution function of the re-emitted particles and that of those particles of the gas located near the wall and moving along the same direction. In this case, in order to inhibit the influence of boundary effects one needs to take very large systems (normal distance between the plates much larger than the mean free path), which is not practical from a computational point of view. To overcome this difficulty, an alternative possibility is to assume that both baths are out of equilibrium in a state close to that of the actual gas. Since such a state is not known "a priori," a simple choice is to assume that the state of the baths is described by the BGK solution of the planar Couette flow, Eq. (5.123). Although the boundary effects are still unavoidable, it turns out that the above mismatch between re-emitted and gas particles is much smaller (Montanero et al., 1994; Montanero et al., 2000).

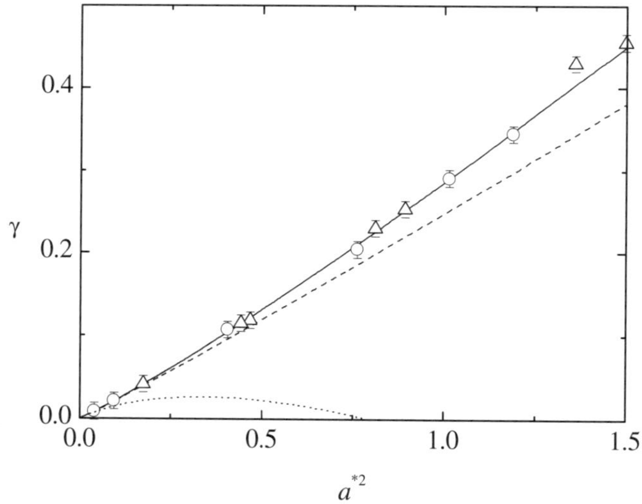

Figure 5.15. Thermal curvature parameter γ as a function of a^{*2}. The symbols are simulation data for Maxwell molecules (circles) and for hard spheres (triangles), while the lines are the theoretical predictions given by the ES kinetic model (solid line), the BGK kinetic model (dashed line), and Grad's method (dotted line).

Let us compare the simulation results for (three-dimensional) hard spheres and Maxwell molecules obtained from the DSMC method (Montanero et al., 2000) with the (universal) predictions of Grad's method and the BGK and ES kinetic models. The simulations show the existence of a bulk domain where the pressure is practically constant, the local values of the reduced shear rate $a^* = \nu^{-1}\partial u_x/\partial y$ are also constant, and the temperature profile is consistent with Eq. (5.18), from which the thermal curvature parameter γ can be measured. Its shear-rate dependence is shown in Fig. 5.15. It is observed that the simulation data for both interactions seem to lie on a common curve. This indicates that, as predicted by the models, the transport properties are hardly sensitive to the interaction potential, provided that the quantities are conveniently scaled. While the kinetic models exhibit a good quantitative (in the case of the ES model) or qualitative (BGK model) agreement, Grad's method fails, except for very small shear rates. This is not surprising in view of Table 5.1, since $\gamma(a^*)$ behaves as $\gamma(a^*) \approx \frac{1}{5}a^{*2}\left[1+\left(F_\eta^{(2)}-F_\kappa^{(2)}\right)a^{*2}\right]$ for small a^* and Grad's method gives a wrong estimate for $F_\kappa^{(2)}$, leading to a negative value for the difference $F_\eta^{(2)} - F_\kappa^{(2)}$, in contrast to what happens in the Boltzmann solution.

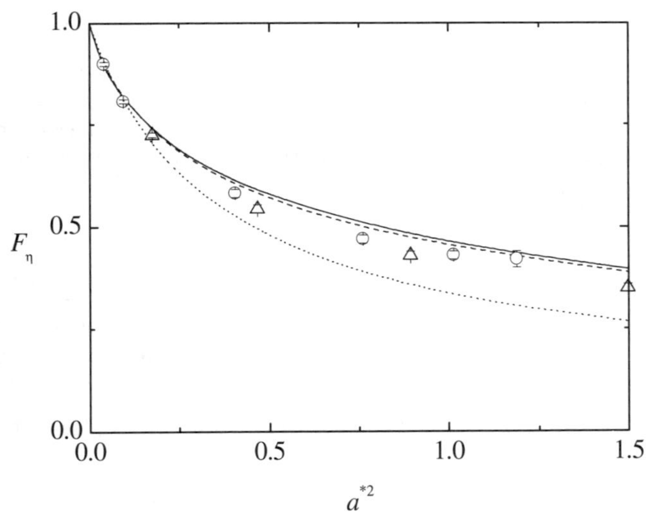

Figure 5.16. Viscosity function F_η versus a^{*2}. The symbols are simulation data for Maxwell molecules (circles) and for hard spheres (triangles), while the lines are the theoretical predictions given by the ES kinetic model (solid line), the BGK kinetic model (dashed line), and Grad's method (dotted line).

The most important transport coefficient is the nonlinear viscosity represented by the function $F_\eta(a^*)$. This is plotted in Fig. 5.16. As is apparent, the three theories retain the qualitative trends of the simulation data, namely the decrease of F_η with increasing a^* (shear-thinning effect). In general, however, the kinetic models (especially the BGK model) have a better agreement than Grad's method. It is also interesting to remark that a slight influence of the interaction potential seems to exist, the shear-thinning effect being a little bit more significant for hard spheres than for Maxwell molecules.

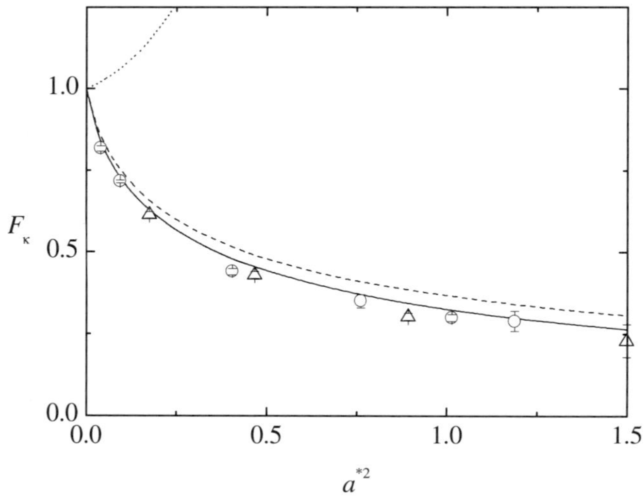

Figure 5.17. Thermal conductivity function F_κ versus a^{*2}. The symbols are simulation data for Maxwell molecules (circles) and for hard spheres (triangles), while the lines are the theoretical predictions given by the ES kinetic model (solid line), the BGK kinetic model (dashed line), and Grad's method (dotted line).

The nonlinear thermal conductivity $F_\kappa(a^*)$ is plotted in Fig. 5.17. Clearly, Grad's solution does not capture even the qualitative shape of F_κ, as expected from Table 5.1. This failure is also present in the two-dimensional case of hard disks (Montanero and Garzó, 1998). As before, the kinetic models present a good agreement, especially in the case of the ES model. A slight sensitivity to the interaction potential is again observed.

The viscometric functions $\Psi_{1,2}(a^*)$, which were not evaluated from Grad's method by Risso and Cordero (1997, 1998), are well described by the kinetic models, as shown in Figs. 5.18 and 5.19. At a quantitative level, the agreement is better in the case of Maxwell molecules. In fact, the viscometric functions, especially the second one, exhibit a certain influence of the potential. The friction function $F_\mu(a^*) = -(P_{xy}/P_{yy})/a^*$ is plotted in Fig. 5.20, showing an agreement between the theories and the simulation data similar to the one found in Fig. 5.16 for the viscosity function.

The comparison with simulation results for the visco-thermal function Φ defined by Eq. (5.22) provides a stringent test of the theories, since it is a

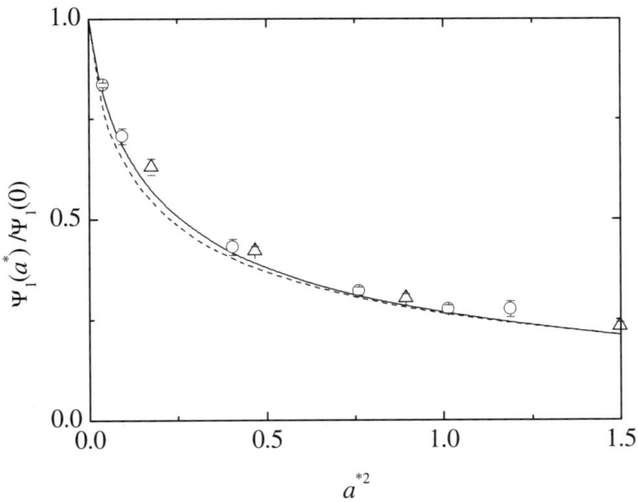

Figure 5.18. First viscometric function $\Psi_1(a^*)$, relative to the Burnett value $\Psi_1(0)$, versus a^{*2}. The symbols are simulation data for Maxwell molecules (circles) and for hard spheres (triangles), while the lines are the theoretical predictions given by the ES kinetic model (solid line) and the BGK kinetic model (dashed line).

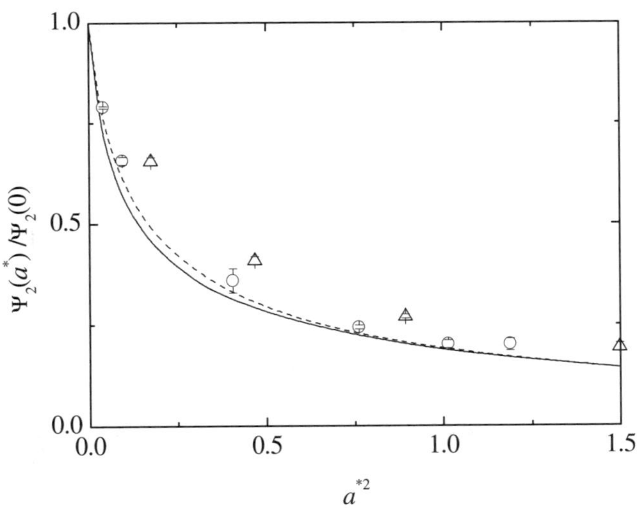

Figure 5.19. Second viscometric function $\Psi_2(a^*)$, relative to the Burnett value $\Psi_2(0)$, versus a^{*2}. The symbols are simulation data for Maxwell molecules (circles) and for hard spheres (triangles), while the lines are the theoretical predictions given by the ES kinetic model (solid line) and the BGK kinetic model (dashed line).

generalization of a Burnett coefficient that measures the component of the heat

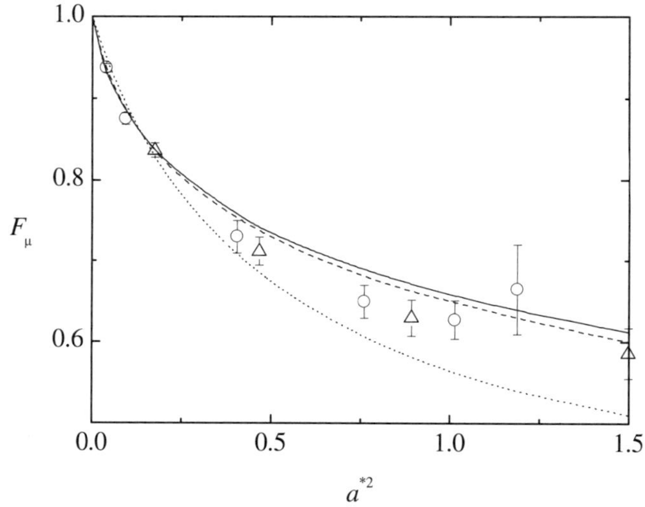

Figure 5.20. Friction function F_μ versus a^{*2}. The symbols are simulation data for Maxwell molecules (circles) and for hard spheres (triangles), while the lines are the theoretical predictions given by the ES kinetic model (solid line), the BGK kinetic model (dashed line), and Grad's method (dotted line).

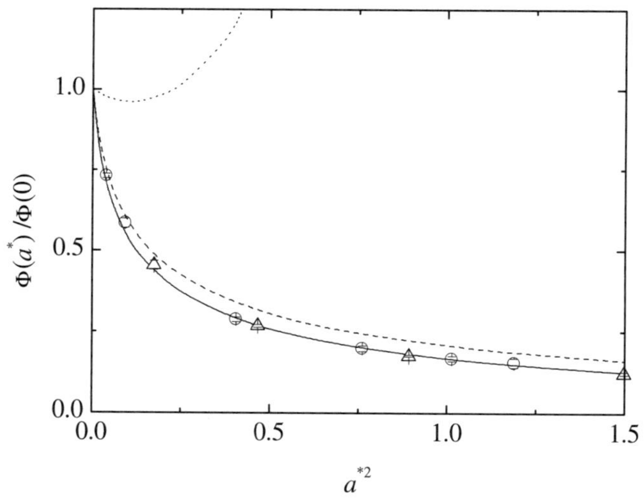

Figure 5.21. Visco-thermal function $\Phi(a^*)$, relative to the Burnett value $\Phi(0)$, versus a^{*2}. The symbols are simulation data for Maxwell molecules (circles) and for hard spheres (triangles), while the lines are the theoretical predictions given by the ES kinetic model (solid line), the BGK kinetic model (dashed line), and Grad's method (dotted line).

flux orthogonal to the thermal gradient. Figure 5.21 shows that, as happened with the thermal conductivity function $F_\kappa(a^*)$, Grad's method gives a wrong prediction for the shear-rate dependence of $\Phi(a^*)$. On the other hand, the kinetic models describe fairly well the nonlinear behavior of this function. In the case of the ES model, the agreement with the simulation data is practically perfect. Note that Fig. 5.21 focuses on the shear-rate dependence of the ratio $\Phi(a^*)/\Phi(0)$, so the failure of the BGK model to give the correct value of the Burnett coefficient $\Phi^{(0)} = \Phi(0) = -\frac{7}{2}$ is not relevant in Fig. 5.21.

The comparisons carried out in Figs. 5.14–5.21 show that the kinetic models give a fairly good description of the generalized transport coefficients. However, Grad's method yields a shear viscosity in qualitative agreement with the simulations but it dramatically fails for the coefficients measuring the heat flux. This is basically due to the truncation scheme of Grad's thirteen-moment method at the level of the heat flux. Grad's method would possibly give a much better prediction for the heat flux if higher-degree moments were retained. The physical idea behind a kinetic model is quite different, since it consists of replacing the true Boltzmann collision operator by a simple relaxation term but otherwise all the velocity moments are taken into account. As a consequence, while in Grad's method one has to solve a closed set of coupled differential equations for the moments, in the case of the kinetic model one gets the velocity distribution function and determines the fluxes from it.

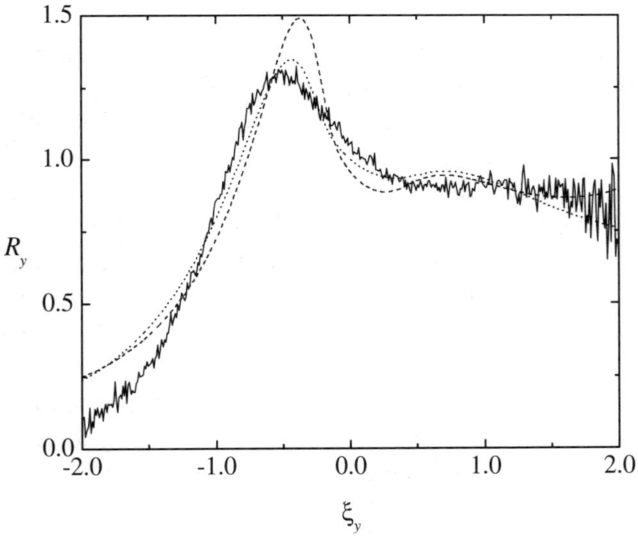

Figure 5.22. Marginal velocity distribution function, relative to the local equilibrium distribution, for Maxwell molecules in the case $a^* = 0.636$ and $\epsilon = -0.272$. The solid line is obtained from a DSMC simulation, while the dashed line is the theoretical ES distribution and the dotted line is the theoretical BGK distribution.

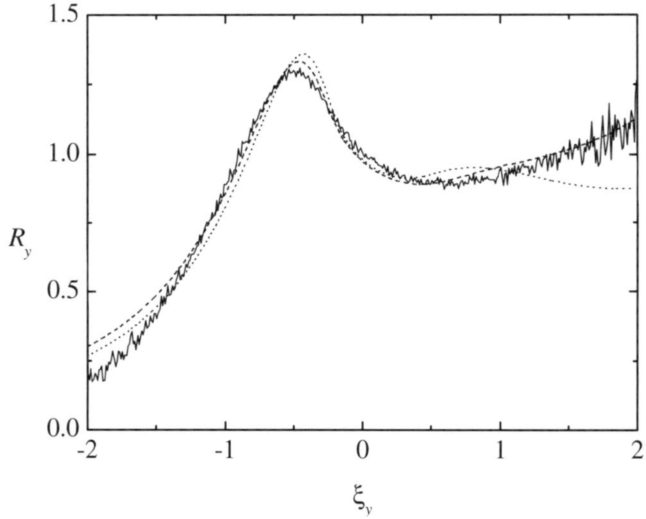

Figure 5.23. Marginal velocity distribution function, relative to the local equilibrium distribution, for hard spheres in the case $a^* = 0.419$ and $\epsilon = -0.195$. The solid line is obtained from a DSMC simulation, while the dashed line is the theoretical ES distribution and the dotted line is the theoretical BGK distribution.

So far, we have compared the simulation values of the transport properties with the theoretical predictions. Nevertheless, a more complete information about the state of the system is provided by the velocity distribution function. In the case of the BGK model, the distribution function is given by Eq. (5.123), while the ES distribution function can be obtained from Eq. (5.137) (Garzó and López de Haro, 1997; Montanero and Garzó, 1998; Montanero and Garzó, 2000). Since the reduced distribution function $f^*(\boldsymbol{\xi})$ depends on the three components of the reduced peculiar velocity $\boldsymbol{\xi}$, for practical purposes it is useful to introduce the *marginal* distribution $f_y^*(\xi_y)$ defined by Eq. (3.18). This distribution can be obtained in simulations by means of a histogram method considering those particles inside a small layer located in the bulk region (Montanero et al., 2000). It is not legitimate to consider particles belonging to different layers because, although the reduced shear rate a^* is a constant parameter along the bulk part of the system, the reduced thermal gradient ϵ is a local parameter. The ratio $R_y(\xi_y) \equiv f_y^*(\xi_y)/[\pi^{-1/2}\exp(-\xi_y^2)]$ is plotted in Fig. 5.22 for Maxwell molecules in the case $a^* = 0.636$ and $\epsilon = -0.272$. Both theories capture the main features of the actual distribution. While the BGK distribution exhibits a better agreement near the maximum (around $\xi_y \simeq -0.5$), the ES distribution seems to describe better the region $\xi_y \gtrsim 1$. The case of hard spheres is illustrated in Fig. 5.23 for $a^* = 0.419$ and $\epsilon = -0.195$. In this case, the ES model shows a superiority over the BGK model both near the maximum and for large positive velocities.

In summary, comparison with simulation results for the transport coefficients, as well as for the velocity distribution function, indicates that the BGK and ES kinetic models succeed in capturing the main trends found in simulation experiments carried out in the Couette flow far from equilibrium. In the ES kinetic model the reference distribution function appearing in the collision operator is an anisotropic Gaussian parameterized by the pressure tensor. This allows the model to give the correct Prandtl number $\text{Pr} = \frac{2}{3}$ but at the expense of complicating its mathematical structure. In the case of the BGK model, however, the reference distribution is that of local equilibrium but the model leads to $\text{Pr} = 1$. The agreement of the ES model with simulation is generally better than that of the BGK model, especially in the case of F_κ and Φ. In spite of this, it is fair to say that the performance of the BGK model is quite good, given its simplicity relative to that of the ES model. As happens in the USF problem (cf. Chapter 3), the results reported in this Section show the usefulness of kinetic models to analyze nonlinear transport phenomena in the Couette flow problem.

APPENDIX 5.A: Generating function and consistency conditions

The consistency conditions for the assumed forms of the hydrodynamic fields are Eqs. (5.92)–(5.94). Evaluation of the left-hand sides of these equations requires the explicit form for the distribution function in terms of these fields, Eq. (5.91). In that equation, the operator ∂_s is taken at constant $\mathbf{v}$. However, it is convenient to refer the velocities of the particles to the Lagrangian frame $\mathbf{V} = \mathbf{v} - \mathbf{u}$ and take the spatial derivative at constant $\mathbf{V}$. Therefore,

$$\left(\frac{\partial}{\partial s}\right)_{\mathbf{v}} = \left(\frac{\partial}{\partial s}\right)_{\mathbf{V}} - a^* \frac{\partial}{\partial V_x}, \tag{5.A.1}$$

where we have taken into account that $u_x = a^* s$. Henceforth we will use the symbol ∂_s to mean the spatial derivative at constant $\mathbf{V}$. Equation (5.91) can then be rewritten as

$$f(s, \mathbf{V}) = \int_0^\infty dt\, e^{-t} e^{-tV_y \partial_s} e^{a^* t V_y \partial_{V_x}} \Theta\left(T(s)\right) f_L(s, \mathbf{V}), \tag{5.A.2}$$

where we have dropped the subscript H.

The velocity integrals for the consistency conditions and for the fluxes are all low-degree moments of this distribution function. They can be obtained from appropriate derivatives of the generating function (or Fourier transform in velocity space) (Ernst, 1981; Bobylev, 1988)

$$G(s, \mathbf{k}) = \int d\mathbf{V} e^{i\mathbf{k}\cdot\mathbf{V}} f(s, \mathbf{V}). \tag{5.A.3}$$

Making use of the solution (5.A.2), the generating function becomes

$$\begin{aligned} G(s, \mathbf{k}) &= \int_0^\infty dt\, e^{-t} \int d\mathbf{V}\, e^{i\mathbf{k}\cdot\mathbf{V}} e^{a^* t V_y \partial_{V_x}} e^{-tV_y \partial_s} f_L(s, \mathbf{V}) \\ &= \int_0^\infty dt\, e^{-t} \int d\mathbf{V}\, e^{i\mathbf{k}'(t)\cdot\mathbf{V}} e^{-tV_y \partial_s} f_L(s, \mathbf{V}), \end{aligned} \tag{5.A.4}$$

where

$$\mathbf{k}'(t) = e^{-a^* t k_x \partial_{k_y}} \mathbf{k} \tag{5.A.5}$$

In the last step of Eq. (5.A.4) we have used the general property

$$\int d\mathbf{V}\, X_1(\mathbf{V}) e^{a^* t V_y \partial_{V_x}} X_2(\mathbf{V}) = \int d\mathbf{V}\, X_2(\mathbf{V}) e^{-a^* t V_y \partial_{V_x}} X_1(\mathbf{V}). \tag{5.A.6}$$

Integration over $\mathbf{V}_\perp \equiv \mathbf{V} - V_y \widehat{\mathbf{y}}$ yields

$$G(s, \mathbf{k}) = G_+(s, \mathbf{k}) + G_-(s, \mathbf{k}), \tag{5.A.7}$$

where

$$G_\pm(s, \mathbf{k}) = \frac{2p}{m} \int_0^\infty dV_y \int_0^\infty dt\, e^{-t} e^{\pm \mathrm{i} k'_y(t) V_y} e^{\mp t V_y \partial_s} e^{-V_y^2/v_0^2(s)} F(s, \mathbf{k}'_\perp(t)), \tag{5.A.8}$$

with $v_0(s) \equiv \sqrt{2k_B T(s)/m}$, $\mathbf{k}'_\perp \equiv \mathbf{k}' - k'_y \widehat{\mathbf{y}}$, and

$$F(s, \mathbf{k}'_\perp(t)) = \Theta(v_0^2(s)) v_0^{-3}(s) \pi^{-1/2} \exp\left[-\frac{1}{4} {k'_\perp}^2(t) v_0^2(s)\right]. \tag{5.A.9}$$

We will focus on $G_+(s, \mathbf{k})$ since $G_-(s, \mathbf{k})$ can be evaluated as $G_-(s, \mathbf{k}) = G_+(-s, -\mathbf{k})$.

Next, define the change of variables $z = V_y t$ in the t integration, so that the function $G_+(s, \mathbf{k})$ becomes

$$\begin{aligned} G_+(s, \mathbf{k}) &= \frac{2p}{m} \int_0^\infty dz \int_0^\infty dV_y\, V_y^{-1} e^{-z/V_y} e^{\mathrm{i} k'_y(z/V_y) V_y} \\ &\quad \times e^{-V_y^2/v_0^2(s-z)} F(s-z, \mathbf{k}'_\perp(z/V_y)) \\ &= \frac{2p}{m} \int_0^\infty dz \int_0^\infty du\, e^{-u^2} u^{-1} e^{-t} e^{\mathrm{i} k'_y(t) u v_0(s-z)} \\ &\quad \times F(s-z, \mathbf{k}'_\perp(t)), \end{aligned} \tag{5.A.10}$$

where a change of variables $V_y \to u v_0(s-z)$ in the V_y integral has been made in the last step. Accordingly, the variable t appearing in Eq. (5.A.10) is $t(z, u) = z/u v_0(s-z)$. Next, for the z integral change variables to

$$w = \frac{z}{v_0(s-z)}. \tag{5.A.11}$$

Since $v_0^2(s)$ is a quadratic function [cf. Eq. (5.55)], Eq. (5.A.11) gives rise to a quadratic equation:

$$\left(1 + 2\gamma w^2\right) z^2 - 4\gamma w^2 s z - w^2 v_0^2(s) = 0. \tag{5.A.12}$$

As a consequence, z as a function of w is $z = z_+(s, w)$, where $z_+(s, w)$ is the positive root of (5.A.12). It can be easily shown that

$$\frac{dz}{dw} = v_0^{-2}(s) \left(\frac{z_+}{w}\right)^3 \left[1 - \frac{z_+}{2} \partial_s \ln T(s)\right]^{-1}. \tag{5.A.13}$$

The generating function is now

$$\begin{aligned} G_+(s, \mathbf{k}) &= \frac{2p}{m v_0^2(s) \pi^{1/2}} \int_0^\infty dw \int_0^\infty du\, e^{-u^2} u^{-1} e^{-w/u} \\ &\quad \times \exp\left[\mathrm{i} k'_y(w/u) z_+(w) u/w - {k'_\perp}^2(w/u)(z_+(w)/2w)^2\right] \\ &\quad \times \left[1 - \frac{z_+(w)}{2} \partial_s \ln T(s)\right]^{-1}. \end{aligned} \tag{5.A.14}$$

Finally, a change of variables in the w integration to $t = w/u$ leads to the explicit result

$$G_\pm(s,\mathbf{k}) = \frac{p}{k_B T(s)\pi^{1/2}} \int_0^\infty du\, e^{-u^2} \int_0^\infty dt\, e^{-t} A_\pm(s,\mathbf{k},u,t), \tag{5.A.15}$$

where

$$\begin{aligned} A_\pm(s,\mathbf{k},u,t) &= \left[1 - \frac{z_\pm(s,ut)}{2}\partial_s \ln T(s)\right]^{-1} \\ &\quad \times \exp\left[\mathrm{i}\left(k_y - a^* t k_x\right)\frac{z_\pm(s,ut)}{t} - \frac{k_\perp^2}{4}\frac{z_\pm^2(s,ut)}{u^2t^2}\right]. \end{aligned} \tag{5.A.16}$$

In these equations we have used the property $G_-(s,\mathbf{k}) = G_+(-s,-\mathbf{k})$ to get G_- from G_+. Here $z_-(s,w) = -z_+(-s,w)$ is the negative root of the quadratic equation (5.A.12). Real roots require

$$\left(4\gamma s w^2\right)^2 + 4\left(1 + 2\gamma w^2\right) 2w^2 \frac{k_B T(s)}{m} \geq 0, \tag{5.A.17}$$

which is satisfied for positive $T(s)$ and positive γ.

Once the generating function is known, all the velocity moments can be explicitly evaluated by expanding $G(s,\mathbf{k})$ in powers of $\mathbf{k}$. In particular, we can check the consistency of the assumed hydrodynamic profiles. The consistency condition for the density, Eq. (5.92), becomes

$$\begin{aligned} n(s) &= G(s,\mathbf{k}=\mathbf{0}) \\ &= \frac{p}{k_B T(s)\pi^{1/2}} \int_0^\infty du\, e^{-u^2} \int_0^\infty dt\, e^{-t} \left\{\left[1 - \frac{z_+}{2}\partial_s \ln T(s)\right]^{-1}\right. \\ &\quad \left. + \left[1 - \frac{z_-}{2}\partial_s \ln T(s)\right]^{-1}\right\} \\ &= \frac{p}{k_B T(s)}. \end{aligned} \tag{5.A.18}$$

Use has been made of the identity

$$\left[1 - \frac{z_+}{2}\partial_s \ln T(s)\right]^{-1} + \left[1 - \frac{z_-}{2}\partial_s \ln T(s)\right]^{-1} = 2, \tag{5.A.19}$$

which follows from the explicit forms for the roots of (5.A.12). The result (5.A.18) is consistent with the required equation of state defining p.

The consistency conditions for the flow velocity components are

$$\left[\partial_{k_i} G(s,\mathbf{k})\right]_{\mathbf{k}=\mathbf{0}} = 0. \tag{5.A.20}$$

This is clearly satisfied for the z-component. The x- and y-components are proportional to integrals including the term

$$z_+\left[1 - \frac{z_+}{2}\partial_s \ln T(s)\right]^{-1} + z_-\left[1 - \frac{z_-}{2}\partial_s \ln T(s)\right]^{-1}, \tag{5.A.21}$$

which identically vanishes.

Finally, the consistency condition for the temperature, Eq. (5.94) is

$$3p = -m\left[\partial_{\mathbf{k}}^2 G(s,\mathbf{k})\right]_{\mathbf{k}=\mathbf{0}}. \tag{5.A.22}$$

Direct evaluation gives

$$\begin{aligned} 1 &= \frac{m}{3k_BT(s)\pi^{1/2}}\int_0^\infty du\, e^{-u^2}\int_0^\infty dt\, e^{-t}\left[1+\left(1+a^{*2}t^2\right)u^2\right](ut)^{-2} \\ &\times\left\{z_+^2\left[1-\frac{z_+}{2}\partial_s\ln T(s)\right]^{-1}+z_-^2\left[1-\frac{z_-}{2}\partial_s\ln T(s)\right]^{-1}\right\}. \end{aligned} \tag{5.A.23}$$

Using the identity

$$z_+^2\left[1-\frac{z_+}{2}\partial_s\ln T(s)\right]^{-1}+z_-^2\left[1-\frac{z_-}{2}\partial_s\ln T(s)\right]^{-1} = \frac{4k_BT(s)}{m}\frac{w^2}{1+2\gamma w^2}, \tag{5.A.24}$$

one gets the final result

$$1=\frac{4}{3\pi^{1/2}}\int_0^\infty du\, e^{-u^2}\int_0^\infty dt\, e^{-t}\left[1+\left(1+a^{*2}t^2\right)u^2\right]\left(1+2\gamma u^2t^2\right)^{-1}. \tag{5.A.25}$$

This condition is enforced by using it as the equation giving the dependence of γ on the shear rate a^*. From a practical point of view, Eq. (5.A.25) gives a^{*2} in terms of γ:

$$a^{*2}=\frac{\frac{3}{4}\sqrt{\pi}-\int_0^\infty du\, e^{-u^2}\int_0^\infty dt\, e^{-t}\left(1+u^2\right)\left(1+2\gamma u^2t^2\right)^{-1}}{\int_0^\infty du\, e^{-u^2}\int_0^\infty dt\, e^{-t}u^2t^2\left(1+2\gamma u^2t^2\right)^{-1}}. \tag{5.A.26}$$

To express the right-hand side of Eq. (5.A.26) in a more compact form, it is convenient to introduce the mathematical functions

$$F_r(x)=\left(\frac{\partial}{\partial x}x\right)^r F_0(x), \quad x\geq 0, \tag{5.A.27}$$

where

$$F_0(x)=\frac{4}{\sqrt{\pi}}\int_0^\infty du\, u^2e^{-u^2}\int_0^\infty dt\, t\, e^{-t}\left(1+2xu^2t^2\right)^{-1}. \tag{5.A.28}$$

In particular,

$$\begin{aligned} F_1(x) &= \frac{2}{\sqrt{\pi}}\int_0^\infty du\, u^2e^{-u^2}\int_0^\infty dt\, t^2e^{-t}\left(1+2xu^2t^2\right)^{-1} \\ &= \frac{4}{\sqrt{\pi}}\int_0^\infty du\, u^2e^{-u^2}\int_0^\infty dt\, t\, e^{-t}\left(1+2xu^2t^2\right)^{-2} \\ &= \frac{1}{2x}-\frac{1}{\sqrt{\pi}x}\int_0^\infty du\, e^{-u^2}\int_0^\infty dt\, e^{-t}\left(1+2xu^2t^2\right)^{-1}, \end{aligned} \tag{5.A.29}$$

$$F_2(x)=\frac{2}{\sqrt{\pi}}\int_0^\infty du\, u^2e^{-u^2}\int_0^\infty dt\, t^2e^{-t}\left(1+2xu^2t^2\right)^{-2}, \tag{5.A.30}$$

$$F_3(x)=\frac{1}{\sqrt{\pi}}\int_0^\infty du\, u^2(2u^2-1)e^{-u^2}\int_0^\infty dt\, t^2e^{-t}\left(1+2xu^2t^2\right)^{-2}. \tag{5.A.31}$$

It can be easily seen that $F_3(x)$ can be expressed as a combination of $F_0(x)$, $F_1(x)$, and $F_2(x)$:

$$F_3(x)=\frac{1}{8x}\left[1-F_0(x)\right]-F_2(x)-\frac{1}{4}F_1(x). \tag{5.A.32}$$

From here, one can express the remaining functions $F_r(x)$ with $r \geq 4$ as combinations of $F_0(x)$, $F_1(x)$, and $F_2(x)$. Let us now obtain an alternative integral representation for $F_0(x)$. Using the identity

$$\frac{4}{\sqrt{\pi}} \int_0^\infty du\, u^2 e^{-u^2} (1 + 2xu^2t^2)^{-1} = \int_0^\infty du\, ue^{-u} e^{-xu^2t^2/2} \tag{5.A.33}$$

in Eq. (5.A.28), one gets

$$F_0(x) = \int_0^\infty dt \int_0^\infty du\, tue^{-(t+u)} e^{-xu^2t^2/2}. \tag{5.A.34}$$

Now we make the change $u \to v = ut$ and use the Laplace transform property (Abramowitz and Stegun, 1972)

$$\int_0^\infty dt\, e^{-st} \frac{1}{2t} e^{-k^2/4t} = K_0(k\sqrt{s}), \tag{5.A.35}$$

where $K_0(x)$ is the zeroth-order modified Bessel function (Abramowitz and Stegun, 1972; Gradshteyn and Ryzhik, 1980) to get

$$F_0(x) = \frac{2}{x} \int_0^\infty dt\, te^{-t^2/2} K_0(2x^{-1/4}t^{1/2}). \tag{5.A.36}$$

Using Eqs. (5.A.29) and (5.A.30), Eq. (5.A.26) can be rewritten as

$$a^{*2} = \gamma \left[3 + 2\frac{F_2(\gamma)}{F_1(\gamma)} \right]. \tag{5.A.37}$$

This is an implicit equation giving γ as a function of a^*. In order to get the asymptotic behaviors of γ for small and large shear rates, we need to study the asymptotic behaviors of the functions $F_r(x)$. Let us start from the "seed" function $F_0(x)$. For small x, we expand $\exp(-xu^2t^2/2)$ in Eq. (5.A.34) and then perform the integrations of the t and u variables. The result is

$$F_0(x) = \sum_{k=0}^\infty (2k+1)!(2k+1)!!(-x)^k, \tag{5.A.38}$$

where $(2k+1)!! = (2k+1)!/2^k k!$. This is an alternating series whose general term is unbounded. As a consequence, the series power expansion (5.A.38) is divergent. However, the numerical evaluation of $F_0(x)$, as given by Eqs. (5.A.28), (5.A.34), or (5.A.36), shows that the series representation (5.A.38) is asymptotic. Consequently, $F_0(x) \approx 1 - 18x$ in the limit $x \to 0^+$. The divergent character of the series (5.A.38) implies that $x = 0$ is a singular point of $F_0(x)$, as is apparent from Eq. (5.A.36). To obtain the behavior of $F_0(x)$ for large x, we use the series expansion of $K_0(x)$ (Abramowitz and Stegun, 1972):

$$K_0(x) = \sum_{k=0}^\infty \frac{(x^2/4)^k}{(k!)^2} \left[\psi(k+1) - \ln\frac{x}{2} \right], \tag{5.A.39}$$

where $\psi(k)$ is the dilemma function, which satisfies the recurrence relation

$$\psi(k+1) = \psi(k) + k^{-1} \tag{5.A.40}$$

with

$$\psi(1) = -\gamma_E, \quad \psi\left(\frac{1}{2}\right) = -2\ln 2 - \gamma_E. \tag{5.A.41}$$

Here, $\gamma_E = 0.57721\ldots$ is the Euler constant. Substitution of Eq. (5.A.39) into Eq. (5.A.36) yields

$$F_0(x) = \sum_{k=0}^{\infty} x^{-1-k/2} c_k^{(0)} \left(\frac{1}{2} \ln x + d_k^{(0)} \right), \tag{5.A.42}$$

where

$$c_k^{(0)} = \frac{\Gamma(k/2+1)}{(k!)^2} 2^{k/2}, \quad d_k^{(0)} = 2\psi(k+1) - \frac{1}{2}\psi(k/2+1) - \frac{1}{2}\ln 2. \tag{5.A.43}$$

The series (5.A.42) converges for all $x > 0$, but is especially useful for large x. In that case,

$$F_0(x) \approx \frac{x^{-1}}{2} \left[\ln x - (\ln 2 + 3\gamma_E) \right]. \tag{5.A.44}$$

The asymptotic behaviors for $F_r(x)$ follow from the definition (5.A.27) and Eqs. (5.A.38) and (5.A.42):

$$F_r(x) = \sum_{k=0}^{\infty} (k+1)^r (2k+1)!(2k+1)!!(-x)^k, \tag{5.A.45}$$

$$F_r(x) = \sum_{k=0}^{\infty} x^{-1-k/2} c_k^{(r)} \left(\frac{1}{2} \ln x + d_k^{(r)} \right), \tag{5.A.46}$$

where $c_0^{(r)} = \delta_{r0}$, $d_0^{(r)} = d_0^{(0)} \delta_{r0} + \frac{1}{2}\delta_{r1}$, and

$$c_k^{(r)} = c_k^{(0)} \left(-\frac{k}{2} \right)^r, \quad d_k^{(r)} = d_k^{(0)} - \frac{r}{k}, \quad k \geq 1. \tag{5.A.47}$$

Using the asymptotic behavior of $F_r(x)$ for small x in Eq. (5.A.37), we get the asymptotic expansion of the parameter γ for small shear rates:

$$\gamma(a^*) = \frac{1}{5} a^{*2} \left[1 + \frac{72}{25} a^{*2} - \frac{84\,672}{625} a^{*4} + \frac{51\,668\,928}{3125} a^{*6} + \mathcal{O}(a^{*8}) \right]. \tag{5.A.48}$$

In the opposite limit of large shear rates, we have

$$\gamma(a^*) \approx \frac{1}{3} a^{*2}, \tag{5.A.49}$$

where we have taken into account that $\lim_{x\to\infty} F_2(x)/F_1(x) = 0$.

APPENDIX 5.B: Calculation of the fluxes

The momentum and heat fluxes can also be obtained from the generating function as

$$P_{ij} = -m \left[\partial_{k_i} \partial_{k_j} G(s, \mathbf{k}) \right]_{\mathbf{k}=\mathbf{0}}, \tag{5.B.1}$$

$$\mathbf{q}(s) = \mathrm{i} \frac{1}{2} m \left[\partial_{\mathbf{k}} \partial_{\mathbf{k}}^2 G(s, \mathbf{k}) \right]_{\mathbf{k}=\mathbf{0}}. \tag{5.B.2}$$

Following similar steps as in the consistency condition for the temperature, the nonzero elements of the pressure tensor can be evaluated. They are given by

$$P_{xx} = \frac{2p}{\pi^{1/2}} \int_0^{\infty} du\, e^{-u^2} \int_0^{\infty} dt\, e^{-t} \left[1 + 2\left(a^* t u\right)^2 \right] \left(1 + 2\gamma u^2 t^2 \right)^{-1}, \tag{5.B.3}$$

$$P_{yy} = \frac{4p}{\pi^{1/2}} \int_0^\infty du\, e^{-u^2} u^2 \int_0^\infty dt\, e^{-t} \left(1 + 2\gamma u^2 t^2\right)^{-1}, \tag{5.B.4}$$

$$P_{zz} = \frac{2p}{\pi^{1/2}} \int_0^\infty du\, e^{-u^2} \int_0^\infty dt\, e^{-t} \left(1 + 2\gamma u^2 t^2\right)^{-1}, \tag{5.B.5}$$

$$P_{xy} = -\frac{4a^* p}{\pi^{1/2}} \int_0^\infty du\, e^{-u^2} u^2 \int_0^\infty dt\, e^{-t} t \left(1 + 2\gamma u^2 t^2\right)^{-1}. \tag{5.B.6}$$

It is readily verified that the relationship $p = \frac{1}{3}\left(P_{xx} + P_{yy} + P_{zz}\right)$ yields again the consistency condition for the temperature, Eq. (5.A.25).

Let us consider now the heat flux vector. Note that, although the temperature gradient is only directed along the y-direction, the presence of the shear flow induces a nonzero x-component of the heat flux. Thus the nonzero components are q_x and q_y. Using (5.B.2), q_y is found to be

$$\begin{aligned} q_y(s) &= \frac{mp}{2k_B T(s)\pi^{1/2}} \int_0^\infty du\, e^{-u^2} u^{-2} \int_0^\infty dt\, e^{-t} t^{-3} \left[1 + \left(1 + a^{*2} t^2\right) u^2\right] \\ &\quad \times \left\{ z_+^3 \left[1 - \frac{z_+}{2} \partial_s \ln T(s)\right]^{-1} + z_-^3 \left[1 - \frac{z_-}{2} \partial_s \ln T(s)\right]^{-1} \right\} \\ &= \frac{8\gamma p s}{\pi^{1/2}} \int_0^\infty du\, e^{-u^2} u^2 \int_0^\infty dt\, e^{-t} \frac{1 + \left(1 + a^{*2} t^2\right) u^2}{\left(1 + 2\gamma u^2 t^2\right)^2} t, \end{aligned} \tag{5.B.7}$$

where use has been made of the identity

$$z_+^3 \left[1 - \frac{z_+}{2} \partial_s \ln T(s)\right]^{-1} + z_-^3 \left[1 - \frac{z_-}{2} \partial_s \ln T(s)\right]^{-1} = 16\gamma s \frac{k_B T(s)}{m} \frac{(ut)^4}{(1 + 2\gamma u^2 t^2)^2}. \tag{5.B.8}$$

This can be written as

$$\begin{aligned} q_y &= -\frac{2ps}{\pi^{1/2}} \int_0^\infty du\, e^{-u^2} \int_0^\infty dt\, e^{-t} \left[1 + \left(1 + a^{*2} t^2\right) u^2\right] \frac{\partial}{\partial t} \left(1 + 2\,\gamma u^2 t^2\right)^{-1} \\ &= \frac{4a^{*2} ps}{\pi^{1/2}} \int_0^\infty du\, e^{-u^2} u^2 \int_0^\infty dt\, e^{-t} t \left(1 + 2\gamma u^2 t^2\right)^{-1}. \end{aligned} \tag{5.B.9}$$

An integration by parts in the t integral and the consistency condition (5.A.25) have been used to obtain the last equality. Comparison of this result with that for P_{xy} above shows

$$q_y = \frac{k_B}{2m\gamma} a^* P_{xy} \partial_s T. \tag{5.B.10}$$

This result implies that $\partial q_y / \partial s = -a^* P_{xy}$, in agreement with the energy conservation law (5.3).

In a similar way $q_x(s)$ is calculated as

$$\begin{aligned} q_x(s) &= -\frac{a^* mp}{2k_B T(s)\pi^{1/2}} \int_0^\infty du\, e^{-u^2} u^{-2} \int_0^\infty dt\, e^{-t} t^{-2} \left[2 + \left(1 + a^{*2} t^2\right) u^2\right] \\ &\quad \times \left\{ z_+^3 \left[1 - \frac{z_+}{2} \partial_s \ln T(s)\right]^{-1} + z_-^3 \left[1 - \frac{z_-}{2} \partial_s \ln T(s)\right]^{-1} \right\} \\ &= \frac{4p}{m\pi^{1/2}} a^* \int_0^\infty du\, e^{-u^2} u^2 \int_0^\infty dt\, e^{-t} t^2 \left[2 + \left(1 + a^{*2} t^2\right) u^2\right] \\ &\quad \times \left(1 + 2\gamma u^2 t^2\right)^{-2} \partial_s T. \end{aligned} \tag{5.B.11}$$

As before, the integrals in the expressions for the fluxes can be written in terms of the functions $F_r(x)$. After simple algebra one gets

$$\frac{P_{xx}}{p} = 1 + 4\gamma \left[F_1(\gamma) + F_2(\gamma)\right], \tag{5.B.12}$$

$$\frac{P_{yy}}{p} = 1 - 2\gamma \left[F_1(\gamma) + 2F_2(\gamma)\right], \tag{5.B.13}$$

$$\frac{P_{zz}}{p} = 1 - 2\gamma F_1(\gamma), \tag{5.B.14}$$

$$\frac{P_{xy}}{p} = -a^* F_0(\gamma), \tag{5.B.15}$$

$$q_y = -\frac{k_B p}{2m} \frac{a^{*2}}{\gamma} F_0(\gamma) \partial_s T, \tag{5.B.16}$$

$$q_x = \frac{k_B p}{2m} \left\{ \frac{1}{2\gamma} \left[1 - F_0(\gamma)\right] - F_1(\gamma) + 6F_2(\gamma) + \frac{2a^{*2}}{\gamma} \left[F_1(\gamma) - F_2(\gamma)\right] \right\} a^* \partial_s T. \tag{5.B.17}$$

From Eqs. (5.B.12)–(5.B.17) it is straightforward to get the expressions (5.100)–(5.104) for the generalized transport coefficients.

Chapter 6

PLANAR COUETTE FLOW IN A MIXTURE

1. Introduction

In parallel with what we did in Chapter 4 for the USF, we now extend the description of the Couette flow to the case of a multi-component system made of N species. Needless to say, the problem is more realistic but much more complicated than that of the USF, as already seen in the one-component case. A detailed description from the Boltzmann equation is enormously difficult, even for Maxwell molecules. Consequently, our analysis will be made from a kinetic model for the mixture, namely the Gross–Krook (GK) model (Gross and Krook, 1956). As will be seen, the model admits an exact hydrodynamic solution characterized by constant mole fractions, the absence of mutual diffusion (i.e all the species move with the same flow velocity $\mathbf{u}_s = \mathbf{u}$), a constant pressure, a quasi-linear velocity profile, and a quasi-parabolic temperature profile. Here the prefix "quasi" means that the profiles are measured with respect to a scaled space variable $s(y)$ defined similarly as in the single-gas case. The fact that, in spite of the existence of a thermal gradient, there is no mass flux means that the Soret effect (thermal diffusion) is absent in the GK model, even in the nonlinear regime. Again, the nonlinear momentum and energy transport are described by the five generalized transport coefficients (5.15), (5.17), (5.20)–(5.22). These quantities depend not only on the reduced shear rate a^* but also on the parameters of the mixture (composition, masses, force constants), what makes the problem much more involved than in the case of a single gas. Apart from the transport properties of the mixture as a whole, the kinetic theory provides information on partial contributions associated with each species. In particular, since the system is out of equilibrium, energy equipartition is broken down and this is measured by partial kinetic temperatures different for each species.

From the general description, one can particularize to the tracer limit case, namely a binary mixture where one of the species has a negligible concentration. In this limit the solvent species is not affected by the presence of the impurities and the effect of self-collisions among the latter can be neglected. Under these conditions, an interesting problem is the study of tracer *diffusion* generated by a weak concentration gradient. This diffusion process is characterized by a diffusion tensor instead of a scalar. A complementary problem is the analysis of the *mobility* of the tracer particles due to the action of an external field. This can be understood as an electric field if the impurities are electrically charged or, alternatively, as a gravity field if the tracer particles are massive "mesoscopic" corpuscles. While there exists a simple relationship (Einstein relation) between the diffusion and mobility coefficients when the bath is in equilibrium (McLennan, 1989), the generalized diffusion and mobility tensors are unrelated when the bath is under Couette flow.

2. Kinetic model description for a mixture

Let us consider an N-component mixture in the low-density regime in steady Couette flow. The mixture of gases is enclosed between two parallel plates at $y = \pm L$ in relative motion along the x-axis and kept, in general, at different temperatures (cf. Fig. 5.1). The resulting flow velocity is directed along the x-axis and, from symmetry, it is expected that the hydrodynamic fields vary only in the y-direction. We assume that the mole fractions $\{x_s = n_s/n, s = 1, \ldots N\}$ are constant, so that mutual diffusion is absent. Therefore, transport of momentum and energy are the only relevant transport phenomena occurring in the system. Our goal is to provide a detailed description of such state in terms of the imposed gradients and the mechanical parameters of the mixture (masses, concentrations, and sizes).

In principle, the appropriate theoretical description should be performed by means of the set of N coupled Boltzmann equations for the velocity distribution functions $f_s(y, \mathbf{v})$ $(s = 1, \cdots, N)$ [cf. Eq. (1.128)]. However, the intricacy of the Boltzmann collision operators hinders an exact analysis of the *nonlinear* Couette flow problem, even for a single gas, as shown in Section 5.3. This prompts the use of a kinetic model which preserves the most important qualitative features of macroscopic transport. Specifically, as done in Section 4.3 for the case of the USF, we consider a generalization to repulsive potentials (Garzó and López de Haro, 1995) of the well-known Gross–Krook (GK) model (Gross and Krook, 1956), originally proposed for Maxwell molecules. As said in Section 4.3, the general drawback of the GK model in the limit of particles mechanically equivalent does not appear in the steady Couette flow because the partial temperatures coincide with the global temperature in the above limit. In the geometry of the steady Couette flow, the GK model reads

$$v_y \frac{\partial}{\partial y} f_s = \sum_{r=1}^{N} J_{sr}^{\mathrm{GK}}[f_s, f_r], \tag{6.1}$$

where the operator $J_{sr}^{\mathrm{GK}}[f_s, f_r]$ is defined by Eqs. (4.55)–(4.57) and use has been made of the assumption $\mathbf{u}_s = \mathbf{u}$. We recall that in Eq. (4.57) the partial temperatures T_s are defined as $T_s = p_s/n_s k_B$, where n_s is the partial number density and $p_s = \frac{1}{3}\mathrm{tr}\,\mathsf{P}_s$ is the partial hydrostatic pressure, P_s being the partial pressure tensor defined by Eq. (4.9). The influence of the interaction law enters through the effective collision frequencies ν_{sr}. For repulsive potentials of the form $\phi_{sr} = K_{sr} r^w$, the collision frequencies are given by Eq. (1.198). At equilibrium, $T_s = T$ and so $\nu_{sr} = \nu_{sr}^{(0)}$ [cf. Eq. (4.67)]. It is convenient to define an *average* collision frequency

$$\nu' = \frac{1}{N(N-1)} \sum_{s=1}^{N} \sum_{r \neq s} \frac{\nu_{sr}^{(0)}}{x_r}. \tag{6.2}$$

In the binary case, $\nu' = \nu_{12}^{(0)}/x_2 = \nu_{21}^{(0)}/x_1$, which coincides with the collision frequency defined in Eq. (4.69).

Although simple in structure, the set of Eqs. (6.1) is highly nonlinear due to the coupling between f_{sr} and the hydrodynamic fields and, in general, it is necessary to employ numerical methods. However, as in Chapter 5, we look for a *normal* solution, i.e. one for which the dependence of $f_s(y, \mathbf{v})$ on y occurs only through n_s, $\mathbf{u}$, and T. The next step is to *guess* the spatial form of n_s (or, alternatively, that of the total number density n and the $N-1$ independent mole fractions x_s), $\mathbf{u} = u_x \widehat{\mathbf{x}}$, and T. Based on the results derived for a single gas, we assume that the total hydrostatic pressure $p = \sum_s p_s = n k_B T$ is constant, and the velocity and temperature profiles are

$$\frac{\partial u_x}{\partial s} = a^* = \mathrm{const}, \quad ds = \nu'(y) dy, \tag{6.3}$$

$$\frac{\partial^2 T}{\partial s^2} = -\frac{2\overline{m}}{k_B} \gamma(a^*) = \mathrm{const}, \tag{6.4}$$

where $\overline{m}$ is an *arbitrary* constant with dimensions of a mass and the thermal curvature parameter $\gamma(a^*)$ is a dimensionless function of the reduced shear rate a^* and the parameters of the system, to be determined by consistency. For simplicity, we take $\overline{m}$ as the average mass $\overline{m} = \sum_s x_s m_s$. Equations (6.3) and (6.4) are the generalizations of Eqs. (5.54) and (5.55) to the multi-component case. The knowledge of p and T gives the total number density n but is not enough to determine the mole fractions. As said before, we assume that the mole fractions are uniform. From a hydrodynamic point of view, the state of the system is completely characterized by the variables n_s, u_x, and T. However, from a kinetic point of view, in order to solve Eq. (6.1) we need to guess the spatial dependence of the partial temperatures T_s. Since $T = \sum_s x_s T_s$, we still have $N-1$ independent variables, e.g. the temperature ratios $\chi_s \equiv T_s/T$, $s = 1, \ldots N-1$. The simplest guess, to be confirmed later, is that the temperature ratios χ_s are uniform. As happens with the thermal curvature parameter $\gamma(a^*)$, the temperature ratios $\chi_s(a^*)$ are expected to be nonlinear functions of the shear rate and the other parameters of the mixture.

In terms of the variable $s(y)$, the GK model (6.1) becomes

$$\left(1 + v_y \frac{\nu'}{\nu_s} \frac{\partial}{\partial s}\right) f_s = \sum_{r=1}^{N} \frac{\nu_{sr}}{\nu_s} f_{sr}, \tag{6.5}$$

where

$$\nu_s = \sum_{r=1}^{N} \nu_{sr} \tag{6.6}$$

is the total collision frequency for a particle of species s. The hydrodynamic solution to Eq. (6.5) is

$$f_s(s, \mathbf{v}) = \sum_{r=1}^{N} \frac{\nu_{sr}}{\nu'} \int_0^\infty dt\, e^{-\nu_s t/\nu'} e^{-t v_y \partial_s} \Theta(T(s)) f_{sr}(s, \mathbf{v}), \tag{6.7}$$

where $\partial_s \equiv \partial/\partial s$ and the action of the operator $\exp(-t v_y \partial_s)$ is described by Eq. (5.89) This is the generalization to mixtures of the solution (5.91) to the BGK equation. It still remains to prove that the distribution (6.7)

is consistent with the profiles (6.3) and (6.4), along with uniform pressure, mole fractions, and temperature ratios. The consistency conditions are

$$\int d\mathbf{v}\, \{1, \mathbf{V}, V^2\} f_s = \{n_s, \mathbf{0}, 3\frac{p_s}{m_s}\}. \tag{6.8}$$

The mathematical steps are similar to those made in the case of a single gas in Appendix 5.A. As in the single-gas case, while the consistency conditions for the partial densities and flow velocities are satisfied with independence of γ, the condition for the partial temperatures implies that (Garzó, 2001; Marín and Garzó, 2002)

$$\frac{3}{2}\chi_s \nu_s = \sum_{r=1}^{N} \chi_{sr} \nu_{sr} \left\{\frac{3}{2} + \frac{\nu'^2}{\nu_s^2} a^{*2} F_1(\gamma_{sr}) - \gamma_{sr} \left[2F_2(\gamma_{sr}) + 3F_1(\gamma_{sr})\right]\right\}, \tag{6.9}$$

where

$$\chi_{sr} \equiv \frac{T_{sr}}{T} = \chi_s + 2\frac{m_s m_r}{(m_s + m_r)^2}(\chi_r - \chi_s), \tag{6.10}$$

$$\gamma_{sr} \equiv \frac{\overline{m}\nu'^2}{m_s \nu_s^2}\chi_{sr}\gamma. \tag{6.11}$$

The mathematical functions $F_0(x)$, $F_1(x)$, ...are defined in Appendix 5.A. Equation (6.9) constitutes a coupled set of N nonlinear equations for the unknowns γ and the $N-1$ independent temperature ratios $\chi_s = T_s/T$. Its solution depends, in general, on $\frac{1}{2}(N^2+5N-4)$ independent parameters: the reduced shear rate a^*, the $N-1$ mole fractions, the $N-1$ mass ratios and the $\frac{1}{2}N(N+1)-1$ force constant ratios. From Eq. (6.9) it is possible to get an equation formally giving a^{*2} in terms of χ_s and γ. Multiplying both sides of Eq. (6.9) by x_s and summing over s, one gets

$$a^{*2} = \gamma\overline{m}\frac{\sum_{s,r}(x_s\nu_{sr}\chi_{sr}^2/m_s\nu_s^2)\left[2F_2(\gamma_{sr}) + 3F_1(\gamma_{sr})\right]}{\sum_{s,r}(x_s\nu_{sr}\chi_{sr}/\nu_s^2)F_1(\gamma_{sr})}, \tag{6.12}$$

where we have taken into account that

$$\sum_{s,r} x_s\nu_{sr}(\chi_{sr} - \chi_s) = 2\sum_{s,r} x_s\nu_{sr}\frac{m_s m_r}{(m_s+m_r)^2}\chi_r = 0, \tag{6.13}$$

since $x_s\nu_{sr} = x_r\nu_{rs}$. Equation (6.12) generalizes Eq. (5.95) to the case of a mixture of N species. As a matter of fact, in the special case of a mixture of particles mechanically equivalent ($m_s = m = \overline{m}$, $K_{sr} = K$), one has $\nu_{sr} = x_r\nu'$ and $\nu_s = \nu'$. Consequently, the temperature ratios are $\chi_s = 1$, so that $\chi_{sr} = 1$ and $\gamma_{sr} = \gamma$, and Eq. (6.13) reduces to that of the single gas, Eq. (5.95). In the general case, Eq. (6.12) allows one to get the dominant behavior of the thermal curvature parameter for small shear rates. In that limit, $\chi_s \to 1$, $\gamma_{sr} \to 0$, so that

$$\gamma \approx \frac{1}{5\overline{m}}\frac{\sum_s x_s/\nu_s^{(0)}}{\sum_s x_s/m_s\nu_s^{(0)}}a^{*2}. \tag{6.14}$$

Here, $\nu_s^{(0)} = \sum_r \nu_{sr}^{(0)}$ is the collision frequency of species s at equilibrium. Let us consider now the opposite limit, namely $a^* \to \infty$. In the single-gas case, the thermal curvature parameter behaves as $\gamma \sim a^{*2}$ in that limit, according to Eq. (5.99). Let us assume (to be confirmed by consistency) that the same behavior occurs in the multi-component case. Taking into account that $F_1(x) \approx 1/2x$ and $F_2(x) \ll F_1(x)$ for large x [cf. Eq. (5.A.46)], Eq. (6.9) yields

$$\chi_s \approx \frac{m_s}{\overline{m}}\frac{a^{*2}}{3\gamma}. \tag{6.15}$$

Finally, the constraint $\sum_s x_s\chi_s = 1$ implies that

$$\gamma(a^*) \approx \frac{1}{3}a^{*2}, \quad a^* \gg 1 \tag{6.16}$$

and

$$\lim_{a^*\to\infty} \frac{T_s}{T} = \frac{m_s}{\overline{m}}. \tag{6.17}$$

Equation (6.16) not only confirms that $\gamma \sim a^{*2}$ for large shear rates but also shows that the coefficient is exactly the same as in the single-gas case, provided that $\overline{m}$ is chosen as the average mass. Equation (6.17) deserves some comments. The fact that $T_s \neq T$ represents a breakdown of energy equipartition in a nonequilibrium state. As will be illustrated later for a binary mixture, the temperature ratio T_s/T varies monotonically from 1 (limit of vanishing shear rate) to $m_s/\overline{m}$ (limit of large shear rate) as the shear rate increases. This means that the particles with a mass larger (smaller) than the average mass have a larger (smaller) mean kinetic energy. As a consequence, the differences among the mean square velocities $\langle v^2\rangle_s = 3k_BT_s/m_s$ tend to decrease as the shear rate increases, so that all the particles reach the same mean square velocity $3k_BT/\overline{m}$ for $a^* \to \infty$.

In principle, the presence of a thermal gradient in a multi-component gas induces a mass flux. In the linear regime, this is the well-known Soret or thermal diffusion effect (de Groot and Mazur, 1984) characterized by the coefficients L_{sq} of Eq. (1.143). However, in the special case of Maxwell molecules the associated thermal diffusion coefficients vanish exactly (Chapman and Cowling, 1970), i.e. $L_{sq} = 0$. The fact that the GK kinetic model is consistent with a common flow velocity to all species, i.e. $\mathbf{u}_s = \mathbf{u}$, implies the absence of thermal diffusion, even in the nonlinear regime. This is not surprising since the GK model is inspired in the properties of the Boltzmann collision operator for Maxwell molecules, as discussed in Section 1.7.

Once we have shown the consistency of the assumed profiles, the distributions (6.7) are explicitly known. From them the momentum and heat fluxes can be obtained, by following steps similar to those of Appendix 5.B (Garzó, 2001; Marín and Garzó, 2002). Let us consider first the rheological properties. A nonlinear shear viscosity η and a viscosity function F_η can be defined by Eqs. (5.12) and (5.15), respectively. Here, however, as done in the case of the USF for mixtures, we define $F_\eta(a^*)$ by

$$P_{xy} = -pF_\eta(a^*)\frac{1}{\nu'}\frac{\partial u_x}{\partial y}. \tag{6.18}$$

Since the Navier–Stokes shear viscosity η_0 differs from p/ν' [cf. Eq. (1.215)], the definition (6.18) implies that $F_\eta(0) = \eta_0/(p/\nu') \neq 1$, except in the trivial case of mechanically equivalent particles. The viscosity function $F_\eta(a^*)$ can be written as

$$F_\eta(a^*) = \sum_{s=1}^{N} x_s F_{\eta,s}(a^*), \tag{6.19}$$

where $F_{\eta,s}(a^*)$ is the partial contribution associated with species s. Its expression is (Garzó, 2001; Marín and Garzó, 2002)

$$F_{\eta,s}(a^*) = \frac{\nu'}{\nu_s^2} \sum_{r=1}^{N} \nu_{sr}\chi_{sr}F_0(\gamma_{sr}). \tag{6.20}$$

For vanishing shear rate, $\gamma_{sr} \to 0$ and $\chi_{sr} \to 1$, so that

$$F_{\eta,s}(0) = \frac{\nu'}{\nu_s^{(0)}} \tag{6.21}$$

and the Navier–Stokes shear viscosity η_0 of a mixture described by the GK model is recovered (Gross and Krook, 1956; Goldman and Sirovich, 1967; Garzó et al., 1989). Of course, this Navier–Stokes coefficient coincides with the one obtained in the USF, Eq. (4.73). In the limit $a^* \to \infty$, one has

$$F_{\eta,s}(a^*) \approx 3\frac{m_s\nu_s}{\overline{m}\nu'} a^{*-2} \ln a^*, \tag{6.22}$$

where we have made use of Eq. (6.17) and the asymptotic behavior (5.A.44). Equation (6.22) generalizes Eq. (5.110) to the case of a multi-component gas.

The first viscometric function, defined by Eq. (5.20), can be written as

$$\Psi_1(a^*) = \sum_{s=1}^{N} x_s \Psi_{1,s}(a^*), \tag{6.23}$$

where

$$\Psi_{1,s}(a^*) = -\frac{2}{a^{*2}} \sum_{r=1}^{N} \frac{\nu_{sr}}{\nu_s}\chi_{sr} \left[2\gamma_{sr}F_2(\gamma_{sr}) + \frac{\nu'^2}{\nu_s^2}F_1(\gamma_{sr})a^{*2}\right]. \tag{6.24}$$

Analogously, the second viscometric function, Eq. (5.21), is

$$\Psi_2(a^*) = \sum_{s=1}^{N} x_s \Psi_{2,s}(a^*), \tag{6.25}$$

where

$$\Psi_{2,s}(a^*) = \frac{4}{a^{*2}} \sum_{r=1}^{N} \frac{\nu_{sr}}{\nu_s}\chi_{sr}\gamma_{sr}F_2(\gamma_{sr}). \tag{6.26}$$

In the limit $a^* \to 0$, Eq. (6.11) yields $\gamma_{sr}/a^{*2} \to (\overline{m}\nu'^2/m_s\nu_s^2)\gamma/a^{*2}$, where γ/a^{*2} is given by Eq. (6.14). Thus, when $a^* \to 0$ the partial viscometric functions tend to

$$\Psi_{1,s}(0) = -2\frac{\nu'^2}{\nu_s^{(0)2}}\left(1+\frac{2}{5m_s}\frac{\sum_r x_r/\nu_r^{(0)}}{\sum_r x_r/m_r\nu_r^{(0)}}\right), \tag{6.27}$$

$$\Psi_{2,s}(0) = \frac{4}{5m_s}\frac{\nu'^2}{\nu_s^{(0)2}}\frac{\sum_r x_r/\nu_r^{(0)}}{\sum_r x_r/m_r\nu_r^{(0)}}. \tag{6.28}$$

These values of the viscometric functions at zero shear rate represent Burnett-order coefficients. As in the single-gas case, these Burnett coefficients differ from those obtained in the USF, Eq. (4.77).

Now we focus on the heat flux. Its component parallel to the thermal gradient defines a generalized thermal conductivity coefficient κ through Eq. (5.13). An associated thermal conductivity function F_κ can be defined by Eq. (5.17). Nevertheless, as in the case of the viscosity function $F_\eta(a^*)$, it is preferable to define $F_\kappa(a^*)$ by the relation

$$q_y = -\frac{5k_B}{2\overline{m}}\frac{p}{\nu'}F_\kappa(a^*)\frac{\partial T}{\partial y}. \tag{6.29}$$

Of course, $F_\kappa(0) = \kappa_0/(5k_Bp/2\overline{m}\nu') \neq 1$. This choice guarantees that the relationship (5.19) also applies to a multi-component gas, as can be checked from Eqs. (5.3), (6.4), and (6.18). The thermal conductivity function can be expressed as

$$F_\kappa(a^*) = \sum_{s=1}^{N} x_s F_{\kappa,s}(a^*), \tag{6.30}$$

where the contribution due to species s is (Marín and Garzó, 2002)

$$\begin{aligned} F_{\kappa,s}(a^*) \quad = \quad & \frac{\overline{m}\nu'}{5m_s\nu_s^2}\sum_{r=1}^{N}\nu_{sr}\chi_{sr}^2\Bigg[3F_1(\gamma_{sr})+2F_2(\gamma_{sr}) \\ & +\frac{\nu'^2}{\nu_s^2}a^{*2}\frac{F_0(\gamma_{sr})-F_1(\gamma_{sr})}{\gamma_{sr}}\Bigg]. \end{aligned} \tag{6.31}$$

For vanishing shear rate, the (partial) thermal conductivity function becomes

$$F_{\kappa,s}(0) = \frac{\overline{m}\nu'}{m_s\nu_s^{(0)}}, \tag{6.32}$$

and the Navier–Stokes thermal conductivity of a mixture described by the GK model, Eq. (1.216), is recovered (Gross and Krook, 1956; Garzó et al., 1989).

The coupling between shearing motion and the thermal gradient in the Couette flow generates a heat flux component orthogonal to the thermal gradient but parallel to the flow direction. This x-component of the heat flux defines the visco-thermal function $\Phi(a^*)$ through an equation similar to Eq. (5.22), namely

$$q_x = -\frac{5k_B}{2\overline{m}}\frac{p}{\nu'^2}\Phi(a^*)\frac{\partial u_x}{\partial y}\frac{\partial T}{\partial y}. \tag{6.33}$$

The result can be written as

$$\Phi(a^*) = \sum_{s=1}^{N} x_s \Phi_s(a^*), \tag{6.34}$$

where the contribution due to species s is (Marín and Garzó, 2002)

$$\begin{aligned}\Phi_s(a^*) \quad &= \quad -\frac{2\overline{m}\nu'^2}{5m_s\nu_s^3}\sum_{r=1}^{N}\nu_{sr}\chi_{sr}^2\left[\frac{1-F_0(\gamma_{sr})}{4\gamma_{sr}} - \frac{1}{2}F_1(\gamma_{sr}) + 3F_2(\gamma_{sr})\right.\\ &\quad \left. +\frac{\nu'^2}{\nu_s^2}a^{*2}\frac{F_1(\gamma_{sr}) - F_2(\gamma_{sr})}{\gamma_{sr}}\right].\end{aligned} \tag{6.35}$$

In the limit of small shear rates, the visco-thermal function Φ_s tends to

$$\Phi_s(0) = -\frac{14}{5}\frac{\overline{m}\nu'^2}{m_s\nu_s^{(0)2}}. \tag{6.36}$$

This value represents a Burnett-order coefficient. It is interesting to note that Eqs. (6.21), (6.27), (6.32), and (6.36) imply the simple relationships

$$\Psi_2(0)F_\kappa(0) = -\frac{2}{7}\Phi(0)F_\eta(0), \tag{6.37}$$

$$\Psi_1(0) + \Psi_2(0) = -2\nu'^2\sum_{s=1}^{N}\frac{x_s}{\nu_s^{(0)2}}. \tag{6.38}$$

Of course, in the case of particles mechanically equivalent, Eqs. (6.20), (6.24), (6.26), (6.31), and (6.35) reduce to the results obtained for a single gas in Chapter 5.

Apart from the evaluation of transport properties, the use of the kinetic model allows one to explicitly get the velocity distribution function of each species s, as given by Eq. (6.7). In order to emphasize the hydrodynamic nature of the solution, i.e. the fact that it depends on space through the local values of flow velocity, density, temperature, shear rate, and thermal gradient, it is convenient to define the reduced quantities (4.79). In terms of these quantities Eq. (6.7) becomes

$$f_s^*(\boldsymbol{\xi}) = \sum_{r=1}^{N} \frac{\nu_{sr}}{\nu'} g_{sr}(\boldsymbol{\xi}), \tag{6.39}$$

where

$$\boldsymbol{\xi} = (m_s/2k_BT)^{1/2}\mathbf{V} \tag{6.40}$$

and

$$\begin{aligned} g_{sr}(\boldsymbol{\xi}) &= \frac{\pi^{-3/2}}{|\xi_y|}\chi_{sr}^{-3/2} \int_0^{(m_s/\overline{m})^{1/2}\beta_\pm} du \left[\vartheta\left(\pm\sqrt{\overline{m}/m_s}u\right)\right]^{-5/2} \\ &\quad \times e^{-(\nu_s/\nu')u/|\xi_y|} e^{\pm a^* u \partial/\partial \xi_x} \exp\left[-\frac{\xi^2}{\vartheta\left(\pm\sqrt{\overline{m}/m_s}u\right)\chi_{sr}}\right]. \end{aligned} \tag{6.41}$$

In this equation, the sign $+$ $(-)$ refers to $\xi_y > 0$ ($\xi_y < 0$), the function $\vartheta(u)$ is given by Eq. (5.122), and the parameters $\beta_\pm$ are given by Eq. (5.124). In addition, the dimensionless thermal gradient ϵ is defined by Eq. (5.56), except that the mass m is replaced by the average mass $\overline{m}$ of the mixture. Equations (6.39) and (6.41) generalize Eq. (5.123).

3. Application to the case of a binary mixture

Thus far, all the results hold for an arbitrary number N of species. Of course, the algebraic difficulties increase with N since one has to solve the coupled set of N nonlinear equations (6.9). For the sake of illustration we consider in this Section the case of a binary mixture ($N = 2$). Then the parameter space is five-dimensional: the mass ratio $\mu \equiv m_1/m_2$, the force constant ratios K_{11}/K_{12} and K_{22}/K_{12}, the mole fraction x_1, and the reduced shear rate a^*. The first four parameters define the particular mixture, while the shear rate defines the state of the system. From Eq. (6.9) one gets the following set of two equations:

$$\begin{aligned} &\left(\frac{\nu_1}{\nu_2}\right)^2 \frac{\sum_{r=1}^2 \nu_{2r}\chi_{2r}F_1(\gamma_{2r})}{\sum_{r=1}^2 \nu_{1r}\chi_{1r}F_1(\gamma_{1r})} \\ &= \frac{3\nu_{21}(\chi_2 - \chi_{21}) + 2\sum_{r=1}^2 \nu_{2r}\chi_{2r}\gamma_{2r}\left[2F_2(\gamma_{2r}) + 3F_1(\gamma_{2r})\right]}{3\nu_{12}(\chi_1 - \chi_{12}) + 2\sum_{r=1}^2 \nu_{1r}\chi_{1r}\gamma_{1r}\left[2F_2(\gamma_{1r}) + 3F_1(\gamma_{1r})\right]}, \end{aligned} \tag{6.42}$$

$$a^{*2} = \frac{1}{2}\left(\frac{\nu_1}{\nu'}\right)^2 \frac{3\nu_{12}(\chi_1 - \chi_{12}) + 2\sum_{r=1}^2 \nu_{1r}\chi_{1r}\gamma_{1r}\left[2F_2(\gamma_{1r}) + 3F_1(\gamma_{1r})\right]}{\sum_{r=1}^2 \nu_{1r}\chi_{1r}F_1(\gamma_{1r})}. \tag{6.43}$$

The unknowns in these equations are the temperature ratio $\chi_1 \equiv T_1/T$ (or, equivalently, χ_2) and the thermal curvature parameter γ. Note that $\chi_2 = (1 - x_1\chi_1)/x_2$, χ_{sr} are related to χ_1 through Eq. (6.10), and γ_{sr} are related to γ and χ_1 through Eq. (6.11). In addition, for non-Maxwell molecules, the collision frequencies ν_{sr} also depend on χ_1 [cf. Eq. (1.198)]. As done in the single-gas case, for practical purposes it is convenient to consider γ rather than a^* as an independent variable. Therefore, Eq. (6.42) can be seen as a closed nonlinear equation giving χ_1 in terms of γ (and, of course, of the remaining parameters characterizing the mixture). Once solved, Eq. (6.43) gives directly the shear rate. Although in general Eq. (6.42) must be solved numerically, its solution can be explicitly obtained for small shear rates. In the case of Maxwell molecules, the result is (Garzó, 2001)

$$\chi_1 \approx 1 + \chi_1^{(2)} a^{*2}, \tag{6.44}$$

where

$$\chi_1^{(2)} = \frac{1}{3}\frac{(\mu-1)(\mu+1)^2}{\mu}\frac{\nu_{12}}{\mu x_2\nu_1 + x_1\nu_2}. \tag{6.45}$$

This equation confirms that if $m_1 > m_2$ then $\chi_1 > 1$ (i.e. $T_1 > T_2$). As seen in the multi-component case, Eq. (6.17), $T_1/T_2 \to m_1/m_2$ in the limit of large shear rates. Of course, $\chi_2^{(2)} = -(x_1/x_2)\chi_1^{(2)}$.

In the case of the viscosity function, the first correction to the Navier–Stokes value (6.21) is (Garzó, 2001)

$$F_\eta(a^*) \approx F_\eta(0) + F_\eta^{(2)} a^{*2}, \tag{6.46}$$

where

$$F_\eta^{(2)} = \nu' \sum_{s=1}^{2} \frac{x_s}{\nu_s}\left[\chi_s^{(2)}\left(1 - \frac{2\mu}{(1+\mu)^2}\frac{\nu'}{\nu_s}\right) - 18\left(\frac{\nu'}{\nu_s}\right)^2 \frac{\overline{m}}{m_s}\gamma^{(2)}\right], \tag{6.47}$$

where we have called $\gamma^{(2)}$ the coefficient of a^{*2} in Eq. (6.14). The super-Burnett coefficient $F_\eta^{(2)}$ reduces to $F_\eta^{(2)} = -18/5$ in the case of mechanically equivalent particles, Eq. (5.105), and to $F_\eta^{(2)} = -(18/5)(\nu'/\nu_1)^3$ in the limit $x_2 \to 0$. Apart from these two trivial limits, the coefficient $F_\eta^{(2)}$ has an intricate dependence on the force constant ratios K_{11}/K_{12} and K_{22}/K_{12}, the mole fraction x_1, and the mass ratio $\mu = m_1/m_2$. The corresponding super-Burnett coefficient in the USF problem is given by Eq. (4.75). For that flow, we found that $F_\eta^{(2)}$ could be positive for sufficiently disparate masses. The natural question is whether this effect is also present in the Couette flow, i.e. whether the coefficient $F_\eta^{(2)}$ given by Eq. (6.47) can reach positive values. A careful analysis shows that this is indeed the case for very disparate masses. The change of sign of $F_\eta^{(2)}$

is illustrated in Fig. 6.1, where the coefficient is plotted as a function of the mass ratio μ for equal force constants and three values of the mole fraction. At a given value of the mole fraction x_1, $F_\eta^{(2)}$ is positive for mass ratios smaller than a certain threshold value $\mu_0(x_1) < 1$ (where, without loss of generality, we have taken the species 1 as the lighter one). A plot of $\mu_0(x_1)$ is given by Fig. 6.2, which shows that μ_0 has a maximum value $\mu_0 \simeq 0.0463$ at $x_1 \simeq 0.39$, and goes to 0 both for $x_1 \to 0$ and $x_1 \to 1$. Note that Figs. 6.1 and 6.2 are qualitatively similar to Figs. 4.6 and 4.7. For $\mu < \mu_0(x_1) < 1$ (or, equivalently, $\mu > 1/\mu_0(1-x_1) > 1$), the shear viscosity function starts increasing with the shear rate (*shear-thickening* effect). Since, according to Eq. (6.22) $F_\eta(a^*)$ goes to zero for large shear rates, it is evident that in those cases $F_\eta(a^*)$ reaches a maximum at a certain shear rate and decreases thereafter.

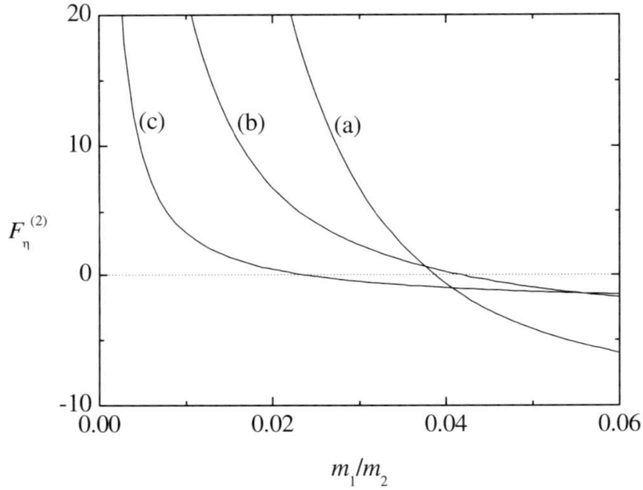

Figure 6.1. Super-Burnett coefficient $F_\eta^{(2)}$ as a function of the mass ratio $\mu = m_1/m_2$ for $K_{11} = K_{12} = K_{22}$ and three values of the mole fraction: (a) $x_1 = 0.25$, (b) $x_1 = 0.5$, and (c) $x_1 = 0.75$.

Let us consider the specific case of a mixture with a mole fraction $x_1 = \frac{1}{6}$, equal force constants ($K_{11} = K_{12} = K_{22}$), and two different values of the mass ratio: $\mu = 0.1$ and $\mu = 10$. We also consider the two extreme cases of Maxwell molecules and hard spheres. In both cases the mass disparity is not enough to give rise to a shear-thickening effect for small shear rates. In Fig. 6.3 we plot the temperature ratio T_1/T_2 as a function of the shear rate. The deviations from the energy equipartition can be weak or strong depending on the strength of the shear rate as well as on the mechanical differences between the species. The fact that $T_1 \neq T_2$ for finite shear rates indicates that the standard one-temperature Chapman–Enskog theory (Chapman and Cowling,

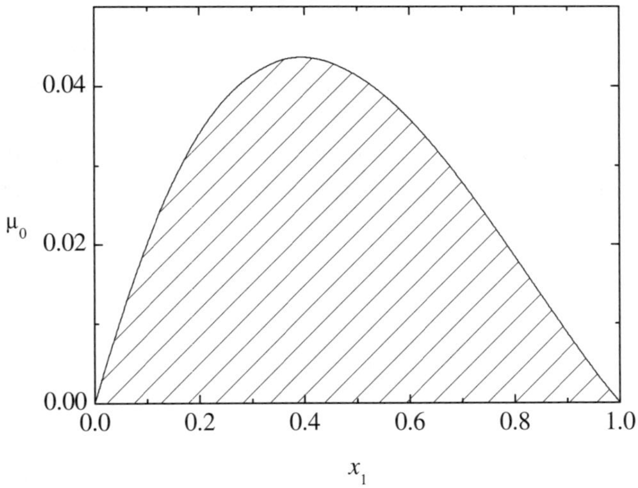

Figure 6.2. Threshold mass ratio $\mu_0(x_1)$ as a function of the mole fraction for $K_{11} = K_{12} = K_{22}$. The shaded region below the curve corresponds to the values of μ and x_1 for which the super-Burnett coefficient $F_\eta^{(2)}$ is positive, so a shear-thickening effect appears for small shear rates.

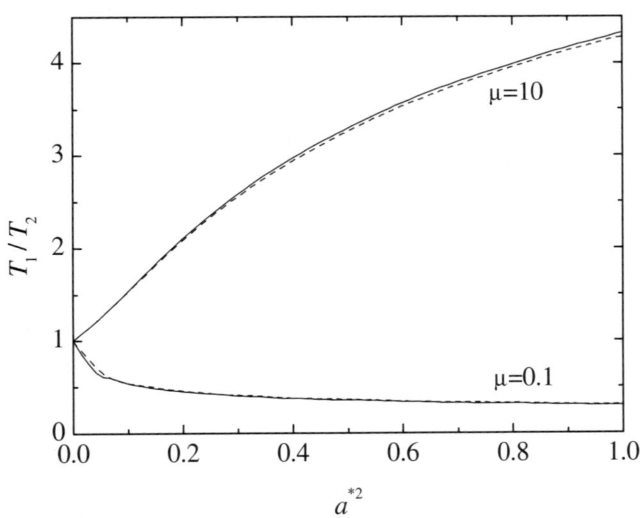

Figure 6.3. Shear-rate dependence of the temperature ratio T_1/T_2 for equal force constants ($K_{11} = K_{12} = K_{22}$), a mole fraction $x_1 = \frac{1}{6}$, and two values of the mass ratio $\mu = m_1/m_2$: $\mu = 10$ and $\mu = 0.1$. The solid lines correspond to Maxwell molecules, while the dashed lines refer to hard spheres.

1970) fails, even for not very disparate masses (Garzó, 2001). This justifies the use of the so-called two-temperature theory (Goebel et al., 1976). With respect to the influence of the interaction potential, Fig. 6.3 shows that the temperature ratio is practically independent of the scattering law considered. In contrast to the non-monotonic behavior of the temperature ratio in the USF (cf. Figs. 4.2 and 4.5), this ratio has a monotonic dependence on the shear rate in the Couette flow.

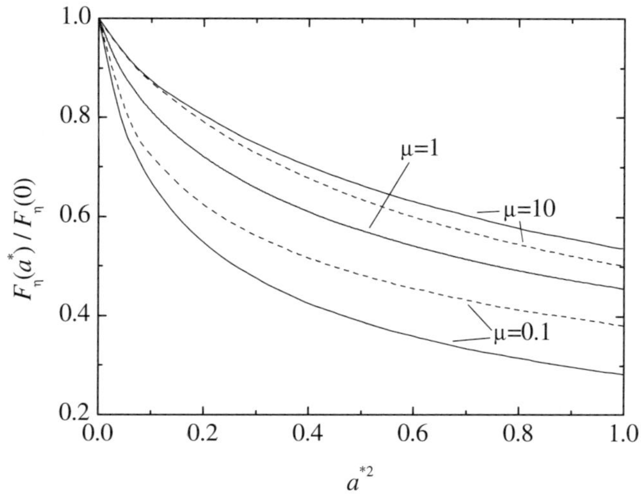

Figure 6.4. Shear-rate dependence of the relative viscosity function $F_\eta(a^*)/F_\eta(0)$ for equal force constants ($K_{11} = K_{12} = K_{22}$), a mole fraction $x_1 = \frac{1}{6}$, and three values of the mass ratio $\mu = m_1/m_2$: $\mu = 10$, $\mu = 1$, and $\mu = 0.1$. The solid lines correspond to Maxwell molecules, while the dashed lines refer to hard spheres.

Once the temperature ratio has been computed, it is worth considering the shear-rate dependence of the generalized transport coefficients. The most important quantity in a sheared flow problem is the shear viscosity function $F_\eta(a^*)$. In Fig. 6.4 we plot the shear viscosity function $F_\eta(a^*)$ normalized to its Navier–Stokes value $F_\eta(0)$. We observe that $F_\eta(a^*)$ decreases as a^* increases (shear-thinning effect). This inhibition of momentum transport is also present in the one-component case, as well as in the USF problem for mixtures [except for very disparate mixtures, as discussed in the paragraph below Eq. (6.47)]. Concerning the dependence on the mass ratio we see that, at a given value of the reduced shear rate $a^* = a/\nu'$, the ratio $F_\eta(a^*)/F_\eta(0)$ increases as the mass of the solute particle increases. In other words, the shear-thinning effect is more significant as the solute particles become lighter. Figure 6.4 also shows that for a given value of a^*, $F_\eta(a^*)/F_\eta(0)$ increases (decreases) as the potential becomes harder when the excess component is heavier (lighter) than the defect

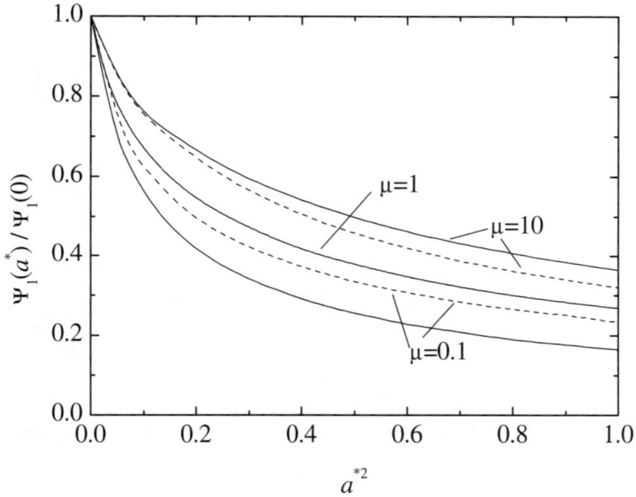

Figure 6.5. Shear-rate dependence of the relative first viscometric function $\Psi_1(a^*)/\Psi_1(0)$ for equal force constants ($K_{11} = K_{12} = K_{22}$), a mole fraction $x_1 = \frac{1}{6}$, and three values of the mass ratio $\mu = m_1/m_2$: $\mu = 10$, $\mu = 1$, and $\mu = 0.1$. The solid lines correspond to Maxwell molecules, while the dashed lines refer to hard spheres.

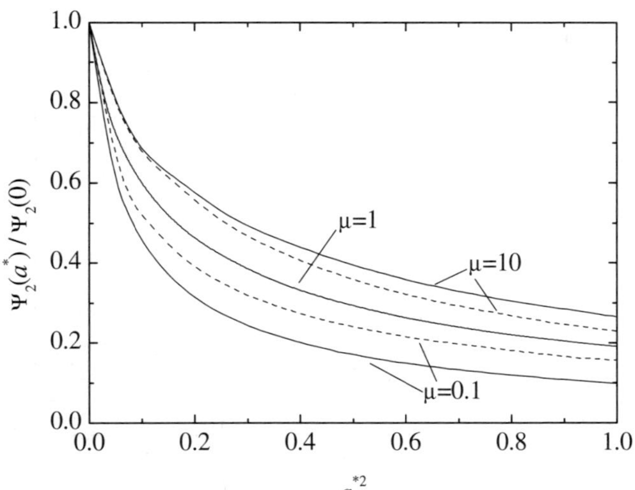

Figure 6.6. Shear-rate dependence of the relative second viscometric function $\Psi_2(a^*)/\Psi_2(0)$ for equal force constants ($K_{11} = K_{12} = K_{22}$), a mole fraction $x_1 = \frac{1}{6}$, and three values of the mass ratio $\mu = m_1/m_2$: $\mu = 10$, $\mu = 1$, and $\mu = 0.1$. The solid lines correspond to Maxwell molecules, while the dashed lines refer to hard spheres.

component. Figures 6.5–6.8 show the remaining reduced transport coefficients

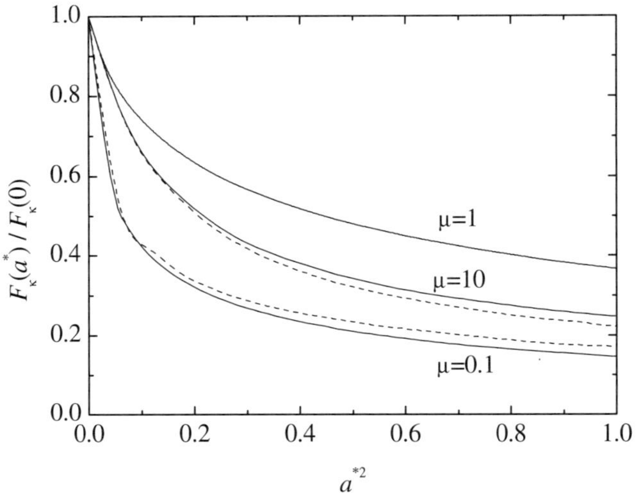

Figure 6.7. Shear-rate dependence of the relative thermal conductivity function $F_\kappa(a^*)/F_\kappa(0)$ for equal force constants ($K_{11} = K_{12} = K_{22}$), a mole fraction $x_1 = \frac{1}{6}$, and three values of the mass ratio $\mu = m_1/m_2$: $\mu = 10$, $\mu = 1$, and $\mu = 0.1$. The solid lines correspond to Maxwell molecules, while the dashed lines refer to hard spheres.

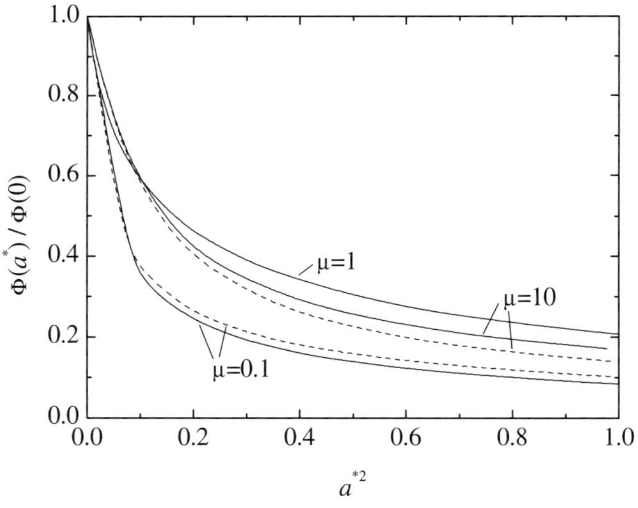

Figure 6.8. Shear-rate dependence of the relative visco-thermal function $\Phi(a^*)/\Phi(0)$ for equal force constants ($K_{11} = K_{12} = K_{22}$), a mole fraction $x_1 = \frac{1}{6}$, and three values of the mass ratio $\mu = m_1/m_2$: $\mu = 10$, $\mu = 1$, and $\mu = 0.1$. The solid lines correspond to Maxwell molecules, while the dashed lines refer to hard spheres.

$\Psi_1(a^*)/\Psi_1(0)$, $\Psi_2(a^*)/\Psi_2(0)$, $F_\kappa(a^*)/F_\kappa(0)$, and $\Phi(a^*)/\Phi(0)$, respectively. In general, we observe that the dependence of these reduced coefficients on the shear rate is very similar to the one found for the shear viscosity, namely, the corresponding transport property decreases as a^* increases and, at a given value of a^*, the coefficients decrease as the mass of the solute particles increases. This effect is more significant for Maxwell molecules than for hard spheres.

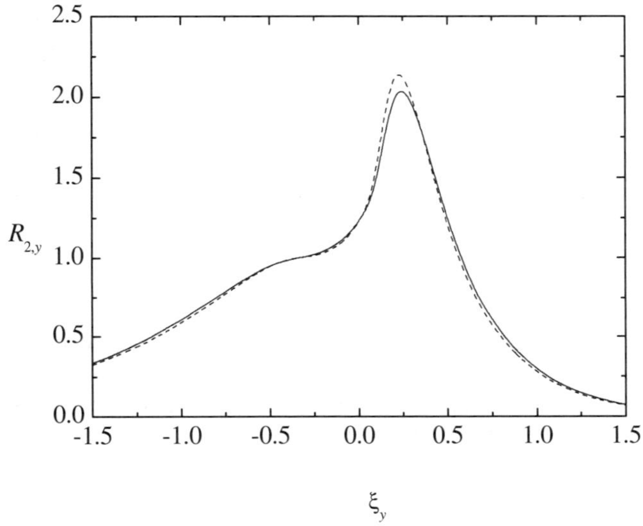

Figure 6.9. Marginal velocity distribution function, relative to the local equilibrium distribution, of the excess component in the case $x_1 = \frac{1}{6}$, $K_{11} = K_{12} = K_{22}$, and $\mu = 0.1$, at a reduced shear rate $a^* = 1$ and a reduced thermal gradient $\epsilon = 0.34$. The solid line refers to Maxwell molecules, while the dashed line refers to hard spheres.

Now we turn our attention to the velocity distributions given by Eqs. (6.39) and (6.41). Since they depend on the three components of $\boldsymbol{\xi}$, it is useful to consider marginal distributions. Their ratios with respect to local equilibrium are given by

$$R_{s,y}(\xi_y) = \frac{\int_{-\infty}^{\infty} d\xi_x \int_{-\infty}^{\infty} d\xi_z \, f_s^*(\boldsymbol{\xi})}{\pi^{-1/2} e^{-\xi_y^2}}, \tag{6.48}$$

with a similar definition of $R_{s,x}(\xi_x)$. Figure 6.9 shows the shape of the distribution $R_{2,y}(\xi_y)$ of the excess component in the case $x_1 = \frac{1}{6}$, $K_{11} = K_{12} = K_{22}$, and $\mu = 0.1$, at a reduced shear rate $a^* = 1$ and a reduced thermal gradient $\epsilon = 0.34$. The same quantity is plotted in Fig. 6.10, but for $\mu = 0.1$ at $a^* = 1$ and $\epsilon = 2$. In general, the distortion from local equilibrium ($R_{2,y} = 1$) is quite important for the two mass ratios considered. We also observe that the qualitative differences between Maxwell molecules and hard spheres are less significant when the mass of the defect particles is smaller than that of the ex-

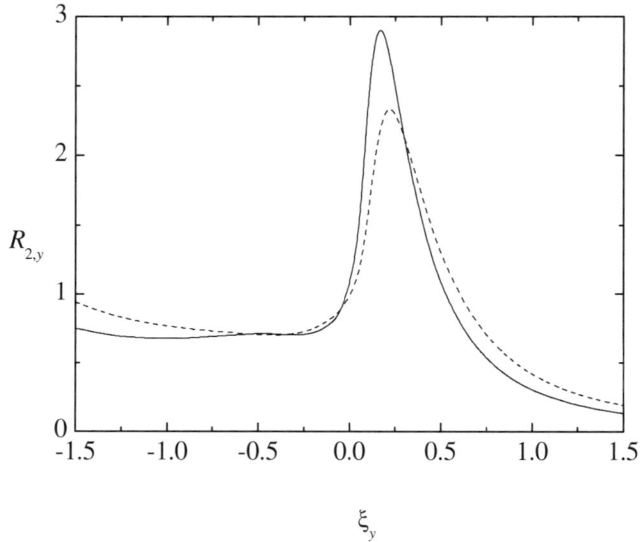

Figure 6.10. Marginal velocity distribution function, relative to the local equilibrium distribution, of the excess component in the case $x_1 = \frac{1}{6}$, $K_{11} = K_{12} = K_{22}$, and $\mu = 10$, at a reduced shear rate $a^* = 1$ and a reduced thermal gradient $\epsilon = 2$. The solid line refers to Maxwell molecules, while the dashed line refers to hard spheres.

cess component ($\mu = 0.1$), especially in the high-velocity region. This can be understood by the following qualitative argument. If the defect particles are much lighter than the excess ones, one expects the velocity distribution of the latter to be hardly influenced by the presence of the former, so the distribution of the excess particles is close to that of a single gas under the same conditions. Since the distribution function of the single gas is universal, it is not surprising that the effect of the interaction potential is quite small in Fig. 6.9. On the other hand, in the case of Fig. 6.10 the small concentration of species 1 is counterbalanced by its larger mass, so the distribution of species 2 is widely affected by the presence of the solute.

Before closing this Section it is worth considering the tracer limit, in which one of the species (say species 1) has a negligible concentration ($x_1 \to 0$). In that case, species 2 is not affected by the presence of the tracer species and so it plays the role of a nonequilibrium bath. Conversely, the state of the tracer particles is mainly governed by their collisions with the bath particles and so one can neglect self-collisions. This limit is then obtained from the general solution by setting $\nu_{11} \to 0$ and $\nu_{21} \to 0$, so $\nu_1 = \nu_{12} = \nu'$ and $\nu_2 = \nu_{22}$. In addition, $\chi_2 = 1$ (since $T_2 \to T$) and $\overline{m} = m_2$. Therefore, Eqs. (6.42) and

(6.43) reduce to

$$\left(\frac{\nu'}{\nu_2}\right)^2 \frac{F_1(\gamma_{22})}{\chi_{12}F_1(\gamma_{12})} = \frac{2\gamma_{22}\left[3F_1(\gamma_{22}) + 2F_2(\gamma_{22})\right]}{3(\chi_1 - \chi_{12}) + 2\chi_{12}\gamma_{12}\left[3F_1(\gamma_{12}) + 2F_2(\gamma_{12})\right]}, \tag{6.49}$$

$$\left(\frac{\nu'}{\nu_2}a^*\right)^2 = \gamma_{22}\frac{3F_1(\gamma_{22}) + 2F_2(\gamma_{22})}{F_1(\gamma_{22})}. \tag{6.50}$$

In the above equations, application of Eq. (6.11) yields $\gamma_{22} = \gamma(\nu'/\nu_2)^2$ and $\gamma_{12} = \gamma\chi_{12}/\mu$. Equation (6.50) gives γ as a function of a^* and, as expected, coincides with the one obtained in the single-gas case, Eq. (5.95), except for the different units taken. Once solved, Eq. (6.49) becomes a nonlinear equation defining the temperature ratio χ_1. Its dependence on the shear rate is qualitatively similar to the one found in the general case (Garzó and Santos, 1993). In particular, if the tracer particles are heavier (lighter) than the bath particles, they have a larger (smaller) temperature. In any case, in contrast to what happens in the USF in the tracer limit (cf. Section 4.4), the temperature ratio remains finite for all shear rates, so a nonequilibrium phase transition is absent in this case.

4. Diffusion and mobility in the tracer limit.

An interesting problem in nonequilibrium statistical mechanics is the description of diffusion of impurities immersed in a bath out of equilibrium. Here we want to characterize the diffusion process when the bath is in the steady Couette flow and a current of impurities is generated by the presence of a weak concentration gradient ∇x_1 and/or a weak external field $\mathbf{E}$ (e.g. gravity or an electric field) acting only on the impurity particles. In the latter situation, for instance, the tracer particles are assumed to have a large mass (sedimentation problem) or to be electrically charged (electric conduction problem).

To fix ideas, let us recall first the situation when the bath is in equilibrium. We will also assume that the tracer particles possess a "charge" q and are subjected to an external "electric" field $\mathbf{E}$. Since the concentration gradient and the external field are assumed to be small, a linear relationship between them and the mass flux $\mathbf{j}_1$ is expected to apply (Woods, 1993):

$$\mathbf{j}_1 = -m_1 n D_0 \nabla x_1 + \frac{m_1}{q}\sigma_0 \mathbf{E}. \tag{6.51}$$

Here, D_0 is the (tracer) diffusion coefficient and σ_0 is the conductivity (or mobility) coefficient. The first contribution to the mass flux in Eq. (6.51) corresponds to Fick's law [cf. Eq. (1.147) in the tracer limit], while the second one represents Ohm's law (if $\mathbf{E}$ is a true electric field). Both coefficients are related through the Einstein relation (Reif, 1965; McLennan, 1989)

$$D_0 = \frac{k_B T}{n_1 q^2}\sigma_0. \tag{6.52}$$

This is not but a consequence of the fluctuation–dissipation theorem (McLennan, 1989), which shows that the equilibrium fluctuations can be probed by applying a weak external field coupled to particles. When the bath is out of equilibrium (in particular, in the steady Couette flow), a linear relationship between the flux and the generalized forces ∇x_1 and $\mathbf{E}$ is still expected (provided that both quantities are weak), but the diffusion and conductivity coefficients must be replaced by tensorial quantities that are nonlinear functions of the nonequilibrium control parameters (e.g. the shear rate). In that case, Fick's and Ohm's laws (6.51) are generalized as

$$\mathbf{j}_1 = -m_1 n \mathsf{D} \cdot \nabla x_1 + \frac{m_1}{q} \boldsymbol{\sigma} \cdot \mathbf{E}, \tag{6.53}$$

where D is the tracer diffusion tensor and $\boldsymbol{\sigma}$ is the conductivity or mobility tensor.[1] The diagonal elements can be seen as generalizations of the equilibrium diffusion (D_0) and conductivity (σ_0) coefficients since they couple the i-th component of the generalized forces to the i-th component of the mass current. On the other hand, the off-diagonal elements measure cross effects not present at equilibrium. The objective here is to get the elements of both tensors when the bath is under Couette flow. These elements will depend on the shear rate and the thermal gradient, as well as on the force constant and mass ratios. As a byproduct, we will find that the usual Einstein relation is no longer valid in this nonequilibrium situation. A similar conclusion was found from the Smoluchowski equation for a sheared suspension of interacting colloidal particles (Szamel et al., 1992).

Let us consider a low-density binary mixture composed of charged particles (of mass m_1, electric charge q, and number density n_1) immersed in a bath constituted by neutral particles (of mass m_2 and number density n_2). We assume that $n_1 \ll n_2$, so that charged particles are in tracer concentration and the interactions of type charged–charged can be neglected in the kinetic equation of f_1. In addition, the smallness of the mole fraction x_1 also allows us to suppose that the state of the neutral gas is not influenced by the tracer motion. Consequently, only interactions of type $1-2$ and $2-2$ will be considered in the description. We assume finally that the mole fraction x_1 is slightly nonuniform and that a weak uniform electric field $\mathbf{E}$ is applied. Both generalized forces generate a current of charged particles, according to Eq. (6.53). The neutral gas is assumed to be under the steady planar Couette flow. Its velocity distribution function f_2 is the solution to the closed Boltzmann equation

$$v_y \frac{\partial}{\partial y} f_2 = J_{22}[f_2, f_2]. \tag{6.54}$$

[1] The tracer diffusion tensor D in Eq. (6.53) is the analog of the mutual diffusion tensor defined in Eq. (4.104).

On the other hand, the velocity distribution function of the tracer particles obeys the Boltzmann–Lorentz equation

$$\frac{\partial}{\partial t} f_1 + \mathbf{v} \cdot \nabla f_1 + \frac{q}{m_1} \mathbf{E} \cdot \frac{\partial}{\partial \mathbf{v}} f_1 = J_{12}[f_1, f_2]. \tag{6.55}$$

At a kinetic level, the mass flux of tracer particles is defined by Eq. (1.138) as

$$\begin{aligned} \mathbf{j}_1 &= m_1 \int d\mathbf{v} \mathbf{V} f_1 \\ &= \rho_1(\mathbf{u}_1 - \mathbf{u}), \end{aligned} \tag{6.56}$$

where $\mathbf{V} = \mathbf{v} - \mathbf{u}$ is the velocity relative to the flow velocity of the mixture, which in the tracer limit coincides with that of the excess neutral component, i.e. $\mathbf{u} = \mathbf{u}_2 \neq \mathbf{u}_1$. The only conservation law for the tracer species corresponds to the conservation of mass. The associated continuity equation is given by Eq. (1.135), that is rewritten here as

$$\frac{\partial x_1}{\partial t} + \frac{1}{m_1 n} \nabla \cdot \mathbf{j}_1 + \mathbf{u} \cdot \nabla x_1 = 0, \tag{6.57}$$

where we have taken into account that $\nabla \cdot \mathbf{u} = \mathbf{u} \cdot \nabla n = 0$ in the Couette flow.

Since the explicit solution of the Boltzmann equation (6.54) for steady Couette flow is not known for arbitrary values of the velocity and thermal gradients, one has to resort to kinetic models. Here we take the BGK model, whose solution for this problem has been widely discussed in Chapter 5. Analogously, the true Boltzmann–Lorentz equation (6.55) is replaced by its Gross–Krook (GK) version [cf. Eq. (1.74)],

$$J_{12}^{\mathrm{GK}} = -\nu_{12}(f_1 - f_{12}), \tag{6.58}$$

where the reference distribution function f_{12} is given by Eq. (1.184). More explicitly,

$$f_{12} = n_1 \left(\frac{m_1}{2\pi k_B T_{12}} \right)^{3/2} \exp\left[-\frac{m_1}{2k_B T_{12}} (\mathbf{v} - \mathbf{u}_{12})^2 \right], \tag{6.59}$$

with

$$\mathbf{u}_{12} = \frac{m_1 \mathbf{u}_1 + m_2 \mathbf{u}}{m_1 + m_2}, \tag{6.60}$$

$$T_{12} = T_1 + \frac{2m_1 m_2}{(m_1 + m_2)^2} (T - T_1) - \frac{m_1}{3k_B} (\mathbf{u}_{12} - \mathbf{u})^2. \tag{6.61}$$

Note that Eqs. (6.59) and (6.61) differ from Eqs. (4.56) and (4.57) since now $\mathbf{u}_1 \neq \mathbf{u}_2 = \mathbf{u}$. For simplicity, we will assume that the particles interact via the Maxwell potential, so the effective collision frequencies ν_{sr} defined by Eq. (1.198) are independent of the temperatures and are given by

$$\nu_{sr} = \text{A} n_r \left(K_{sr} \frac{m_s + m_r}{m_s m_r} \right)^{1/2} , \tag{6.62}$$

where A is a pure number.[2] The extension to more general potentials can be done by following similar steps (Garzó, 1996). The ratio of collision frequencies ν_{12}/ν_{22} is a constant quantity given by

$$\frac{\nu_{12}}{\nu_{22}} = \frac{\nu'}{\nu_2} = \left(\frac{K_{12}}{K_{22}} \frac{1+\mu}{2\mu} \right)^{1/2} , \tag{6.63}$$

where $\mu \equiv m_1/m_2$.

Equation (6.55) is quite difficult to solve, even in the GK model. Nonetheless, as said before, we are interested in the limit of small concentration gradient and external field. In addition, the goal is not to find the solution to Eq. (6.55) for arbitrary initial and boundary conditions, but to get the *normal* solution. To that end, we shall follow a perturbation scheme in the same spirit as in the Chapman–Enskog method (Chapman and Cowling, 1970), analogous to the one followed in the study of mutual diffusion under USF carried out in Section 4.5. Assuming that ∇x_1 and $\mathbf{E}$ are weak (both of order of a bookkeeping uniformity parameter ε), we perform an expansion taking ε as the perturbation parameter. The main feature of this expansion is that the zeroth-order approximation is a nonequilibrium state with arbitrarily large velocity and temperature gradients. As a consequence, the transport coefficients associated with the mass flux will be nonlinear functions of both gradients. Thus, we write the distribution function for the tracer particles in the form

$$f_1 = f_1^{(0)} + \varepsilon f_1^{(1)} + \cdots . \tag{6.64}$$

The reference state $f_1^{(0)}$ is the solution of the GK model for steady Couette flow in the tracer limit, as described at the end of the preceding Section. The fields $\mathbf{u}_{12}$ and T_{12} must also be expanded in a similar way:

$$\mathbf{u}_{12} = \mathbf{u} + \varepsilon \mathbf{u}_{12}^{(1)} + \cdots , \tag{6.65}$$

$$T_{12} = T_{12}^{(0)} + \varepsilon T_{12}^{(1)} + \cdots . \tag{6.66}$$

[2]The choice $\text{A} = 4\pi \times 0.422 \simeq 5.30$ is consistent with Eq. (1.194) and is more convenient for diffusion processes, while the choice $\text{A} = 3 \times 1.3703 \simeq 4.11$ is consistent with Eq. (1.221) and is more appropriate for shearing phenomena.

From Eqs. (6.60) and (6.61),

$$\mathbf{u}_{12}^{(1)} = \frac{\mathbf{j}_1^{(1)}}{n_1(m_1+m_2)}, \tag{6.67}$$

$$T_{12}^{(0)} = T_1^{(0)} + \frac{2m_1m_2}{(m_1+m_2)^2}\left(T - T_1^{(0)}\right), \tag{6.68}$$

$$T_{12}^{(1)} = \frac{m_1^2+m_2^2}{(m_1+m_2)^2} T_1^{(1)}. \tag{6.69}$$

In Eq. (6.68), $T_1^{(0)}$ is obtained from the solution of Eq. (6.49). The expansion of the distribution f_{12} is

$$f_{12} = f_{12}^{(0)} + \varepsilon f_{12}^{(1)} + \cdots, \tag{6.70}$$

where $f_{12}^{(0)}$ is given by Eq. (4.134) with $T_{12} \to T_{12}^{(0)}$ and

$$f_{12}^{(1)} = \left[\frac{1}{n_1 k_B T_{12}^{(0)}} \frac{m_1}{m_1+m_2} \mathbf{V}\cdot\mathbf{j}_1^{(1)} + \left(\frac{m_1 V^2}{2k_B T_{12}^{(0)}} - \frac{3}{2}\right)\frac{T_{12}^{(1)}}{T_{12}^{(0)}}\right] f_{12}^{(0)}. \tag{6.71}$$

As in the usual Chapman–Enskog method, the time derivative is also expanded in powers of ε:

$$\frac{\partial}{\partial t} = \varepsilon \partial_t^{(1)} + \cdots. \tag{6.72}$$

In particular, according to the hydrodynamic equation (6.57), the operator $\partial_t^{(1)}$ acts as

$$\partial_t^{(1)} x_1 = -\mathbf{u}\cdot\nabla x_1, \tag{6.73}$$

where we have neglected the term $\nabla\cdot\mathbf{j}_1^{(1)} \sim \varepsilon^2$. As we will see later, this implies that $\nabla\cdot[(1/\nu')\nabla x_1] \sim \varepsilon^2$. As a consequence of Eq. (6.73),

$$\partial_t^{(1)} f_1^{(0)} = -\left(\mathbf{u}\cdot\nabla \ln x_1\right) f_1^{(0)}, \tag{6.74}$$

since $f_1^{(0)} \propto x_1$. Inserting the expansions in the GK version of the Boltzmann–Lorentz equation (6.55), one gets the following kinetic equation to first order in ε:

$$\begin{aligned}\left(1 + v_y \partial_s\right) f_1^{(1)} &= f_{12}^{(1)} - \frac{1}{\nu'}\left(\mathbf{V}\cdot\nabla\ln x_1 + \frac{q}{m_1}\mathbf{E}\cdot\frac{\partial}{\partial\mathbf{V}}\right) f_1^{(0)} \\ &\equiv \Xi_1^{(1)}.\end{aligned} \tag{6.75}$$

Here, the spatial operator $\partial_s \equiv (1/\nu')\partial/\partial y$ is understood at constant x_1. In other words, the operator ∂_s only "sees" the spatial dependence through the flow velocity $\mathbf{u}$ and the bath temperature T. Equation (6.75) has a structure similar to that of Eq. (5.85), except that the role of f_L is played now by the auxiliary function $\Xi_1^{(1)}$, and its formal solution can be written in a form analogous to Eq. (5.91) as

$$f_1^{(1)}(s,\mathbf{v}) = \int_0^\infty dt\, e^{-t} e^{-tv_y\partial_s} \Xi_1^{(1)}(s,\mathbf{v}). \tag{6.76}$$

Of course, this solution is formal since $\Xi_1^{(1)}$ depends on $f_1^{(1)}$ through the mass flux $\mathbf{j}_1^{(1)}$ and the partial temperature $T_1^{(1)}$. However, assuming (to be confirmed by consistency) that $\partial_s \mathbf{j}_1^{(1)} = \mathbf{0}$ and $\partial_s^2 \left(x_1 T_1^{(1)}/T_1^{(0)}\right) = 0$, the spatial dependence of $\Xi_1^{(1)}$ becomes explicit. The determination of $\mathbf{j}_1^{(1)}$ and $T_1^{(1)}$ is closed by taking appropriate velocity moments in both sides of Eq. (6.76). The algebra involved is quite tedious and the details can be found elsewhere (Garzó and Santos, 1995b; Garzó, 1998). Here we give the structure of the final results. The mass flux $\mathbf{j}_1^{(1)}$ is given by Eq. (6.53), with the tracer diffusion and mobility tensors having the form

$$D_{ij} = D_0 D_{ij}^*(a^*), \quad D_0 = \frac{m_1+m_2}{m_1 m_2}\frac{k_B T}{\nu'}, \tag{6.77}$$

$$\sigma_{ij} = \sigma_0 \sigma_{ij}^*(a^*), \quad \sigma_0 = \frac{n_1 q^2}{k_B T} D_0, \tag{6.78}$$

where the dimensionless tensors $D_{ij}^*(a^*)$ and $\sigma_{ij}^*(a^*)$ are highly nonlinear functions of the shear rate and the force constant and mass ratios (Garzó and Santos, 1995b; Garzó, 1998). The structure of the first-order contribution to the partial temperature is

$$T_1^{(1)} = \left(k_B T \boldsymbol{\Omega}(a^*)\cdot\nabla \ln x_1 + q\boldsymbol{\Omega}'(a^*)\cdot\mathbf{E}\right)\frac{1}{m_1 \nu'^2}\frac{\partial T}{\partial y}, \tag{6.79}$$

where $\boldsymbol{\Omega}(a^*)$ and $\boldsymbol{\Omega}'(a^*)$ are dimensionless vectors. Since $\mathbf{j}_1^{(1)} \sim (1/\nu')\nabla x_1 \sim \varepsilon$, it follows that $\partial_s \mathbf{j}_1^{(1)} \sim \partial_s (1/\nu')\nabla x_1 \sim \varepsilon^2$, so it was legitimate to neglect it in the calculations. Similarly, since $\partial_s^3 T = 0$, Eq. (6.79) leads to $\partial_s^2 \left(x_1 T_1^{(1)}/T\right) \sim \varepsilon^2$, as assumed before. It is worth noting that the diffusion and mobility tensors, as well as the vectors $\boldsymbol{\Omega}$ and $\boldsymbol{\Omega}'$ are independent of the thermal gradient. However, the deviation $T_1^{(1)}$ of the partial temperature of the impurities from the unperturbed value is proportional to the thermal gradient.

The explicit expressions of the elements D_{ij}^* and σ_{ij}^* are very lengthy and will be omitted here. For illustration, we will restrict ourselves to the elements yy and xy, which are the relevant ones when the concentration gradient and the external field are parallel to the y-direction. For small shear rates, the above elements behave as (Garzó and Santos, 1995b; Garzó, 1998)

$$D_{yy}^*(a^*) \approx 1 - \frac{11\mu^2 + 18\mu + 5}{5\mu^2} a^{*2}, \tag{6.80}$$

$$D_{xy}^*(a^*) \approx -(2 + \mu)a^*, \tag{6.81}$$

$$\sigma_{yy}^*(a^*) \approx 1 - (1 + \mu)\frac{5\mu^2 + 18\mu + 5}{15\mu^2} a^{*2}, \tag{6.82}$$

$$\sigma_{xy}^*(a^*) \approx -(1 + \mu)a^*. \tag{6.83}$$

Note that the coefficients appearing in the above expansions are independent of the force constant ratio K_{12}/K_{22}, once the shear rate is nondimensionalized with respect to the collision frequency $\nu' = \nu_{12} = \nu_1$. The breakdown of Einstein's relation ($D_{ij}^* \neq \sigma_{ij}^*$) is clearly apparent in the domain of small shear rates. On the other hand, the shear-rate dependence of both tensors is qualitatively similar in that domain. In particular, D_{yy}^* and σ_{yy}^* decay as the shear rate increases, thus indicating that the shear flow inhibits the mass current parallel to the y-direction. Both cross contributions D_{xy}^* and σ_{xy}^* are negative. This means that if $\partial x_1/\partial y > 0$ and $qE_y < 0$, then the current component along the flow direction is positive.

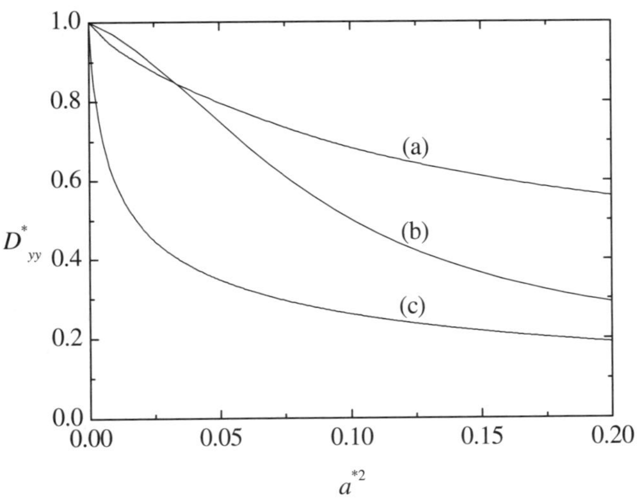

Figure 6.11. Shear-rate dependence of the reduced diagonal element D_{yy}^* of the diffusion tensor for $K_{12} = K_{22}$ and (a) $\mu = 1$, (b) $\mu = 10$, and $\mu = 0.1$.

Figures 6.11 and 6.12 show the shear-rate dependence of D_{yy}^* and σ_{yy}^*, respectively, for $K_{12} = K_{22}$ and a few representative values of the mass ratio

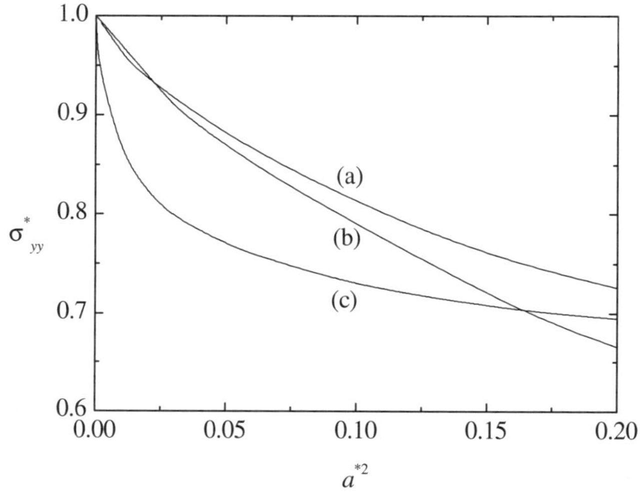

Figure 6.12. Shear-rate dependence of the reduced diagonal element σ_{yy}^* of the conductivity tensor for $K_{12} = K_{22}$ and (a) $\mu = 1$, (b) $\mu = 2$, and (c) $\mu = 0.1$.

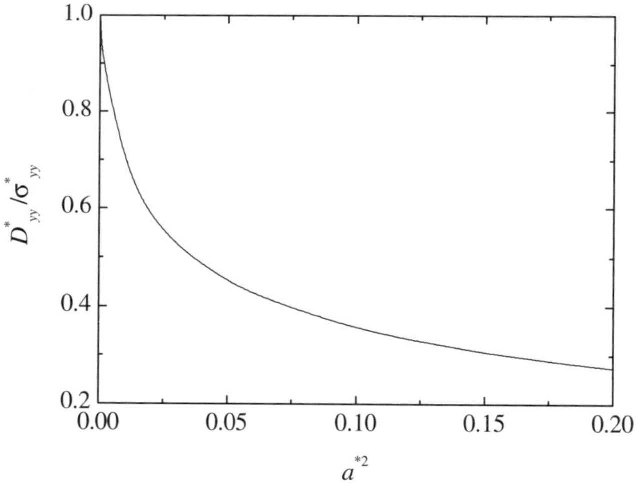

Figure 6.13. Shear-rate dependence of the ratio D_{yy}^*/σ_{yy}^* for $K_{12} = K_{22}$ and $\mu = 0.1$.

μ, including the self-diffusion case $\mu = 1$ (Santos and Garzó, 1992). Both quantities monotonically decrease with the shear rate, as expected from their behaviors for small shear rates. The deviation of the ratio D_{yy}^*/σ_{yy}^* from unity is a measure of the violation of Einstein's relation in the Couette flow. For small shear rates, Eqs. (6.80) and (6.82) yields

$$\frac{D^*_{yy}}{\sigma^*_{yy}} \approx 1 - \frac{10 + 31\mu + 10\mu^2 - 5\mu^3}{15\mu^2} a^{*2}. \tag{6.84}$$

Thus, for small shear rates, the ratio is smaller than 1 if $\mu < 3.780$, while it is larger than 1 if $\mu > 3.780$. The ratio D^*_{yy}/σ^*_{yy} is plotted in Fig. 6.13 for $K_{12} = K_{22}$ and $\mu = 0.1$. We observe that this ratio quickly decays with the shear rate in this situation where the mass of the tracer particles is much smaller than that of the bath particles.

In summary, the diffusion and mobility of impurities or tracer particles immersed in a bath under Couette flow are widely influenced by the value of the shear rate, as well as the impurity/bath mass ratio. The diffusion and mobility processes are characterized by tensorial quantities instead of the scalar coefficients when the bath is in equilibrium. The diagonal elements of the tracer diffusion tensor D and the mobility tensor $\boldsymbol{\sigma}$ show that shearing motion produces an inhibition of the mass current, relative to its equilibrium value. Moreover, there are in general nonzero components of the mass flux orthogonal to the concentration gradient or to the electric field. The latter cross effects are characterized by the off-diagonal elements of D and $\boldsymbol{\sigma}$. Besides, the presence of shearing gives rise to a breakdown of Einstein's relation between the tracer diffusion and mobility quantities at finite shear rates.

Appendix A
List of symbols

Table A.1: List of main symbols used in the book. In general, those symbols which do not have an important meaning and appear only in a few consecutive pages are not included in the list. The third column indicates the pages on which the symbols are introduced.

Symbol	*Meaning*	*Page(s)*
a	Shear rate	56, 217
a^*	Reduced shear rate	63, 170, 217, 273
a_c^*	Critical shear rate for the fourth-degree moments	77
$a_c^{(k)}$	Critical shear rate for the kth-degree moments	81
a, a_{ij}	Rate-of-strain tensor	57, 98
a^*, a_{ij}^*	Reduced rate-of-strain tensor	98
A	Set of hydrodynamic fields	134
A	Numerical constant	44, 181
A_r	Collision integral	50,62
$\mathrm{A}^{(s)}$	Collision integral	27, 181
b	Impact parameter	7
B	Differential cross section	8
B_{sr}	Differential cross section for a mixture	30
c_p	Specific heat at constant pressure	26
c_s	Sound velocity	151
C, C_{ij}	Second-degree velocity polynomial	21
D_0	Mutual diffusion coefficient	32
D_t	Material time derivative	13
D_T	Heat diffusion coefficient	151
D, D_{ij}	Symmetric traceless part of the rate-of-strain tensor	21
D, D_{ij}	Generalized mutual diffusion tensor	203
D, D_{ij}	Generalized tracer diffusion tensor	290
e	Internal energy per unit volume	15, 89, 129

Table A.1: List of symbols (continued).

Symbol	*Meaning*	*Page(s)*
$\widehat{e}$	Effective internal energy per unit volume	130
$\mathbf{E}$	Electric field	289
f	Velocity distribution function	4
f^*	Reduced velocity distribution function	98
f_2	Two-body distribution function	11
f_B	Boundary-layer part of f	225
f_{eq}	Equilibrium distribution function	15
f_H	Hydrodynamic part of f	225
f_{IT}	Information-theory distribution function	53, 90
f_L	Local equilibrium distribution function	16
f_R	Reference distribution function	33
f_s	Velocity distribution function of species s	29
f_s^*	Reduced velocity distribution function of species s	189
f_s^L	Local equilibrium distribution function of species s	45
f_{sr}	Reference distribution functions	40
$f_x^{(+)}$, $f_y^{(+)}$	Marginal velocity distribution functions	98
f_x^*, f_y^*	Marginal velocity distribution functions	99
f_{xy}^*	Marginal velocity distribution function	98
$\mathbf{F}$	External force	4
F_r	Mathematical functions	234, 266
F_η	Viscosity function	66, 218, 276
$F_\eta^{(2)}$	Super-Burnett contribution to F_η	183, 281
$F_{\eta,s}$	Partial contribution of species s to F_η	277
F_κ	Thermal conductivity function	218, 278
$F_{\kappa,s}$	Partial contribution of species s to F_κ	278
F_μ	Friction function	240
$\mathbf{g}$, g	Relative velocity	6
G	Generating function of the velocity moments	263
h	Hydrodynamic length	18, 217
h_T	Hydrodynamic length	227
$\hbar$	Planck constant divided by 2π	4
i	Imaginary unit	64
I	3×3 unit tensor	32
$\mathcal{I}$	Identity matrix	76, 170
$\mathbf{j}_s$	Mass flux of species s	31
J	Boltzmann collision operator	10
J_{sr}	Boltzmann collision operator for a mixture	30
$J_{\mathbf{k}}$, J_{k_1,k_2,k_3}	Collisional moment	226
$\mathbf{J}_q$	Heat flux	31
k_B	Boltzmann constant	4
K	Force constant	7
K_{sr}	Force constants for a mixture	44
ℓ	Mean free path	6, 35

Table A.1: List of symbols (continued).

Symbol	*Meaning*	*Page(s)*
ℓ_σ	Eigenvalues of $\boldsymbol{\mathcal{L}}$	73, 77, 171
L_0	Dufour coefficient	32
L_{qs}	Dufour coefficients	31
$L_r^{(\ell+1/2)}$	Generalized Laguerre (or Sonine) polynomials	23
L_{sq}	Thermal diffusion coefficients	31
L, L_{ij}	Generalized Dufour tensor	203
L, L_{sr}	Phenomenological coefficients	31
$\mathcal{L}$	Linearized Boltzmann collision operator	21
$\boldsymbol{\mathcal{L}}$, $\mathcal{L}_{\sigma\sigma'}$	Matrix for the evolution of moments	75, 169
$\mathcal{L}_{\mathbf{kk}'}$	Matrix for the evolution of moments	80
m	Mass of a particle	4
$\overline{m}$	Average mass in a mixture	273
m_s	Mass of a particle of species s	29
$M_{2\|i}$, $M_{0\|ijk}$	Third-degree velocity moments	50
$M_{4\|0}$, $M_{2\|ij}$, $M_{0\|ijkl}$	Fourth-degree velocity moments	51
$M_{\mathbf{k}}$, $M_{r\ell m}$	Velocity moments	79
$M_{\mathbf{k}}$, M_{k_1,k_2,k_3}	Velocity moments	225
$\boldsymbol{\mathcal{M}}$, $\mathcal{M}_\sigma$	Fourth-degree velocity moments	76
n	Number density	4
$\overline{n}$	Average number density	153
n_s	Number density of species s	29
N	Number of species in a mixture	29
$\mathcal{N}$	Number of particles in a gas	4
P, P_{ij}	Pressure or stress tensor	13, 31
p	Hydrostatic pressure	13
P_s, $P_{s,ij}$	Partial pressure tensor of species s	52, 168
p_s	Partial hydrostatic pressure of species s	52, 168
Pr	Prandtl number	26
P_ℓ	Legendre polynomials	23
q	Interaction parameter for repulsive interactions	26
$\mathbf{q}$	Heat flux	14
$\mathbf{r}$, r_i	Position vector	4
R_y	Marginal distribution f_y^* relative to local equilibrium	99
R_{xy}	Marginal distribution f_{xy}^* relative to local equilibrium	102
s	Entropy per unit volume	14
s_0	Local equilibrium entropy per unit volume	15
s_{ex}	Excess entropy per unit volume	89
$s_{\text{ex,IT}}$	Information-theory excess entropy	90
S	Total entropy	15
$\mathbf{S}$, S_i	Third-degree velocity polynomial	21
t	Time	4
T	Local temperature	4, 29
$\widehat{T}$	Effective temperature	130

Table A.1: List of symbols (continued).

Symbol	*Meaning*	*Page(s)*
$\overline{T}$	Average temperature	153
T_s	Partial temperature of species s	41
T_{sr}	Effective temperatures in f_{sr}	41
$\mathbf{u}$	Flow velocity	4, 29
$\mathbf{u}_s$	Flow velocity of species s	41
$\mathbf{u}_{sr}$	Effective flow velocities in f_{sr}	41
$\mathbf{v}$	Velocity of a particle	4
v_0	Thermal velocity	6
$\mathbf{V}$	Peculiar velocity	13
w	Power exponent of repulsive interactions	7
x_s	Mole fraction of species s	32
$Y_{2\vert i}$, $Y_{0\vert ijk}$	Third-degree velocity polynomials	50
$Y_{4\vert 0}$, $Y_{2\vert ij}$, $Y_{0\vert ijkl}$	Fourth-degree velocity polynomials	51
Y_ℓ^m	Spherical harmonics	22
α	Heating rate, thermostat parameter	59, 118
α^*	Reduced heating rate,	
	reduced thermostat parameter	63, 170
β	Reduced impact parameter	7
γ	Thermal curvature parameter	218, 273
γ_{11}, γ_{22}	Collisional parameters for Maxwell mixtures	33
γ_{sr}	Thermal curvature parameters in a mixture	275
Γ	Sound damping constant	151
ϵ	Reduced thermal gradient	227, 280
$\boldsymbol{\epsilon}$, ϵ_i	Field strength	148
ε	Uniformity parameter	
	in the Chapman–Enskog expansion	19, 134, 204
ζ	Amplitude of a non-conservative force	14, 209
ζ	Effective collision frequency in kinetic models	38
η	Nonlinear shear viscosity	217
η_0	Navier–Stokes shear viscosity	25
η_{ijkl}	Generalized shear viscosity tensor	135
θ	Nonequilibrium thermodynamic temperature	129
θ_4, θ_5	Burnett coefficients	220
Θ	Heaviside's step function	10
κ	Nonlinear thermal conductivity	217
κ_0	Navier–Stokes thermal conductivity	26
κ_{ij}	Generalized thermal conductivity tensor	135
$\widetilde{\kappa}_{ij}$	Generalized thermal conductivity tensor	144
λ_B	Thermal de Broglie wavelength	4
λ_{ij}	Susceptibility tensor	148
$\lambda_{\mathbf{k}}^0$, $\lambda_{r\ell}^0$	Eigenvalues of $\mathcal{L}$	22
$\lambda_{\mathbf{k}}$	Eigenvalues of $\mathcal{L}_{\mathbf{kk}'}$	81
λ_{sr}	Collisional parameters for Maxwell mixtures	52
λ_{sr}'	Collisional parameters for Maxwell mixtures	52
Λ	Operator on velocity space	97

Table A.1: List of symbols (continued).

Symbol	*Meaning*	*Page(s)*
μ	Mass ratio	170, 280
μ_{ij}	Generalized transport coefficients	135
μ_s	Chemical potential per unit mass of species s	32
ν	Effective collision frequency for a single gas	36, 50, 62
ν	Effective collision frequency for a mixture	170
ν'	Effective collision frequency for a mixture	182
ν_1	Effective collision frequency for a single gas	51, 75
ν_{HS}	Collision frequency for hard spheres	34
ν_s	Effective collision frequency of species s	40
ν_{sr}	Effective collision frequency in a mixture	40, 42, 44, 49
$\nu_{sr}^{(0)}$	Effective collision frequency in a mixture in equilibrium	181
ν_η	Effective collision frequency associated with the momentum transport	37
ν_κ	Effective collision frequency associated with the energy transport	37
$\boldsymbol{\xi}$	Reduced peculiar velocity	23, 189
π	Nonequilibrium thermodynamic pressure	129
$\mathbf{\Pi}$, Π_{ij}	Irreversible part of the pressure tensor	14
ϖ	Numerical constant	33
ϖ_2–ϖ_6	Burnett coefficients	219
ρ	Mass density	13, 29
ρ_s	Mass density of species s	29
σ	Range of the interaction potential, hard-sphere diameter	5, 7
σ_0	Mobility or conductivity coefficient	289
$\sigma_s^{(c)}$, $\sigma_s^{(F)}$	Entropy production terms	15
$\sigma_\psi^{(c)}$, $\sigma_\psi^{(F)}$	Production terms of the quantity ψ	12
$\boldsymbol{\sigma}$, σ_{ij}	Generalized mobility or conductivity tensor	290
τ	Mean free time	6
τ_h	Hydrodynamic time	18
τ_p	Period	158
φ_0	Reduced local equilibrium distribution	86, 244
ϕ	Interaction potential	7
ϕ_{sr}	Interaction potential in a mixture	40
Φ	Visco-thermal function	219
$\Phi^{(k)}$	kth-order contribution to f in the Chapman–Enskog expansion	21, 86, 98, 244
Φ_s	Partial contribution of species s to Φ	279
$\boldsymbol{\Phi}_s$	Entropy flux	15
$\boldsymbol{\Phi}_\psi$	Flux of the quantity ψ	12
χ	Scattering angle	6
χ_s	Temperature ratio	274
χ_{sr}	Temperature ratio	275

Table A.1: List of symbols (continued).

Symbol	*Meaning*	*Page(s)*
Ψ_1, Ψ_2	Viscometric functions	66, 219
$\Psi_{1,s}$, $\Psi_{2,s}$	Partial contribution of species s to $\Psi_1$and Ψ_2	277
$\Psi_{\mathbf{k}}$, $\Psi_{r\ell m}$	Eigenfunctions of $\mathcal{L}$	22, 79

References

Abramowitz, M. and Stegun, I. A., editors (1972). *Handbook of Mathematical Functions*. Dover, New York.

Acedo, L., Santos, A., and Bobylev, A. V. (2002). On the derivation of a high-velocity tail from the Boltzmann–Fokker–Planck equation for shear flow. *J. Stat. Phys.*, 109:1027–1050.

Alexander, F. J. and Garcia, A. L. (1997). The direct simulation Monte Carlo method. *Comput. Phys.*, 11:588–593.

Allen, M. P. and Tildesley, D. J. (1987). *Computer Simulation of Liquids*. Clarendon, Oxford.

Alterman, Z., Frankowski, K., and Pekeris, C. L. (1962). Eigenvalues and eigenfunctions of the linearized Boltzmann collision operator for a Maxwell gas and for a gas of rigid spheres. *Astrophys. J. Suppl. Ser.*, 7:291–331.

Anderson, H. L., editor (1989). *A Physicist's Desk Reference*. American Institute of Physics, New York.

Andries, P., Aoki, K., and Perthame, B. (2002). A consistent BGK-type model for gas mixtures. *J. Stat. Phys.*, 106:993–1018.

Andries, P., Tallec, P. L., Perlat, J., and Perthame, B. (2000). The Gaussian-BGK model of Boltzmann equation with small Prandtl number. *Eur. J. Mech. B-Fluids*, 19:813–830.

Aoki, K. and Doi, T. (1994). High-speed vapor flows condensing on a plane condensed phase in the presence of a noncondensable gas. In Shizgal, B. and Weaver, D. P., editors, *Rarefied Gas Dynamics 17*, pages 521–536, Washington DC. AIAA.

Aoki, K., Takata, S., and Kosuge, S. (1998). Vapor flows caused by evaporation and condensation on two parallel plane surfaces: Effect of the presence of a noncondensable gas. *Phys. Fluids*, 10:1519–1533.

Asmolov, E., Makashev, N. K., and Nosik, V. I. (1979). Heat transfer between parallel plates in a gas of Maxwellian molecules. *Sov. Phys. Dokl.*, 24:892–894.

Balescu, R. (1975). *Equilibrium and Non-Equilibrium Statistical Mechanics*. Wiley, New York.

Baranyai, A. (1996). Phase-space compression and entropy for steady states. *J. Chem. Phys.*, 105:7723–7726.

Bear, J. (1971). *Dynamics of Fluids in Porous Media*. Elsevier, Amsterdam.

Bhatnagar, P. L., Gross, E. P., and Krook, M. (1954). A model for collision processes in gases. I. Small amplitude processes in charged and neutral one-component systems. *Phys. Rev.*, 94:511–525.

Bird, G. (1994). *Molecular Gas Dynamics and the Direct Simulation of Gas Flows*. Clarendon, Oxford.

Bird, R. B. (1983). Non-Newtonian behavior of polymeric liquids. *Physica A*, 118:3–16.

Bird, R. B., Armstrong, R. C., and Hassager, O. (1987a). *Dynamics of Polymeric Liquids. Volume 1. Fluid Mechanics*. Wiley, New York.

Bird, R. B., Curtiss, C. F., Armstrong, R. C., and Hassager, O. (1987b). *Dynamics of Polymeric Liquids. Volume 2. Kinetic Theory*. Wiley, New York.

Bobylev, A. V. (1975a). Exact solutions of the Boltzmann equation. *Sov. Phys. Dokl.*, 20:822–824.

Bobylev, A. V. (1975b). Fourier transform method in the theory of the Boltzmann equation for Maxwellian molecules. *Sov. Phys. Dokl.*, 20:820–822.

Bobylev, A. V. (1981). The Chapman–Enskog and Grad methods for solving the Boltzmann equation. *Sov. Phys. Dokl.*, 27:29–31.

Bobylev, A. V. (1984). Exact solutions of the nonlinear Boltzmann equation and of its models. *Fluid Mech.-Sov. Res.*, 13:105–110.

Bobylev, A. V. (1988). The theory of the nonlinear spatially uniform Boltzmann equation for Maxwell molecules. *Sov. Sci. Rev. C. Math. Phys.*, 7:111–233.

Bobylev, A. V. and Spiga, G. (1992). On the relaxation processes in a mixture of Maxwell gases. *Riv. Mat. Univ. Parma*, 1:255–264.

Boltzmann, L. (1995). *Lectures on Gas Theory*. Dover, New York.

Brey, J. J. and Santos, A. (1984). Solution of the BGK model kinetic equation for Very Hard Particle interaction. *J. Stat. Phys.*, 37:123–149.

Brey, J. J. and Santos, A. (1992). Nonequilibrium entropy of a gas. *Phys. Rev. A*, 45:8566–8572.

Brey, J. J., Santos, A., and Dufty, J. W. (1987). Heat and momentum transport far from equilibrium. *Phys. Rev. A*, 36:2842–2849.

Brey, J. J., Santos, A., and Garzó, V. (1993). Analysis of the Evans and Baranyai variational principle in dilute gases. *Phys. Rev. Lett.*, 70:2730–2733.

Brush, S. G. (1965). *Kinetic Theory, Vols. I–III*. Pergamon Press, London.

Brush, S. G. (1976). *The kind of motion we call heat, Vols. I and II*. Nort–Holland, Amsterdam.

Buck, B. and Macauly, V. A. (1991). *Maximum Entropy in Action*. Clarendon Press, Oxford.

Casas-Vázquez, J. and Jou, D. (1994). Nonequilibrium temperature versus local-equilibrium temperature. *Phys. Rev. E*, 49:1040–1048.

Cercignani, C. (1988). *The Boltzmann Equation and Its Applications*. Springer–Verlag, New York.

Cercignani, C. (1989). Kinetic theory with "bounce-back" boundary conditions. *Transp. Theory Stat. Phys.*, 182:125–131.

Cercignani, C. (1990). *Mathematical Methods in Kinetic Theory*. Plenum Press, New York.

Cercignani, C. (1998). *Ludwig Boltzmann. The Man Who Trusted Atoms*. Oxford University Press, Oxford.

Cercignani, C. and Cortesse, S. (1994). Validation of a Monte Carlo simulation of the plane Couette flow a rarefied gas. *J. Stat. Phys.*, 75:817–838.

Cercignani, C., Illner, R., and Pulvirenti, M. (1994). *The Mathematical Theory of Dilute Gases*. Springer–Verlag, New York.

Chapman, S. and Cowling, T. G. (1970). *The Mathematical Theory of Nonuniform Gases*. Cambridge University Press, Cambridge.

Daivis, P. J. and Evans, D. J. (1993). Thermal conductivity of a shearing fluid. *Phys. Rev. E*, 48:1058–1065.

de Groot, S. R. and Mazur, P. (1984). *Nonequilibrium Thermodynamics*. Dover, New York.

Dealy, J. M. (1995). Official nomenclature for material functions describing the response of a viscoelastic fluid to various shearing and extensional deformations. *J. Rheol.*, 39:253–265.

Dorfman, J. R. and van Beijeren, H. (1977). The Kinetic Theory of Gases. In Berne, B. J., editor, *Statistical Mechanics, Part B*, pages 65–179, New York. Plenum.

Duderstadt, J. J. and Martin, W. R. (1979). *Transport Theory*. Wiley, New York.

Dufty, J. W. (1983). Nonlinear diffusion and mode-coupling anomalies. *Phys. Rev. Lett.*, 51:2159–2162.

Dufty, J. W. (1984). Diffusion in shear flow. *Phys. Rev. A*, 30:1465–1476.

Dufty, J. W. (1990). Kinetic theory of fluids far from equilibrium — Exact results. In López de Haro, M. and Varea, C., editors, *Lectures on Thermodynamics and Statistical Mechanics*, pages 166–181, Singapore. World Scientific.

Dufty, J. W., Brey, J. J., and Santos, A. (1986a). Some theoretical aspects of nonequilibrium simulation methods. In Ciccotti, G. and Hoover, W. G., editors, *Molecular-Dynamics Simulation of Statistical-Mechanical Systems*, pages 294–303, Amsterdam. North–Holland.

Dufty, J. W., Santos, A., and Brey, J. J. (1996). Practical kinetic model for hard sphere dynamics. *Phys. Rev. Lett.*, 77:1270–1273.

Dufty, J. W., Santos, A., Brey, J. J., and Rodríguez, R. F. (1986b). Model for nonequilibrium computer simulation methods. *Phys. Rev. A*, 33:459–466.

Ernst, M. H. (1981). Nonlinear model-Boltzmann equations and exact solutions. *Phys. Rep.*, 78:1–171.

Ernst, M. H. and Brito, R. (2002). High-energy tails for inelastic Maxwell models. *Europhys. Lett.*, 58:182–187.

Ernst, M. H. and Hendriks, E. M. (1979). An exactly solvable non-linear Bolzmann equation. *Phys. Lett. A*, 70:183–185.

Erpenbeck, J. (1984). Shear viscosity of the hard-sphere fluid via nonequilibrium molecular dynamics. *Phys. Rev. Lett.*, 52:1333–1335.

Evans, D. J. (1982). Homogeneous NEMD algorithm for thermal conductivity — Application of non-canonical linear response theory. *Phys. Lett. A*, 91:457–460.

Evans, D. J. (1989). On the entropy of nonequilibrium states. *J. Stat. Phys.*, 57:745–758.

Evans, D. J. (1991). Green–Kubo relations for weak vector processes in strongly shearing fluids. *Phys. Rev. A*, 44:3630–3632.

Evans, D. J. and Baranyai, A. (1991). Possible variational principle for steady states far from equilibrium. *Phys. Rev. Lett.*, 67:2597–2600.

Evans, D. J. and Baranyai, A. (1992). The Gaussian thermostast, phase compression and the conjugate pairing rule. *Mol. Phys.*, 67:1209–1216.

Evans, D. J. and Hoover, W. G. (1986). Flows far from equilibrium via molecular dynamics. *Annu. Rev. Fluid Mech.*, 18:243–264.

Evans, D. J. and Morriss, G. P. (1984). Non-Newtonian molecular dynamics. *Comput. Phys. Rep.*, 1:299–343.

Evans, D. J. and Morriss, G. P. (1990). *Statistical Mechanics of Nonequilibrium Liquids*. Academic Press, London.

Ferziger, J. H. and Kaper, G. H. (1972). *Mathematical Theory of Transport Processes in Gases*. North–Holland, Amsterdam.

Frenkel, D. and Smit, B. (1996). *Understanding Molecular Simulations: From Algorithms to Applications*. Academic Press, San Diego.

Galkin, V. S. (1995). Exact solution of the system of equations of the second-order kinetic moments for a two-scale homoenergetic affine monatomic gas flow. *Fluid Dynamics*, 30:467–476.

Garzó, V. (1988). On the derivation of the Burnett hydrodynamic equations from the Hilbert expansion. *Physica A*, 149:551–560.

Garzó, V. (1991). Transport equations from the Liu model. *Phys. Fluids A*, 3:1980–1982.

Garzó, V. (1993). Thermal conductivity of a dilute gas in a thermostatted shear-flow state. *Phys. Rev. E*, 48:3589–3593.

Garzó, V. (1994). Heat flux induced by an external force in a strongly shearing dilute gas. *J. Chem. Phys.*, 101:1423–1430.

Garzó, V. (1995a). Heat transport in a dilute gas under uniform shear flow. *Phys. Rev. E*, 51:3156–3163.

Garzó, V. (1995b). Thermal transport generated by an external force in a sheared dilute gas. *J. Chem. Phys.*, 103:4626–4631.

Garzó, V. (1996). Tracer diffusion under heat and momentum transport for general repulsive interactions. *Physica A*, 234:108–128.

Garzó, V. (1997). Kinetic model for uniform shear flow. *Physica A*, 243:113–128.

Garzó, V. (1998). Electrical conductivity in a dilute gas far from equilibrium. *Phys. Rev. E*, 57:4186–4197.

Garzó, V. (2001). Heat and momentum transport in a multicomponent mixture far from equilibrium. *Physica A*, 289:37–56.

Garzó, V. and López de Haro, M. (1991). Tracer diffusion in shear flow. *Phys. Rev. A*, 44:1397–1400.

Garzó, V. and López de Haro, M. (1992). Kinetic models for diffusion in shear flow. *Phys. Fluids A*, 4:1057–1069.

Garzó, V. and López de Haro, M. (1993). Effect of mass-ratio dependence of the force law for tracer diffusion in shear flow. *Phys. Fluids A*, 5:1059–1061.

Garzó, V. and López de Haro, M. (1994). Kinetic model for heat and momentum transport. *Phys. Fluids*, 6:3787–3794.

Garzó, V. and López de Haro, M. (1995). Tracer diffusion under shear flow for general repulsive interactions. *Phys. Fluids*, 7:478–486.

Garzó, V. and López de Haro, M. (1997). Nonlinear transport for a dilute gas in steady Couette flow. *Phys. Fluids*, 9:776–787.

Garzó, V. and Marín, C. (1998). Nonlinear transport in a dilute binary mixture. *Recent Res. Devel. in Phys. Fluids*, 1:44–60.

Garzó, V. and Santos, A. (1993). Heat and momentum transport in a gaseous dilute solution. *Phys. Rev. E*, 48:256–262.

Garzó, V. and Santos, A. (1995a). Comparison between the Boltzmann and BGK equations for uniform shear flow. *Physica A*, 213:426–434.

Garzó, V. and Santos, A. (1995b). Diffusion in a gaseous dilute solution under heat and momentum transport. *Phys. Rev. E*, 52:4942–4951.

Garzó, V., Santos, A., and Brey, J. J. (1989). A kinetic model for a multicomponent gas. *Phys. Fluids A*, 1:380–383.

Garzó, V., Santos, A., and Brey, J. J. (1990). Influence of nonconservative external forces on self-diffusion in dilute gases. *Physica A*, 163:651–671.

Gillan, M. J. and Dixon, M. (1983). The calculation of thermal conductivies by perturbed molecular dynamics simulation. *J. Phys. C*, 16:869–878.

Goebel, C. J., Harris, S. M., and Johnson, E. A. (1976). Two-temperature disparate-mass gas mixtures: A thirteen moment description. *Phys. Fluids*, 19:627–635.

Goldman, E. and Sirovich, L. (1967). Equations for gas mixtures. *Phys. Fluids*, 10:1928–1940.

Goldstein, H. (1959). *Classical Mechanics*. Addison–Wesley, Massachusetts.

Gómez Ordóñez, J., Brey, J. J., and Santos, A. (1989). Shear-rate dependence of the viscosity for dilute gases. *Phys. Rev. A*, 39:3038–3040.

Gómez Ordóñez, J., Brey, J. J., and Santos, A. (1990). Velocity distribution function of a dilute gas under uniform shear flow: Comparison between a Monte Carlo simulation method and the Bhatnagar–Gross–Krook equation. *Phys. Rev. A*, 41:810–815.

Grad, H. (1949). On the kinetic theory of rarefied gases. *Commun. Pure Appl. Math.*, 2:331–407.

Grad, H. (1963). Asymptotic theory of the Boltzmann equation. *Phys. Fluids*, 6:147–181.

Gradshteyn, I. S. and Ryzhik, I. M. (1980). *Table of Integrals, Series, and Products*. Academic Press, San Diego.

Gross, E. P. and Jackson, E. A. (1959). Kinetic models and the linearized Boltzmann equation. *Phys. Fluids*, 2:432–441.

Gross, E. P. and Krook, M. (1956). Model for collisions processes in gases. Small amplitude oscillations of charged two-component systems. *Phys. Rev.*, 102:593–604.

Hamel, B. B. (1966). Two-fluid hydrodynamic equations for a neutral, disparate-mass, binary mixture. *Phys. Fluids*, 9:12–22.

Hanley, H. J. M., editor (1983). *Nonlinear Fluid Behavior*. North–Holland, Amsterdam.

Hendriks, E. M. and Nieuwenhuizen, T. M. (1982). Solution to the nonlinear Boltzmann equation for Maxwell models for nonisotropic initial conditions. *J. Stat. Phys.*, 29:591–615.

Hirschfelder, J. O., Curtiss, C. F., and Bird, R. B. (1964). *Molecular Theory of Gases and Liquids*. Wiley, New York.

Holway, L. H. (1966). New statistical models for kinetic theory: Methods of construction. *Phys. Fluids*, 9:1658–1673.

Hoover, W. G. (1983). Nonequilibrium molecular dynamics. *Annu. Rev. Phys. Chem.*, 34:103–127.

Hoover, W. G. (1991). *Computational Statistical Mechanics*. Elsevier, Amsterdam.

Ikenberry, E. and Truesdell, C. (1956). On the pressures and the flux of energy in a gas according to Maxwell's kinetic theory, I. *J. Rat. Mech. Anal.*, 5:1–54.

Johnson, E. A. (1978). Effect of cross-sectional mass dependence in the theory of disparate-mass gas mixtures. *Phys. Fluids*, 21:1239–1240.

Jou, D., Casas-Vázquez, J., and Criado-Sancho, M. (2001). *Thermodynamics of Fluids Under Flow*. Springer–Verlag, Berlin.

Jou, D., Casas-Vázquez, J., and Lebon, G. (1992). Extended irreversible thermodynamics: An overview of recent bibliography. *J. Non-Equilib. Thermodyn.*, 17:383–396.

Jou, D., Casas-Vázquez, J., and Lebon, G. (1999). Extended irreversible thermodynamics revisited (1988-98). *Rep. Prog. Phys.*, 62:1035–1142.

Jou, D., Casas-Vázquez, J., and Lebon, G. (2000). *Extended Irreversible Thermodynamics*. Springer–Verlag, Berlin.

Jou, D. and Criado-Sancho, M. (2001). Thermodynamics of dilute gases in shear flow. *Physica A*, 292:75–86.

Kao, M. H., Yodh, A. G., and Pine, D. J. (1993). Observation of Brownian motion on the time scale of hydrodynamic interactions. *Phys. Rev. Lett.*, 70:242–245.

Karlin, I. V., Dukek, G., and Nonnenmacher, T. F. (1997). Invariance principle for extension of hydrodynamics: Nonlinear viscosity. *Phys. Rev. E*, 55:1573–1576.

Karlin, I. V., Dukek, G., and Nonnenmacher, T. F. (1998). Reply to "Comment on 'Invariance principle for extension of hydrodynamics: Nonlinear viscosity' ". *Phys. Rev. E*, 57:3674–3675.

Kawasaki, K. and Gunton, J. D. (1973). Theory of nonlinear transport processes: Nonlinear shear viscosity and normal stress effects. *Phys. Rev. A*, 8:2048–2064.

Kim, C. S., Dufty, J. W., Santos, A., and Brey, J. J. (1989a). Analysis of nonlinear transport in Couette flow. *Phys. Rev. A*, 40:7165–7174.

Kim, C. S., Dufty, J. W., Santos, A., and Brey, J. J. (1989b). Hilbert-class or "normal" solutions for stationary heat flow. *Phys. Rev. A*, 39:328–338.

Kirkpatrick, T. R. and Nieuwoudt, T. C. (1986). Stability analysis of a dense hard-sphere fluid subjected to large shear — Shear-induced ordering. *Phys. Rev. Lett.*, 56:885–888.

Kogan, M. N. (1969). *Rarefied Gas Dynamics*. Plenum Press, New York.

Kourganoff, V. (1963). *Basic Methods in Transfer Problems*. Oxford University Press, Oxford.

Krook, M. and Wu, T. T. (1976). Formation of Maxwellian Tails. *Phys. Rev. Lett.*, 36:1107–1109.

Krook, M. and Wu, T. T. (1977). Exact solutions of the Boltzmann equation. *Phys. Fluids*, 20:1589–1595.

Lee, M. and Dufty, J. W. (1997). Transport far from equilibrium: Uniform shear flow. *Phys. Rev. E*, 56:1733–1745.

Lee, M., Dufty, J. W., Montanero, J. M., Santos, A., and Lutsko, J. F. (1996). Long wavelength instability for uniform shear flow. *Phys. Rev. Lett.*, 76:2702–2705.

Lees, A. W. and Edwards, S. F. (1972). The computer study of transport processes under extreme conditions. *J. Phys. C*, 5:1921–1929.

Liem, S. Y., Brown, D., and Clarke, J. H. R. (1992). Investigation of the homogeneous-shear nonequilibrium-molecular-dynamics method. *Phys. Rev. A*, 45:3706–3713.

Liu, G. (1990). A method for constructing a model for the Boltzmann equation. *Phys. Fluids A*, 2:277–280.

Loose, W. (1988). The constant-temperature constraint in the nonequilibrium molecular dynamics and the non-Newtonian viscosity coefficients of gases. *Phys. Lett. A*, 128:39–44.

Loose, W. and Hess, S. (1987). Velocity distribution function of a streaming gas via nonequilibrium molecular dynamics. *Phys. Rev. Lett.*, 58:2443–2445.

Loose, W. and Hess, S. (1988). Nonequilibrium velocity distribution function of gases: Kinetic theory and molecular dynamics. *Phys. Rev. A*, 37:2099–2111.

Loose, W. and Hess, S. (1989). Rheology of dense model fluids via nonequilibrium molecular dynamics: shear thinning and ordering transition. *Rheologica Acta*, 28:91–101.

Lutsko, J. F. (1996). Molecular chaos, pair correlations, and shear-induced ordering of hard spheres. *Phys. Rev. Lett.*, 77:2225–2228.

Lutsko, J. F. (1997). Approximate solution of the Enskog equation far from equilibrium. *Phys. Rev. Lett.*, 78:243–246.

Lutsko, J. F. (1998). Viscoelastic effects from the Enskog equation for uniform shear flow. *Phys. Rev. E*, 58:434–446.

Lutsko, J. F. and Dufty, J. W. (1987). Possible instability for shear-induced order-disorder transition. *Phys. Rev. Lett.*, 57:2775–2778.

Lutsko, J. F., Dufty, J. W., and Das, S. P. (1989). Fluctuations and dissipation in a fluid under shear: Linear dynamics. *Phys. Rev. A*, 39:1311–1324.

Luzzi, R., Vasconcellos, A. R., and Galvão Ramos, J. (2002). *Predictive Statistical Mechanics. A Nonequilibrium Ensemble Formalism*. Kluwer Academic Publishers, Dordrecht.

Makashev, N. K. and Nosik, V. I. (1981). Steady Couette flow (with heat transfer) of a gas of Maxwellian molecules. *Sov. Phys. Dokl.*, 25:589–591.

Marchetti, M. C. and Dufty, J. W. (1982). Nonlinear irreversible fluxes in a fluid with large shear rate. *Int. J. Quant. Chem.*, 16:83–89.

Marchetti, M. C. and Dufty, J. W. (1983). Tagged particle fluctuations in uniform shear flow. *J. Stat. Phys.*, 32:255–277.

Margeneau, H. and Murphy, G. M. (1956). *The Mathematics of Physics and Chemistry*. Krieger, Huntington, N.Y.

Marín, C. and Garzó, V. (1996). Uniform shear flow in a binary mixture with general repulsive interactions. *Phys. Fluids*, 8:2756–2765.

Marín, C. and Garzó, V. (1997). Kinetic model for momentum transport in a binary mixture. In Shen, C., editor, *Rarefied Gas Dynamics 20*, pages 91–96, Beijing. Peking University Press.

Marín, C. and Garzó, V. (1998). Mutual diffusion in a binary mixture under shear flow. *Phys. Rev. E*, 57:507–513.

Marín, C. and Garzó, V. (1999). Electrical current density in a sheared dilute gas. *Physica A*, 265:508–519.

Marín, C. and Garzó, V. (2000). Transport of mass and energy in a mixture far from equilibrium. *Trends in Stat. Phys.*, 3:127–136.

Marín, C. and Garzó, V. (2002). Exact solution of the Gross–Krook kinetic model for a multicomponent gas in steady Couette flow. *Physica A*, 312:315–341.

Marín, C., Garzó, V., and López de Haro, M. (2000). Shear-rate dependent transport coefficients in a binary mixture of Maxwell molecules. *Phys. Fluids*, 12:717–726.

Marín, C., Garzó, V., and Santos, A. (1994). Nonlinear transport in a dilute binary mixture of mechanically different particles. *J. Stat. Phys.*, 75:797–816.

Marín, C., Garzó, V., and Santos, A. (1995). Transport properties in a binary mixture under shear flow. *Phys. Rev. E*, 52:3812–3820.

Marín, C., Garzó, V., and Santos, A. (1997). Tracer limit in a gas mixture under shear flow with repulsive interactions. *Phys. Rev. E*, 56:2291–2294.

Marín, C., Montanero, J. M., and Garzó, V. (1996a). Kinetic models for diffusion generated by an external force. *Physica A*, 225:235–253.

Marín, C., Montanero, J. M., and Garzó, V. (1996b). Monte Carlo simulation of the Boltzmann equation in the colour conductivity problem for general repulsive interactions. *Mol. Phys.*, 78:1249–1261.

Marín, C., Santos, A., and Garzó, V. (1996c). Non-equilibrium phase transition in a binary mixture. *Europhys. Lett.*, 33:599–604.

McLennan, J. A. (1965). Convergence of the Chapman–Enskog expansion of the linearized Boltzmann equation. *Phys. Fluids*, 8:1580–1584.

McLennan, J. A. (1989). *Introduction to Non-Equilibrium Statistical Mechanics*. Prentice Hall, Englewood Cliffs, N.J.

Mead, L. R. and Papanicolau, N. (1984). Maximum entropy in the problem of moments. *J. Math. Phys.*, 25:2404–2417.

Montanero, J. M., Alauoi, M., Santos, A., and Garzó, V. (1994). Monte Carlo simulation of the Boltzmann equation for steady Fourier flow. *Phys. Rev. A*, 49:367–375.

Montanero, J. M. and Garzó, V. (1998). Nonlinear Couette flow in a dilute gas: Comparison between theory and molecular-dynamics simulation. *Phys. Rev. E*, 58:1836–1842.

Montanero, J. M. and Garzó, V. (2000). Nonlinear Couette flow in dilute gases. *Trends in Stat. Phys.*, 3:105–115.

Montanero, J. M., Garzó, V., and Santos, A. (1997a). High velocity tail in a dilute gas under shear. In Shen, C., editor, *Rarefied Gas Dynamics 20*, pages 113–117, Beijing. Peking University Press.

Montanero, J. M. and Santos, A. (1995). Comparison of the DSMC method with an exact solution of the Boltzmann equation. In Harvey, J. and Lord, G., editors, *Rarefied Gas Dynamics 19*, pages 899–905, Oxford. Oxford University Press.

Montanero, J. M. and Santos, A. (1996). Nonequilibrium entropy of a sheared gas. *Physica A*, 225:7–18.

Montanero, J. M., Santos, A., and Garzó, V. (1995a). Analysis on the stability of the uniform shear flow from a Monte Carlo simulation of the Boltzmann equation. *Phys. Lett. A*, 203:73–76.

Montanero, J. M., Santos, A., and Garzó, V. (1995b). Does the Gaussian thermostat maximize the phase-space compression factor? *J. Stat. Phys.*, 81:989–1005.

Montanero, J. M., Santos, A., and Garzó, V. (1996a). Monte Carlo simulation of the Boltzmann equation for uniform shear flow. *Phys. Fluids*, 8:1981–1983.

Montanero, J. M., Santos, A., and Garzó, V. (1996b). Singular behavior of the velocity moments of a dilute gas under uniform shear flow. *Phys. Rev. E*, 53:1269–1272.

Montanero, J. M., Santos, A., and Garzó, V. (1997b). Distribution function for large velocities of a two-dimensional gas under shear flow. *J. Stat. Phys.*, 88:1165–1181.

Montanero, J. M., Santos, A., and Garzó, V. (2000). Monte Carlo simulation of nonlinear Couette flow in a dilute gas. *Phys. Fluids*, 12:3060–3073.

Montanero, J. M., Santos, A., Lee, M., Dufty, J. W., and Lutsko, J. F. (1998). Stability of uniform shear flow. *Phys. Rev. E*, 57:546–556.

Morse, T. F. (1964). Kinetic model equations for a gas mixture. *Phys. Fluids*, 7:2012–2013.

Naitoh, T. and Ono, S. (1979). The shear viscosity of a hard-sphere fluid via nonequilibrium molecular dynamics. *J. Chem. Phys.*, 70:4515–4523.

Nosik, V. I. (1981). Heat transfer between parallel plates in a mixture of gases of Maxwellian molecules. *Sov. Phys. Dokl.*, 25:495–497.

Nosik, V. I. (1983). Degeneration of the Chapman–Enskog expansion in one-dimensional motions of Maxwellian molecule gases. In Belotserkovskii, O. M., Kogan, M. N., Kutateladze, S. S., and Rebrov, A. K., editors, *Rarefied Gas Dynamics 13*, pages 237–244, New York. Plenum Press.

Oran, E. S., Oh, C. K., and Cybyk, B. Z. (1998). Direct simulation Monte Carlo: Recent advances and applications. *Annu. Rev. Fluid Mech.*, 30:403–441.

Rapaport, D. C. (1995). *The Art of Molecular Dynamics Simulation*. Cambridge University Press, Cambridge.

Reif, R. (1965). *Fundamentals of Statistical and Thermal Physics*. McGraw–Hill, New York.

Résibois, P. and de Leener, M. (1977). *Classical Kinetic Theory of Fluids*. Wiley, New York.

Risso, D. and Cordero, P. (1997). Dilute gas Couette flow: Theory and molecular dynamics simulation. *Phys. Rev. E*, 56:489–498.

Risso, D. and Cordero, P. (1998). Erratum: Dilute gas Couette flow: Theory and molecular dynamics simulation. *Phys. Rev. E*, 57:7365–7366.

Rodríguez, R., Salinas-Rodríguez, E., and Dufty, J. W. (1983). Fokker–Planck and Langevin description of fluctuations in uniform shear flow. *J. Stat. Phys.*, 32:279–298.

Sabbane, M. and Tij, M. (2002). Calculation algorithm for the collisional moments of the Boltzmann equation for Maxwell molecules. *Comp. Phys. Comm.*, 149:19–29.

Sahimi, M. (1993). Flow phenomena in rocks: from continuum models to fractals, percolation, cellular automata, and simulated annealing. *Rev. Mod. Phys.*, 65:1393–1534.

Santos, A. (2000). Nonlinear viscosity and velocity distribution function in a simple longitudinal flow. *Phys. Rev. E*, 62:6597–6607.

Santos, A. and Brey, J. J. (1991a). Far from equilibrium velocity distribution of a dilute gas. *Physica A*, 174:355–390.

Santos, A. and Brey, J. J. (1991b). Velocity distribution function of a gas under shear flow. In Beylich, A. E., editor, *Rarefied Gas Dynamics 17*, pages 101–108, Weinheim. VCH Publisher.

Santos, A., Brey, J. J., and Dufty, J. W. (1986a). Divergence of the Chapman–Enskog expansion. *Phys. Rev. Lett.*, 56:1571–1574.

Santos, A., Brey, J. J., and Garzó, V. (1986b). Kinetic model for steady heat flow. *Phys. Rev. A*, 34:5047–5050.

Santos, A., Brey, J. J., Kim, C. S., and Dufty, J. W. (1989). Velocity distribution for a gas with steady heat flow. *Phys. Rev. A*, 39:320327–3974.

Santos, A. and Garzó, V. (1992). Self-diffusion in a dilute gas under heat and momentum transport. *Phys. Rev. A*, 46:3276–3287.

Santos, A. and Garzó, V. (1995a). Exact moment solution of the Boltzmann equation for uniform shear flow. *Physica A*, 213:409–425.

Santos, A. and Garzó, V. (1995b). Exact non-linear transport from the Boltzmann equation. In Harvey, J. and Lord, G., editors, *Rarefied Gas Dynamics 19*, pages 13–22, Oxford. Oxford University Press.

Santos, A., Garzó, V., and Brey, J. J. (1992). Comparison between the homogeneous-shear and the sliding-boundary methods to produce shear flows. *Phys. Rev. A*, 46:8018–8020.

Santos, A., Garzó, V., and Brey, J. J. (1995). On the validity of a variational principle for far-from-equilibrium steady states. *Europhys. Lett.*, 29:693–698.

Santos, A., Garzó, V., Brey, J. J., and Dufty, J. W. (1993). Singular behavior of shear flow far from equilibrium. *Phys. Rev. Lett.*, 71:3971–3974.

Santos, A., Montanero, J. M., Dufty, J. W., and Brey, J. J. (1998). Kinetic model for the hard-sphere fluid and solid. *Phys. Rev. E*, 57:1644–1660.

Sarman, S. and Evans, D. J. (1992a). Heat flow and mass diffusion in binary Lennard–Jones mixtures. *Phys. Rev. A*, 45:2370–2379.

Sarman, S. and Evans, D. J. (1992b). Heat flow and mass diffusion in binary Lennard–Jones mixtures. II. *Phys. Rev. A*, 46:1960–1966.

Sarman, S., Evans, D. J., and Baranyai, A. (1992). Mutual and self-diffusion in fluids undergoing strong shear. *Phys. Rev. A*, 46:893–902.

Sarman, S., Evans, D. J., and Baranyai, A. (1994). Extremum properties of the Gaussian thermostat. *Physica A*, 208:191–204.

Sirovich, L. (1962). Kinetic modeling of gas mixtures. *Phys. Fluids*, 5:908–918.

Sone, Y. (1991). Asymptotic theory of a steady flow of a rarefied gas past bodies for small Knudsen numbers. In Gatignol, R. and Soubbaramayer, editors, *Advances in Kinetic Theory and Continuum Mechanics*, pages 19–31, Berlin. Springer–Verlag.

Sone, Y. (1997). Continuum gas dynamics in the light of kinetic theory and new features of rarefied gas flows. In Shen, C., editor, *Rarefied Gas Dynamics 20*, pages 3–24, Beijing. Peking University Press.

Sone, Y. (2000). Flows induced by temperature fields in a rarefied gas and their ghost effect on the behavior of a gas in the continuum limit. *Annu. Rev. Fluid Mech.*, 32:779–811.

Sone, Y. (2002). *Kinetic Theory and Fluid Dynamics*. Birkhäuser, Boston.

Sone, Y., Aoki, K., and Doi, T. (1992). Kinetic theory analysis of gas flows condensing on a plane condensed phase: case of a mixture of a vapor and a noncondensable gas. *Transp. Theory Stat. Phys.*, 21:297–328.

Sone, Y. and Onishi, Y. (1979). Flow of rarefied gas over plane wall. *J. Phys. Soc. Jpn.*, 47:572–672.

Sone, Y., Takata, S., and Ohwada, T. (1990). Numerical analysis of the plane Couette flow of a rarefied gas on the basis of the linearized Boltzmann equation for hard-sphere molecules. *Eur. J. Mech. B/Fluids*, 9:273–2888.

Sone, Y. and Yamamoto, K. (1970). Flow of rarefied gas over plane wall. *J. Phys. Soc. Jpn.*, 29:495–508.

Stanley, H. (1971). *Introduction to Phase Transitions and Critical Phenomena*. Oxford University Press, Oxford.

Szamel, G., Bławzdziewicz, J., and Leegwater, J. A. (1992). Self-diffusion in sheared suspensions: Violation of the Einstein relation. *Phys. Rev. A*, 45:R2173–R2176.

Tehver, R., Toigo, F., Koplik, J., and Banavar, J. R. (1998). Thermal walls in computer simulations. *Phys. Rev. E*, 57:R17–R20.

Tij, M. and Santos, A. (1995). Combined heat and momentum transport in a dilute gas. *Phys. Fluids*, 7:2858–2866.

Trozzi, C. and Ciccotti, G. (1984). Stationary nonequilibrium states by molecular dynamics. II. Newton's law. *Phys. Rev. A*, 29:916925–128.

Truesdell, C. (1956). On the pressures and the flux of energy in a gas according to Maxwell's kinetic theory, II. *J. Rat. Mech. Anal.*, 5:55–128.

Truesdell, C. and Muncaster, R. G. (1980). *Fundamentals of Maxwell's Kinetic Theory of a Simple Monatomic Gas*. Academic Press, New York.

van Beijeren, H. and Dorfman, J. R. (2000). On thermostats and entropy production. *Physica A*, 279:21–29.

van Beijeren, H. and Ernst, M. H. (1973). The modified Enskog equation. *Physica*, 68:437–456.

van Zhon, R. (1999). Kinetic approach to the Gaussian thermostat in a dilute sheared gas in the thermodynamic limit. *Phys. Rev. E*, 60:4158–4163.

Wadsworth, D. C. (1993). Slip effects in a confined rarefied gas. I: Temperature slip. *Phys. Fluids A*, 5:1831–1839.

Weinert, U. (1982). Multi-temperature generalized moment method in Boltzmann transport theory. *Phys. Rep.*, 91:297–399.

Welander, P. (1954). On the temperature jump in a rarefied gas. *Arkiv Fysik*, 7:507–553.

Woods, L. C. (1993). *An Introduction to the Kinetic Theory of Gases and Magnetoplasmas*. Oxford University Press, Oxford.

Zhu, J. X., Durian, D. J., Müller, J., Weitz, D. A., and Pine, D. J. (1992). Scaling of transient hydrodynamic interactions in concentrated suspensions. *Phys. Rev. Lett.*, 68:2559–2562.

Zwanzig, R. (1979). Nonlinear shear viscosity of a gas. *J. Chem. Phys.*, 71:4416–4420.

Index